Study Guide

Peter S. Shenkin
COLUMBIA UNIVERSITY

Chemistry
experiment and theory

Bernice G. Segal
BARNARD COLLEGE
COLUMBIA UNIVERSITY

John Wiley & Sons
NEW YORK CHICHESTER BRISBANE TORONTO SINGAPORE

*This book is dedicated to all those who taught me chemistry,
but especially to Mr. Lawrence Finbelstein, of the Bronx High
School of Science, Prof. Donald W. Slocum, of Southern Illinois
University, and Prof. Walter J. Kauzmann, of Princeton University.
There are bits and pieces of all of them in this book.*

*Of the many who helped this Study Guide come into being,
I would like to particularly thank Bernice G. Segal, the text
author, for asking me to write it, Dennis Sawicki and his staff
at John Wiley & Sons, for their infinite patience, Alan Crosswell,
of Columbia University Center for Computing Activities, for help
in producing the manuscript on their Imogen laser printer, and
Cristina Healey, who appeared in the roles of amanuensis, muse
and fury, abd who shared the agony as well as the joy.*

Copyright © 1985 by John Wiley & Sons, Inc.

All rights reserved.

Reproduction or translation of any part of
this work beyond that permitted by Sections
107 or 108 of the 1976 United States Copyright
Act without the permission of the copyright
owner is unlawful. Requests for permission or
further information should be addressed to the
Permissions Department, John Wiley & Sons, Inc.

ISBN 0 471 80972 1

Printed in the United States of America

10 9 8 7 6 5 4 3 2 1

CONTENTS

Chapter 0.	Introduction	0-1
Chapter 1.	**Atomic Structure and Reactivity**	
	Scope	1-1
	Questions	1-11
	Answers	1-19
Chapter 2.	**Introduction to the Periodic Table and to Inorganic Nomenclature**	
	Scope	2-1
	Questions	2-9
	Answers	2-14
Chapter 3.	**The Gas Laws and Stoichiometry Involving Gases**	
	Scope	3-1
	Questions	3-12
	Answers	3-17
Chapter 4.	**The Properties of Gases and the Kinetic-Molecular Theory**	
	Scope	4-1
	Questions	4-14
	Answers	4-21
Chapter 5.	**Intermolecular Forces, Condensed Phases, and Changes of Phase**	
	Scope	5-1
	Questions	5-14
	Answers	5-20
Chapter 6.	**Properties of Dilute Solutions**	
	Scope	6-1
	Questions	6-16
	Answers	6-24
Chapter 7.	**Aqueous Solutions and Ionic Reactions**	
	Scope	7-1
	Questions	7-12
	Answers	7-17
Chapter 8.	**Introduction to the Law of Chemical Equilibrium**	
	Scope	8-1
	Questions	8-18
	Answers	8-24
Chapter 9.	**Acids, Bases and Salts**	
	Scope	9-1
	Questions	9-18
	Answers	9-23
Chapter 10.	**Buffer Solutions and Acid-Base Titrations**	
	Scope	10-1
	Questions	10-20
	Answers	10-26

Chapter 11.	**Equilibria Involving Slightly Soluble Electrolytes**	
	Scope	11-1
	Questions	11-13
	Answers	11-18
Chapter 12.	**Atomic Structure, Atomic Spectra and the Introduction of the Quantum Concept**	
	Scope	12-1
	Questions	12-10
	Perspective: The Schroedinger Equation	12-15
	Answers	12-23
Chapter 13.	**The Electronic Structure of Atoms, the Periodic Table, and Periodic Properties**	
	Scope	13-1
	Questions	13-17
	Answers	13-26
Chapter 14.	**The Nature of the Chemical Bond**	
	Scope	14-1
	Questions	14-12
	Answers	14-18
Chapter 15.	**Oxidation States and Oxidation-Reduction Reactions**	
	Scope	15-1
	Questions	15-12
	Perspective: Descriptive Redox Chemistry; Acidity of the Oxides	15-20
	Answers	15-28
Chapter 16.	**Electrochemistry**	
	Scope	16-1
	Questions	16-23
	Answers	16-34
Chapter 17.	**Energy Enthalpy and Thermochemistry**	
	Scope	17-1
	Questions	17-17
	Answers	18-22
Chapter 18.	**Entropy, Free Energy, and Equilibrium**	
	Scope	18-1
	Questions	18-23
	Perspective: Dependence of the Free Energy on Concentration; K_{eq}	18-31
	Answers	18-35
Chapter 19.	**Chemical Kinetics**	
	Scope	19-1
	Questions	19-20
	Perspective: The Steady State Approximation; Michaelis-Menten Kinetics	19-27
	Answers	19-35
Chapter 20.	**Coordination Compounds**	
	Scope	20-1
	Questions	20-12
	Answers	20-24
Chapter 21.	**Properties and Structures of Metallic and Ionic Crystalline Solids**	
	Scope	21-1
	Questions	21-12
	Perspective: Crystal Structures and the Close Packing of Spheres	21-17
	Answers	21-24

Chapter 22.	**Radioactivity and Nuclear Chemistry**	
	Scope	22-1
	Questions	22-10
	Answers	22-16
Chapter 23.	**Introduction to Organic Chemistry**	
	Scope	23-1
	Questions	23-14
	Answers	23-22

INTRODUCTION

With the exception of this **Introduction**, all the chapters of this study guide correspond in name and number, as well as subject matter, with those of Segal's *Chemistry, Experiment and Theory*. The study guide is your companion to the text, and its purpose is to make the text easier to use and understand. This **Introduction** is your companion to the study guide itself. The next few pages describe what is in it, and contain hints and advice about how to use it.

The text combines a presentation of chemistry as a coherent science with an emphasis on problem-solving techniques. Although you may be able to memorize methods for solving particular kinds of problems, you will not usually find it possible to solve a new kind of problem unless you understand the underlying science. If you feel you understand the principles, but still cannot solve the problems, it may mean that you don't understand what you thought you did. Another possibility is that you need to practice your algebra and arithmetic (most college freshmen do!). The Study Guide should help you both to understand the principles and to solve problems, and especially to help you pinpoint exactly where your difficulties lie.

Each Study Guide chapter has a section called the **Scope**. This section amounts to my own running (and occasionally rambling) commentary on the corresponding text chapter. Its purpose is to clarify chemical principles. Sometimes I expand the text's explanation of a difficult topic, or give additional examples or an alternative explanation; sometimes I give a lengthy analysis of a text example; sometimes I pass lightly over a topic that I feel will probably be clear or obvious. Often I go beyond the text, introducing a definition or concept or experimental result that I think makes a principle easier to understand, and occasionally I introduce a new application of a principle, in order to show its usefulness. When I do talk about something that is not in the text, I usually say so.

Each chapter also has a set of **Questions and Answers**. These are keyed to the text, section by section, and the majority are of the short-answer type. Where a numerical question is given, the numbers are chosen, where possible, so as to make computation easy; you should not need a calculator to do these, and in fact, you should avoid using one. This will give you practice in doing simple calculations in your head, or on the back of the proverbial envelope. In those few cases where it has not been possible to make up numerical questions which compute easily, the answer is given in algebraic as well as numerical form. For the harder questions, the answer also contains a line or so of explanation.

The **Questions** and **Answers** are designed to test your understanding. If you read a text section, then do the corresponding Study Guide questions, it should be immediately obvious what you do and do not understand. You should be aware that a lot of the material in the text is subtle, and some of the questions, though demanding only one-word answers, will require some thought, and maybe even some writing. (I have put together the **Questions** section in "workbook" format, so you will have room to do your writing in the Study Guide, should you wish to.) On the other hand, I have not written even a single question "just to make it hard." If there is one you get wrong, or cannot do, assume it is because there is something you don't understand, and and that the same confusion will come back to haunt you on other problems and exams. You should go back to the corresponding text section and/or Study Guide **Scope** and try to make sense of whatever it is. Often you still will not be able to figure it out. It is very helpful in most cases to discuss the material with your fellow students. If this does not help either, you should make sure you take it up with your teaching assistant or instructor. It is very useful to get into the habit of making a list of your own "unanswered questions" concerning the course material. The more explicit these are, the more useful they will be. You can then use this list when you approach your instructor for guidance, and often the very process of writing a cogent question will lead you to the answer.

Starting at the middle of the Study Guide, several chapters have an additional section called a **Perspective**. This section either introduces an entirely new subject, not discussed in the text, or else goes over material discussed in the text from a very different viewpoint. You can view the **Perspectives** as "special topics," and some instructors may wish to use some of them.

Enough advice for now; you'll find plenty more as you go along in both the text and the Study Guide! Bernice Segal and I both hope you find chemistry to be as much fun as we do, but even in the extremely unlikely event that you don't, we hope that the text and Study guide help you learn what you need to know of it.

CHAPTER 1: ATOMIC STRUCTURE AND STOICHIOMETRY

Scope

The general pedagogical approach of the text is to introduce important topics twice: once to confer familiarity with the fundamental idea, and then again later to discuss the subject in depth. Chapter 1 really amounts to a brief introduction to the basic ideas of chemistry: what an atom is, what a molecule is, what a chemical reaction is. All of these notions will be considerably elaborated in later chapters, and are introduced here in only their simplest forms.

Matter and Atoms

We now know that *matter* (that which has mass), exists on the earth in the form of either *chemical compounds* or *chemical elements*, which in turn are combinations of *atoms*. Atoms consist of negatively electrically charged *electrons* surrounding a positively electrically charged *nucleus*. The nucleus is made up of *neutrons*, which have mass but no charge, and *protons*, which have charge as well as mass. There are also other particles in the nucleus, and in addition protons and neutrons can be described in terms of still more elementary particles, called *quarks* and *gluons*. How far this procedure of subdivision can go — or even if it has an end — is not known. *Chemistry*, however, is concerned with how atoms interact with each other. In an *ordinary chemical interaction*, nuclei do not break up; they simply form different combinations with each other, and the electrons associated with the atoms alter the paths they travel to correspond to the new positions. On a fundamental physical level, virtually all of chemistry can be accounted for by the *Coulombic forces* that attract the negatively charged electrons to positively charged nuclei, and that repel electrons from other electrons and nuclei from other nuclei. It turns out, however, that objects as *microscopic* as electrons and nuclei obey physical laws somewhat altered from those we are accustomed to in the *macroscopic* world. The microscopic behaviors are embodied in *quantum mechanics*, which is discussed in Chapters 12-14.

Coulombic forces are not the only forces in the universe; there is also *gravity*, by which any objects which have mass are mutually attracted, and also the *strong* and *weak nuclear interactions*, which are responsible for holding nuclei and subatomic particles together. There may be other, yet undiscovered, forces as well, but only the Coulombic interaction is important in chemistry.

Quantum mechanics reveals that electrons do not travel in fixed paths or orbits; they move so fast, and are in a sense so hard to pin down, that the electron is often depicted as a diffuse blur in space, either a spherical blur, as in Figure 1.1, or some other kind of blur, as you will see in Chapter 13. It is interesting to note historically that after it became clear (in 1911) that an atom does in fact consist of a small, massive, positively charged nucleus surrounded by negative electrons, the first attempts to describe the motion of the electrons around the nucleus did involve fixed orbits; this was the Bohr model of the atom, described in Chapter 12, but we now know that this picture is fundamentally wrong.

Even though an atomic nucleus can be split, it loses its chemical identity when this occurs. Thus, as the text states, if the nucleus of a copper atom is split, it is no longer copper. The intact nucleus is the smallest entity that retains its chemical identity.

A proton and a neutron have almost identical mass, approximately equal to 1amu (*atomic mass unit*). Each electron has a mass of about 1/1836 amu. In most chemical calculations we ignore the mass of the electrons, and we also ignore the difference between the mass of a proton and that of a neutron. Thus a hydrogen atom, with a single proton and a single electron, has an *atomic mass number* of one, whereas a helium atom, with two protons, two neutrons and two electrons, has an atomic mass number of four. The *atomic number* is the number of protons in the nucleus, and is one for hydrogen and two for helium. An isolated atom is electrically neutral, and must have as many electrons as protons, one for hydrogen and two for helium.

The electrical charge on a chemical species is given in units of *electronic charge*; that is, the charge on an electron. A species obtained by taking away or adding electrons to a neutral atom is called an *ion*; thus if one electron is taken from hydrogen, we have the naked proton, denoted H^+; if one is added, instead, we have the *hydride ion*, designated H^-. A neutral hydrogen atom is designated H. Other elements are given one- or two-letter symbols, the first one of which is capitalized; thus helium is He; take away one electron and we have the helium ion, He^+, with a charge of +1.

Chapter 1   Atomic Structure and Stoichiometry

The symbol for an element always designates a nucleus with a given number of protons, the atomic number, but for the same atomic number, the atomic mass can vary, meaning the number of neutrons can vary. When this happens, the kind of chemical interactions that take place hardly change at all. This is because there is no change in the Coulombic interaction between the electrons and the nuclei, because neutrons have no charge, and we said earlier that chemistry is based on Coulombic forces. Thus a change in the number of neutrons in the nucleus does not change the element. If we add a neutron to the proton in hydrogen's nucleus, we get an *isotope* of hydrogen known as deuterium, but deuterium is still hydrogen. When you look up the atomic mass of an element in a table, the value given is the average over the naturally occurring isotopic composition; for example, the atomic mass of H is 1.0079, indicating that most hydrogen has only a proton in the nucleus. Isotopes are sometimes symbolized $^A_Z X$, where X is the symbol for the element, A is the atomic mass and Z the atomic number; thus deuterium is $^2_1 H$, and ordinary hydrogen is $^1_1 H$.

You also need to know that usually only the outermost electrons (called *valence electrons*) in an atom take part in chemical interactions. This is because the inner electrons (the *core electrons*) are so highly attracted to the positively charged nucleus that they are not affected much by neighboring nuclei. The valence electrons are sometimes indicated by dots, so that sodium, whose symbol is Na, is sometimes represented as Na·, to indicate that it has one valence electron.

The mass of an isotope in atomic mass units is not identical to the number of protons plus neutrons, for three reasons. The first reason is that a proton and a neutron have slightly different masses. The second reason is that the masses of protons and neutrons vary a bit from nucleus to nucleus; this will be explained in Chapter 22. The third reason has to do with the way an atomic mass unit is defined (see below), but the number of protons plus neutrons is always very close to the weight in atomic mass units; thus $^{28}_{14}$Si (the most common isotope of silicon) has a mass of 27.97693 amu. As discussed in Example 1.1, the atomic weight of Si is given in tables as 28.09 amu, which is the weighted average of the masses of the naturally occurring isotopes. By *definition*, an atom of $^{12}_6 C$ has a mass of 12 amu, so that the amu contains the mass of an electron, as well as the mass of a proton and neutron of this isotope.

The Mole Concept

A *mole* of a substance is *Avogadro's number* of atoms or molecules of the substance. Since we have not yet discussed molecules, we will use atoms for our description. If one atom of hydrogen weighs, on average, 1.008 amu (we need to say "on average" because of isotopic composition) and one atom of carbon weighs, on average, 12.011 amu, then, on average, a carbon atom weighs 12.011/1.008=11.916 times as much as a hydrogen atom. Two carbon atoms, or a million, or *any* number, will always weigh 11.916 times as much as the same number of hydrogen atoms. Now suppose I take enough carbon atoms to weigh 12.011g. (This will certainly be a large number!) The same number of hydrogen atoms will then weigh 12.011g/11.916=1.008g. The weight, in grams, of a substance numerically equal to the atomic (or molecular) weight of the substance is a mole of that substance. Thus a mole of C is 12.011g of C, and a mole of H is 1.008g of H. We already showed that these two must contain the same number of atoms; a mole of *any* element contains the same number of atoms. This number is 6.02×10^{23}, and is called *Avogadro's number*; in our text it is symbolized N_A. The mole concept is one of the most important ideas in chemistry, as will become apparent even before you finish this chapter.

Chemical Compounds and Chemical Formulas

Molecules are combinations of atoms in fixed ratios. For example, a molecule of water contains one atom of oxygen and two of hydrogen, and is symbolized H_2O. This symbol is the *chemical formula* for water; the subscript following the symbol for an element in such a formula indicates the number of that kind of atom in the formula. As another example, P_2O_5 (phosphorus pentoxide) contains two phosphorus and five oxygen atoms in each molecule. Not all substances are made up of molecules, however. A molecule is actually an independent entity containing the number of atoms given by the formula. One mole of water contains N_A of H_2O molecules.

A chemical *element* is a substance that consists of only one kind of atom, but certain elements exist as molecules, rather than isolated atoms. Thus nitrogen gas contains only nitrogen atoms, but these are combined into molecules with the formula N_2. Similarly for hydrogen (H_2), oxygen (O_2), fluorine (F_2) and chlorine (Cl_2).

The *molecular weight* of a substance is the sum of the atomic weights of the atoms in the molecule; thus the molecular weight of H_2O is approximately $2 \times 1 + 1 \times 16 = 18$, and that of P_2O_5 is approximately $2 \times 31 + 5 \times 16 = 142$. Also, it should be clear that a mole of P_2O_5 contains two moles of P atoms and five moles of O atoms, and that 1/3 mole of P_2O_5 contains 2/3 mole of P atoms and 5/3 mole of O atoms. Furthermore, knowing the atomic weights and the molecular formula allows us to calculate the percentage by weight of the constituent elements: one mole of P_2O_5 weighs about 142g, and contains about $5 \times 16 = 80$g of O; so the fraction by weight O in P_2O_5 is $80/142 \approx 0.563$; multiplying by 100 gives 56.3 as the *weight percent* O in P_2O_5.

We said that an element is composed of a single kind of atom. A *chemical compound* is the general term for substances composed of more than one kind of atom chemically bonded together. We said earlier that not all substances are molecular; as one example, most elements are not; for example, the gas helium exists as individual He atoms, not as molecules. There are, however, compounds which are non-molecular. An example is ordinary table salt, (NaCl) sodium chloride. It is still true that, as the formula indicates, a mole of NaCl (23g+35g = 58g) contains one mole each of sodium and chlorine atoms, but these are not bonded together into discrete molecules; rather the sodium exists as Na^+ *cations* (a cation is a positively charged ion) and the chlorine as Cl^- *anions* (negatively charged ions), called chloride ions. These are distributed in an extended *crystal lattice* (Figure 1.3). There is no inidication in the chemical formula that NaCl is an ionic solid rather than an assembly of discrete molecules; you need to know some chemistry to know that, and by the time you finish this course you will understand why NaCl is an ionic solid whereas H_2O is a molecular liquid. But even though an NaCl crystal doesn't have discrete molecules, we still define a mole of sodium chloride to be the weight in grams of the composition in the formula. The weight in grams corresponding to a chemical formula (e.g., 58g for NaCl) is sometimes called a *gram-formula weight*; that of an atom (such as 12g for carbon) is sometimes called a *gram-atomic weight*, and that of a molecule (such as 18g for H_2O) is sometimes called (you guessed it!) a *gram-molecular weight*. In any case, they each correspond to a mole of the substance.

Empirical and Molecular Formulas

Given an unknown compound, very often the first step in analysing its composition is to break it apart into its elements and determine how much of each element is present. For example, suppose we have an unknown substance, and 15g of it is found to contain 12g of C and 3g of H. This corresponds to about 1 mole of C and 3 moles of H. Thus we might be tempted to say that the chemical formula is CH_3. In fact, there is no stable substance with this molecular composition; the gas ethane, however, has the formula C_2H_6, which is also consistent with the data: ethane has a molecular weight of 30, so 15g is half a mole. One mole of C_2H_6 contains 2 moles C and 6 moles H, so a half a mole contains 1 mole C and 3 moles H. Just from knowing the elemental composition, however, it is not possible to know whether the substance has formula CH_3 or C_2H_6 or C_3H_9..., or even whether the substance is molecular or ionic! So, until further data is available, one must content oneself with the *empirical formula* ("empirical" means "from experimental evidence"), which is the *simplest* formula consistent with the data, in this case, CH_3[*]. Incidentally, if we had data from another type of experiment that told us that the molecular weight of our compound was 30, then we would in fact know that the *molecular formula*, which gives the composition of the molecule, must consist of double the empirical formula, or C_2H_6.

Chemical Equations and Stoichiometry

In chemistry we are interested not only in what substances are made of, but also in how they react, or transform into one another. These transformations, or *chemical reactions*, sometimes take place just upon mixing chemicals, but often something else, such as heat, or pressure, or a spark, is necessary. Remember that when substances transform chemically, the atomic nuclei remain intact; they simply move from their starting arrangements (*reactants*) to their final arrangements (*products*). This is symbolized by a *chemical equation*, such as the following for the reaction of hydrogen and oxygen to

---

[*] There is a general scientific principle called *Occam's razor*, named after the fourteenth century English scholastic, William of Occam. Occam's razor states that until futher evidence is available, one should always adopt the simplest hypothesis consistent with the data.

Chapter 1 — Atomic Structure and Stoichiometry

make water.

$$2H_2 + O_2 \rightarrow 2H_2O$$

What this says is that two hydrogen molecules (H_2) and one oxygen molecule (O_2) form two water molecules (H_2O).

Note that all the ingredients are given in the forms in which they actually occur: hydrogen and oxygen as H_2 and O_2 molecules, for example. Note also that the equation is *balanced*; this requires that there be the same number of each kind of atom on both sides of the reaction — four H's and two O's in this example. The numbers in *front* of the reactants and products are called *stoichiometric coefficients*; the stoichiometric coefficients of H_2 and of H_2O are both 2; that of O_2 is 1 (which is not written explicitly). *Balancing an equation* means figuring out the stoichiometric coefficients, given just the reactants and products; for example, the *unbalanced* equation for the above reaction is

$$H_2 + O_2 \rightarrow H_2O$$

Some chemical equations are hard to balance, and for these detailed methods can be used (Chapter 15); however, most that you encounter in this course will be simple. For this reaction, note that since oxygen starts out as O_2, we must have two O's on the right. This allows the guess that the stoichiometric coefficient of H_2O is 2. Then there are four H's on the right, and the way to achieve this on the left is to give H_2 a stoichiometric coefficient of 2 also. Since the result balances, the guess of 2 for H_2O was a good one.

If two H_2 molecules and one O_2 molecule combine to form two H_2O molecules, then two hundred H_2 molecules must combine with one hundred O_2 molecules to give two hundred H_2O molecules, and $2N_A$ H_2 molecules must combine with N_A O_2 molecules to give $2N_A$ H_2O molecules, which is another way of saying that two *moles* of H_2 combine with one *mole* of O_2 to form two *moles* of water. Thus, stoichiometric coefficients can be understood to pertain to moles as well as atoms. And, in fact, the stoichiometric coefficients can be used to determine how much of each reactant is needed to form a given amount of product; for example, suppose I want to make 9g of water from the elements. Since H_2O has a molecular weight of 18, 9g is half a mole. From the stoichiometric coefficients it is clear that I will need the same number of moles of H_2 as the number of moles of water I want, so I need 1/2 mole of H_2. The molecular weight of H_2 is 2, so I need (1/2)2g = 1g H_2. Then the stoichiometric coefficients tell me I need *half* this number of moles of O_2, or (1/2)(1/2 mole) = 1/4 mole O_2. The molecular weight of O_2 is 2×16 = 32, so I need (1/4)(32g) = 8g of O_2. Note that the mass of the reactants (1g H_2+8g O_2) must be equal to the mass of the products (9g H_2O).

In fact, since stoichiometric coefficients can be understood as pertaining to moles, the following are also balanced equations for the above reaction

$$H_2 + \frac{1}{2}O_2 \rightarrow H_2O$$

$$4H_2 + 2O_2 \rightarrow 4H_2O$$

When a balanced equation is requested, however, you are usually expected to give the form in which the stoichiometric coefficients are the smallest possible non-fractional numbers.

Sometimes ions appear in chemical reactions. In this event, it is important that the equation be balanced with regard to charge as well as atoms, as in the text example

$$2Al(s) + 6H^+(aq) \rightarrow 2Al^{3+}(aq) + 3H_2(g)$$

Note that there is a charge of +6 on both sides of the equation. Note also that the state of the substance may also be indicated, (g) for gas, (l) for a pure liquid, (aq) for aqueous solution and (s) for solid. Incidentally, this equation, which describes the reaction of aluminum metal with acid (an acid consists of hydrogen ions in water) should not be understood to mean that the solution itself has a positive charge. It is just that the negative ions which blance the charge can be omitted from the chemical equation, since they do not participate in the reaction; they are *spectators*. Thus, if the negative ion is Cl^-, the reaction could be written

$$2Al(s) + 6H^+(aq) + 6Cl^-(aq) \rightarrow 2Al^{3+}(aq) + 3H_2(g) + 6Cl^-(aq)$$

No additional information is communicated by including the chlorides in the equation, so it is customary to leave them out.

It is important to understand one more thing about chemical equations. They do not say anything about how the reaction occurs on a detailed, step-by-step molecular level. For example, the equation for the formation of water does *not* say that two H_2 molecules and one O_2 molecule collide, and two H_2O molecules come flying out. The situation is far more complicated than this, and how reactions actually do take place is discussed in Chapter 19. The chemical equations simply tell you what the reactants are and what the products are, and in what molar proportions.

Limiting Reagent and Reaction Yield

Now suppose I have two moles each of H_2 and O_2, and these react to form water. After the reaction is complete, what will I have? From the balanced equation, the two moles of H_2 can only react with a single mole of O_2. These form 2 moles of H_2O. So 1 mole O_2 will be left over, and will still be present when the reaction is complete. We say that H_2 is the *limiting reagent*; adding more H_2 would allow more H_2O to be formed, whereas adding more O_2 would not.

The 2 moles of H_2O which we expect to get out of the above reaction is called the *theoretical yield*; it is the maximum that could be obtained. Often in practice the theoretical yield is not achieved, however, due to side reactions, purification problems or the fact that some reactions just do not go to completion (you will see why in Chapters 17 and 18). Suppose for the sake of argument that after the experiment is over we only obtain 30g of water. We expected 36g (2 moles×18g/mole = 36g). The *percent yield* is then (30g×100%/36g) = 83%. This is the percent of the maximum theoretical yield which was actually achieved under whatever conditions were used.

Postscript

Although the symbols we use for chemical substances and chemical equations may seem simple on the surface, they actually in and of themselves summarize knowledge and understanding that was obtained over hundreds (some would say thousands) of years of experimentation. Just the fact that we write H_2O for water means that we accept that there are atoms, and that they combine in fixed ratios to form compounds. The fact that we pay attention to balancing an equation like

$$2H_2 + O_2 \rightarrow 2H_2O$$

means that we accept that atoms maintain their own identity when chemicals react; they are not created or destroyed, or changed into other kinds of atoms. The fact that we can use such an equation to calculate how much H_2 and O_2 will be needed to form a given amount of H_2O means that we not only accept the idea that different atoms have different weights, but also means that we know what the ratios of these weights are. And the fact that we know that 18g of H_2O contains 6.02×10^{23} molecules means that we know not just the weight ratios, but also the absolute weights of the atoms. Thus there is a lot of science to be understood just in the very notation that chemists use, so do not be surprised if it takes hard work to be comfortable with some of the "simple" problems which you may be asked to do at this stage. In the coming chapters you will come to understand better where some of these ideas come from.

Chapter 1　　　　　　　　　　　　　　　　　　　　　　Atomic Structure and Stoichiometry

Questions

1.1 Subatomic particles and the nuclear atom,
1.2 Isotopes and
1.3 The atomic weight scale

Note: Consult a periodic table or table of atomics weights when necessary.

1. Which particles of chemical interest are in the nucleus?

2.* The nucleus takes up *most of the mass* and *most of the space* of the atom.

3. The valence electrons are *closer* to the nucleus than the core electrons.

4. Nitrogen has five valence electrons.

   (a) Give a dot chemical symbol for a nitrogen atom.

   (b) How many electrons, protons and neutrons does the most common isotope of nitrogen atom have?

5. Give the isotopic symbol (such as ^1_1H for hydrogen) for each of the following

   (a) The isotope of iron with A=56

   (b) The isotope of chlorine with 18 neutrons

---

* In this sort of question, write "true" if the words in italics are true; otherwise, change them to make the statement true.

Chapter 1   Atomic Structure and Stoichiometry

6. The atomic weight of fluorine is 18.9948. Give the isotopic symbol for the most common isotope.

7. Suppose some element has an isotope with A=30 that is 90% abundant, and another isotope with A=31 that is 10% abundant. What is the atomic weight of A?

8. For the isotope $^{7}_{3}Li$, the atomic mass number is _____, the atomic number is _____, and a single atom contains _____ protons, _____ neutrons and _____ electrons.

1.4 Avogadro's number and the concept of a mole

Note: In these and subsequent questions, express your answers in terms of N_A, where appropriate.

1. How much does Avogadro's number of neutrons weigh (to one significant figure)?

2. The atomic weight of carbon is about 12. How many grams does a single carbon atom weigh?

3. The mass of a single atom of a particular isotope of boron is 11.009305 amu. How much would a mole of these atoms weigh?

4. How many neutrons are there in a mole of $^{7}_{3}Li$?

5. A flask contains equal weights of nitrogen atoms (atomic weight 14), lithium atoms (atomic weight 7) and silicon atoms (atomic weight 28).

    (a) How many Li atoms are there for each N atom?

Chapter 1   Atomic Structure and Stoichiometry

(b) How many Si atoms are there for each N atom?

(c) How many Li atoms are there for each Si atom?

1.5 Molecular formulas and molecular weights for discrete molecules

Note: Look up atomic weights and atomic numbers for these and subsequent questions which require them.

1. What is the molecular weight of each of the following compounds?

    (a) CO_2 (carbon dioxide)

    (b) Nitrous oxide

    (c) Water

    (d) $C_2H_4O_2$ (acetic acid)

2. Chlorine occurs as the diatomic gas Cl_2. In one mole of this gas

    (a) How many molecules are there?

    (b) How many atoms are there?

    (c) How many grams are there?

    (d) How many electrons are there?

3. In 1/2 mole of CO_2

    (a) How many moles of C are there?

Chapter 1   Atomic Structure and Stoichiometry

(b) How many moles of O are there?

(c) How many moles of O_2 are there?

(d) What is the mole ratio of O to C?

(e) If instead of 1/2 mole I had specified 1/4 mole of CO_2, which of the above would have changed?

1.6 The significance of "molecular" formulas and "molecular weights" for ionic crystalline solids

1. Write the symbols for the cation and the anion for each of the following ionic species:

    (a) KCl

    (b) Li_2O

    (c) FeO

    (d) Fe_2O_3

    (e) NaCl

    (f) $CaCl_2$

    (g) Na_3PO_4

2. In one mole of Li_2O

    (a) How many cations are there?

    (b) How many anions are there?

    (c) How many ions total are there?

Chapter 1 Atomic Structure and Stoichiometry

(d) How many Li_2O molecules are there?

1.7 Distinction between the terms molecular weight, formula weight and gram atomic weight

1. Which of the following terms includes the other two?

   _____ Gram atomic weight

   _____ Gram formula weight

   _____ Gram molecular weight

1.8 Empirical formulas

1. A substance gives, on decomposition, 0.60 moles of Fe and 0.60 moles of S. What is its empirical formula?

2. A substance gives, on decompositon, 0.60 moles of Fe and 0.90 moles of S. What is its empirical formula?

3. A substance gives, on decomposition, 0.60 moles of Fe and 0.30 moles of O_2. What is its empirical formula?

4. A substance gives, on decomposition, 0.60 moles of Fe and 0.45 moles of O_2. What is its empirical formula?

5. A substance gives, on decomposition, 1.2g of C and 1.6g of O_2. What is its empirical formula?

Chapter 1 Atomic Structure and Stoichiometry

6. The deuterated form of sulfuric acid, D_2SO_4 (where D stands for ^2_1H) has a molecular weight of 100. What is the percent by weight of each of the following in the compound?

   (a) D

   (b) S

   (c) O

1.9 The significance of balanced chemical equations

1. Balance the following equations

   (a) $H_2 + O_2 \rightarrow H_2O$

   (b) H_2O_2(hydrogen peroxide) $\rightarrow$ oxygen + hydrogen

   (c) $Na + H_2O \rightarrow Na^+ + OH^- + H_2$

   (d) $Ca(OH)_2(s) + HCl(aq) \rightarrow Ca^{2+}(aq) + Cl^-(aq) + H_2O(l)$

2. For the reaction

   $$2ClO^-(aq) \rightarrow Cl^-(aq) + ClO_2^-(aq)$$

   suppose I start with 1 mole of ClO^-.

   (a) Suppose 1/2 mole of ClO^- reacts. How many moles will there then be of

   i. ClO^-

   ii. Cl^-

   iii. ClO_2^-

   iv. total ions

   (b) Suppose instead of 1/2 mole of ClO^- reacting, I tell you that x moles react. How many moles

Chapter 1   Atomic Structure and Stoichiometry

will there then be of

    i. ClO^-

    ii. Cl^-

    iii. ClO_2^-

    iv. total ions

1.10 Stoichiometric calculations and the empirical formula

1. $\dfrac{\text{mass in grams}}{\text{formula weight}} = \underline{\qquad}$

2. (a) The units of formula weight are _____.

   (b) The units of number of moles are _____.

   (c) The units of mass in grams are _____.

3. The two sides of a chemical equation must contain (true or false)

   _____(a) the same number of atoms

   _____(b) the same number of moles

   _____(c) the same number of grams

   _____(d) the same overall charge

4. Carbon can react with oxygen to form CO_2 (carbon dioxide), or CO (carbon monoxide), or a mixture of the two. Suppose 12g of C is completely reacted with oxygen, and the gaseous products are found to weigh 32g. How many moles of each gas are in the mixture?

1.11 The limiting reagent and the yield of product

1. For the reaction

$$2NO + O_2 \rightarrow 2NO_2$$

suppose I have one-half mole of each reactant.

(a) Which reactant is limiting?

(b) How much of what do I have to add to make the other reactant limiting?

(c) In the original situation, what is the theoretical yield, in grams?

(d) If the theoretical yield is achieved, how many moles of gas, total, are there at the end of the experiment?

(e) In fact, suppose 15g of NO_2 is formed. What is the percent yield?

Chapter 1 — Atomic Structure and Stoichiometry

Answers

1.1 Subatomic particles and the nuclear atom
1.2 Isotopes and
1.3 The atomic weight scale

1. proton and neutron
2. (true); almost none of the space
3. further
4. (a) $:\ddot{N}\cdot$
   (b) 7e, 7p, 7n
5. (a) $^{56}_{26}Fe$
   (b) $^{35}_{17}Cl$
6. $^{19}_{9}F$
7. 30.1
8. 7, 3, 3, 4, 3

1.4 Avogadro's number and the concept of a mole

1. 1g
2. $(12/N_A)$ grams
3. 11.009305g
4. $4N_A$
5. (a) 2
   (b) 1/2
   (c) 4

1.5 Molecular formulas and molecular weights for discrete molecules

1. (a) 44
   (b) 44
   (c) 18
   (d) 60
2. (a) N_A
   (b) $2N_A$
   (c) about 71
   (d) $(2)(17)N_A = 34N_A$
3. (a) 1/2
   (b) 1
   (c) 1/2
   (d) 2
   (e) (a), (b) and (c)

1.6 The significance of molecular weights for discrete molecules

1.

	Cation	Anion
(a)	K^+	Cl^-
(b)	Li^+	O^{2-}
(c)	Fe^{2+}	O^{2-}

Chapter 1 — Atomic Structure and Stoichiometry

(d) Fe^{3+} O^{2-}
(e) Na^+ Cl^-
(f) Ca^{2+} Cl^-
(g) Na^+ PO_4^{3-}

2. (a) $2N_A$
   (b) N_A
   (c) $3N_A$
   (d) none!

1.7 Distinction between the terms molecular weight, formula weight, and gram atomic weight

1. gram formula weight

1.8 Empirical formulas

1. Fe S
2. Fe_2S_3
3. FeO
4. Fe_2O_3
5. CO
6. (a) 4%
   (b) 32%
   (c) 64%

1.9 The significance of balanced chemical equations

1. (a) $2H_2 + O_2 \to 2H_2O$
   (b) $H_2O_2 \to H_2 + O_2$
   (c) $2Na + 2H_2O \to 2Na^+ + 2OH^- + H_2$
   (d) $Ca(OH)_2(s) + 2HCl(aq) \to Ca^{2+}(aq) + 2Cl^-(aq) + 2H_2O(l)$

2. (a)
       i. 1/2
       ii. 1/4
       iii. 1/4
       iv. 1
   (b)
       i. 1-x
       ii. x/2
       iii. x/2
       iv. 1

1.10 Stoichiometric calculations and the empirical formula

1. Number of moles
2. (a) g/mole
   (b) (unitless)
   (c) g
3. (a) true
   (b) false
   (c) true
   (d) true
4. Let x = g CO, y = g CO_2
   $x + y = 32$

Chapter 1   Atomic Structure and Stoichiometry

$$\frac{x}{28} + \frac{y}{44} = 1$$

Second equation says (moles CO + moles CO_2 = 1) which must be true because we started with 1 mole C. Result is x=21g=3/4 mole CO, so that there is 1/4 mole CO_2 present. *Quick way to do it*: if all C reacted to form CO, product would weigh 28g. If all formed CO_2, product would weigh 44g. Mixture weighs 32g, which is 1/4 the way from all CO to all CO_2, so it must contain 3/4 CO and 1/4 CO_2.

1.11 The limiting reagent and the yield of product

1   (a) NO
      (b) If add more than 1/2 mole additional NO, then O_2 will be limiting.
      (c) 23g
      (d) 0 mole NO+1/2 mole NO_2+1/4 mole O_2=3/4 mole
      (e) 65%

CHAPTER 2: INTRODUCTION TO THE PERIODIC TABLE AND INORGANIC NOMENCLATURE

Scope

The material summarized in Chapter 1 included concepts common to all of chemistry: atoms, compounds, molecules, ions, stoichiometry. We noted that nothing in the formula H_2O indicates that water is molecular, and that nothing in the formula NaCl indicates that table salt is ionic. Also, when we gave the equation

$$2H_2 + O_2 \rightarrow 2H_2O$$

we said nothing about why this occurs. If hydrogen and oxygen are placed in a vessel and a spark is passed through the mixture, water is formed; if water is placed in a vessel and a spark is passed through it, H_2 and O_2 are not formed, even though

$$2H_2O \rightarrow 2H_2 + O_2$$

is a perfectly good balanced chemical equation.

This chapter introduces the part of chemistry which seeks to understand the differences between different substances: why some reactions take place, whereas others do not. The last chapter introduced properties that all substances have. Throughout your study of chemistry you will see an interplay between these two aspects of chemistry, and you will learn a great deal about both of them.

The Periodic Table

Much of what is sometimes called *descriptive chemistry* is embodied in the *periodic table of the elements* (Figure 2.1). It turns out that when the elements are laid out in this manner, in order of increasing atomic number, elements that lie in the same *column*, or *group*, have many similar properties. You will see in Chapter 12 that the particular shape of the periodic table is due to the way in which electrons bind to the nuclei, which is predicted by quantum mechanics; the periodic table, however, in one form or another, has been around since at least the late nineteenth century, and is usually attributed to the Russian chemist Dmitri Mendele'ev, who proposed a version of it in 1869. Its underlying rationale based on the ideas of quantum mechanics only became clear in the 1920's.

The evidence that first led chemists to the idea of the periodic table is exemplified by Figure 2.2, which shows *atomic volume* (*molar volume* is actually the more correct term) of the elements as a function of atomic number. Note that sodium (Na), potassium (K), rubidium (Rb) and cesium (Cs) exhibit higher atomic volumes than their neighbors. Now look at Figure 2.1, and note that these lie under each other on the periodic chart. It appears that starting with any one of these elements and moving to the right on the chart (in the same *row* or *period*), atomic volumes decrease, go through a minimum, then increase again; for example, Figure 2.2 shows that, starting with sodium and going toward the right, atomic volume decreases with magnesium, reaches a minimum with aluminum, then increases steadily through neon. Note also that hydrogen and lithium, which are at the top of the first column of the chart, do *not* have atomic volumes at peaks of Figure 2.2. This points out another aspect of periodic properties, and of the use of the periodic chart to predict chemistry: periodic laws are rarely rigorous laws; they are usually "rules of thumb," which have exceptions. Not all periodic laws are easy to explain, and when they can be explained, one still has to account for the exceptions! An explanation of Figure 2.2 is given in Chapter 13 of the study guide.

The periodic chart can be divided into sections, as exhibited in Figure 2.1, and summarized in the following table.

Sections of the Periodic Chart

Representative Elements	First two and last six columns (groups IA-VIIA, and O)
Active Metals	First two columns
Inert bases	Last column
Nonmetals	Upper-right triangle of groups IIIA-VIIA
Representative metals	lower-left triangle of groups IIA-VIIA
Transition Metals	Middle portion of chart (groups IB-VIIIB)
Inner Transition Metals	Two rows at bottom

Chapter 2 The Periodic Table and Inorganic Nomenclature

Lanthanides First row at bottom (Z=58-71)

Actinides Second row at bottom (Z=90-103)

Outer Transition Metals Remainder of transition elements

Recall that we said earlier that substances <...... (column) share similar properties. We also said that these properties are due to the way the electrons bind to their nuclei. For the representative elements, the number of valence electrons is given by the Roman numeral of the group; thus Na and K have one valence electron, and F and Cl have seven. The exception to this rule is the inert gases, group O, which have eight valence electrons.

In groups IIIA-VIA as one goes down in a group one passes from non-metals to metals (Figure 2.1); thus lead (Pb) is in the same group as carbon. Clearly, not all the properties of elements within the same group act the same! In fact, there are some properties that do not follow vertical lines on the periodic chart, but which instead follow lines parallel to the division between metals and non-metals within the representative elements. For example, carbon and iodine (I) lie on such a line, and though they are usually thought of as non-metals, both have crystal forms (graphite, for carbon) which exhibit metallic luster.

The Active Metals

Metals have the following properties:

— they are malleable and ductile

— they are good conductors of heat and electricity

— they shine (they exhibit "metallic luster")

The reason for some of these is discussed in Chapter 21, but they all have to do ultimately with the fact that an atom of a metal holds its valence electrons only loosely. This fact also allows metal atoms to give up one or more valence electrons to become cations. Thus, in ionic substances, such as NaCl, it is the metal which has the positive charge.

Chapter 2   **The Periodic Table and Inorganic Nomenclature**

The *active metals* are those in the first two columns of the periodic chart, and consist of the *alkali metals* (IA) and the *alkaline earth metals* (IIA). These have one and two valence electrons, respectively, and almost always exist in compounds as ions with +1 and +2 charge, respectively, so that lithium chloride is LiCl, an ionic substance where Li exists as Li^+ and Cl exists as Cl^-, and calcium chloride is $CaCl_2$, where Ca exists as Ca^{2+}. The reason these metals are called "active" is that they very readily give up their valence electrons, and the pure metals react readily — sometimes violently — with oxygen or even with water, according to the following typical reactions

$$2Li + O_2 \rightarrow 2LiO$$

$$2Li(s) + 2H_2O(l) \rightarrow 2Li^+(aq) + 2OH^-(aq) + H_2(g)$$

Because these metals react so readily with commonly available substances, they are not found free in nature, but are always found in compounds, in which they are in their ionic forms.

The Noble Gases

These are also called *rare* or *inert* gases; the terms "noble" and "inert" derive from the fact that these gases have very little tendency to combine with other elements. The use of the term "noble" to indicate this seems to evince a rather Victorian attitude toward chemical combination on the part of the early chemists, who would probably have been shocked and dismayed to learn that xenon and krypton do in fact form compounds with fluorine and oxygen.

The Nonmetals

Nonmetals not only hold onto their electrons more tightly than metals; some of them — the ones to the right of the periodic chart — have a tendency to attract additional electrons and become anions. Thus, as we have pointed out, Cl and F (which are *halogens*) tend to exist as the Cl^- (chloride) and F^- (fluoride) ions in ionic compounds. O and S (which are *chalcogens*) have a tendency to exist as O^{2-} (oxide) and S^{2-} (sulfide) ions. All of these participate in many molecular compounds as well, however; for example, HCl (hydrogen chloride, or hydrochloric acid) and H_2O are both molecular, rather than ionic. In addition, oxygen, in particular, also forms the ions O_2^{2-} (peroxide) and O_2^- (superoxide).

The halogens are, for the most part, found in the combined state; when isolated in pure form, they exist as diatomic molecules (F_2, Cl_2, etc.). Oxygen (a chalcogen) also exists as O_2, but also as the reactive gas O_3, ozone, which is an *allotrope* (a different form) of the element.

Metallic character is not an all-or-nothing thing; we mentioned that graphite resembles a metal in some ways (luster, good conductor of heat and electricity), but not in others (malleability and ductility). The elements bordering the zig-zag line separating the metals and non-metals in the periodic table are *semimetals*, and exhibit some metallic properties.

Hydrogen

Hydrogen can lose an electron to form the cation H^+, or gain one to form the hydride anion, H^-. Actually, H^+ is a naked proton, and is never found uncombined; for example, in aqueous solution $H^+(aq)$ is sometimes written $H_3O^+(aq)$, the result of the reaction

$$H^+ + H_2O \rightarrow H_3O^+$$

to indicate that it is combined with the surrounding molecules. In the free state hydrogen exists as $H_2(g)$. Almost all of the mass in the universe exists in the form of H atoms, even though hydrogen comprises only about 1% by weight of the earth's crust.

The Transition Metals

The outer transition elements all have rather similar chemistry, from a gross point of view. They are all metals, for example, and many are commonly found both combined with other elements (in ores) and also in the free state; thus they are not as reactive as the active metals. On closer examination, however, the subtle distinctions between the behaviors of the transition metals have vast implications for both biological and industrial processes. For example, hemoglobin contains iron bound to long organic chain molecules (proteins), and no other metal bound to the same protein succeeds in performing the oxygen-carrying function that the iron in hemoglobin can carry out. The same sort of thing is true in industrial processes, where *catalysts* (agents added in small quantities to speed up a reaction) are often composed of specific combinations of transition metals with other substances.

Ionic and Molecular Compounds

Ionic solids have high melting points; this is because the Coulombic forces holding adjacent ions together are very strong. Molecular solids, on the other hand, have low melting points. Each molecule is neutral, and the forces holding adjacent neutral molecules together are much weaker than the forces attracting two ions of unlike charge.

Solutions of ionic compounds in water conduct electricity. This is because the ions can move toward positive and negative electrodes immersed in the solution. Molecular substances usually have low conductivity in water solutions; notable exceptions to this are the acids, such as HCl, which are molecules, but which dissociate into ions — H^+ plus Cl^-, in this case — when placed in water.

The fact that the crystal forces are stronger in ionic than in molecular solids also makes ionic solids hard and brittle, compared to molecular solids. You should be aware that there are two additional kinds of solids: metallic solids and network solids. A metallic solid can be viewed as an array of metal ions sharing a common "sea" of electrons; it is the mobility of these electrons that makes metals good conductors of electricity. In network solids the bonds between the atoms are of the same sort as the bonds between atoms within a molecule, but the bonds form a network that encompasses the entire crystal, which can therefore be viewed as one big molecule. Network solids are very hard and melt very high, but the melt does not conduct electricity, and these solids tend not to dissolve in common solvents. One (relatively!) common example is diamond, an allotrope of carbon.

Inorganic Nomenclature

Detailed rules for naming inorganic compounds are given in the text; only the most useful and simple ones are summarized here.

Cations are usually named directly after the parent metal; Na^+ is the sodium ion. When a metal can form more than one kind of ion, the ion is named after the metal, with the charge given in Roman numerals following the name; thus Fe^{2+} is the iron(III) ion, and Fe^{3+} is the iron(III) ion. An older naming system still in use is that when a metal can form two ions, the ion name has as its root the Latin name of the element, "fer" for iron, followed by "-ic" for the ion with the higher charge, or "-ous" for the ion with the lower charge. Thus Fe^{3+} is the ferric ion, and Fe^{2+} is the ferrous ion; Cu^{2+} is the cupric ion, and Cu^+ is the cuprous ion. Of course, presented with Fe^{2+} and Cu^{2+}, you have to know that iron also forms Fe^{3+} and that copper also forms Cu^+ in order to know what names to use.

You need to memorize the names and formulas of the following polyatomic ions: NH_4^+ (ammonium ion), H_3O^+ (hydronium ion), Hg_2^{2+} (mercurous ion).

Names of monatomic anions are derived from the element names, with "-ide" at the end: F^- is fluoride, H^- is hydride, O^{2-} is oxide, and so on. *Oxyanions*, which consist of a central atom bonded to oxygens, have names ending with "-ite" or "-ate," the latter being reserved for the form with more oxygens, if it exists: SO_4^{2-} is sulfate, and SO_3^{2-} is sulfite. Other examples are given in Table 2.4. Some especially important examples which must be memorized are OH^- (hydroxide), CN^- (cyanide), O_2^{2-} (peroxide), MnO_4^- (permanganate), CO_3^{2-} (carbonate), NO_3^- (nitrate), SO_4^{2-} (sulfate) and PO_4^{3-} (phosphate).

An ionic compound is named by giving first the cation name then the anion name, as in $Ca_3(PO_4)_2$, calcium phosphate. Furthermore, from the name calcium phosphate it is possible to construct the above formula provided you know the charges on the ions. The key to doing this is the fact that the overall substance must be electrically neutral. Since calcium ion has a charge of +2 and since phosphate has a charge of -3, the way to achieve neutrality most simply is to have three Ca^{2+}'s for each two PO_4^{3-}'s.

Most molecular compounds known are *organic compounds*, meaning that they contain carbon. These are named in special ways (Chapter 23). Simple inorganic molecular compounds are named by adding the prefixes di-, tri-, tetra, pent-, etc. to the atom names, depending on how many of the particular kind of atom the molecule contains. Examples are P_2O_5, diphosphorus pentoxide, SF_4, sulfur tetrafluoride and N_2O, dinitrogen oxide. Some of these also have older names, as shown in Example 2.3. In both writing the formula and making up the name, the element that is closer to the metals on the periodic chart is given first, so that IF_5, iodine pentafluoride, not F_5I, pentafluorine iodide, is the correct description of this compound.

Nonstoichiometric Compounds

These are compounds whose formula can not be given in a form like $A_m B_n$, with m and n small integers. For example, what we usually write as "FeO" (ferrous oxide) actually has the composition range $Fe_{.84-.94}O$. In pure FeO there would be one Fe^{2+} ion for each O^{2-} ion. In real "FeO" some of the iron ions are iron(III), rather than iron(II). In order to maintain charge neutrality, some iron sites in the crystal are vacant. Such a *vacancy* is one kind of crystal *defect*; you will learn about others in Chapter 21. To illustrate this, suppose we start with a sample of pure FeO which contains exactly 100 Fe^{2+} ions and exactly 100 O^{2-} ions. Suppose now that two Fe^{2+} ions are changed to Fe^{3+}. In order to still have two hundred positive charges to balance the two hundred negative charges from the oxygens, we would have to get rid of one Fe^{2+} ion, so that the sample would contain 100 oxygens and 99 irons, and have the formula $Fe_{.99}O$.

Non-stoichiometric compounds are also called *berthollides*, after the French chemist Claude Louis Berthollet, who refused to believe that chemical compounds were defined by fixed ratios of elements. In fact, in Berthollet's time methods of chemical analysis were crude enough so that this hypothesis (usually attributed to Dalton) was not obvious. As chemical analysis improved, it was applied mostly to simple substances, such as pure gases, which in fact *did* turn out to exhibit fixed elemental ratios, and our entire system of chemical symbols and names embodies this concept; a formula like $Fe_{.9}O$ still raises eyebrows. Berthollides are by far the exception in the chemical world, and it is interesting to speculate on how differently the science of chemistry would have evolved had their existance been proved in the early days of the formulation of the atomic theory.

Chapter 2   The Periodic Table and Inorganic Nomenclature

Questions

Note: You may consult a periodic chart.

2.1 The sections of the periodic table

1. (a) A row of the periodic chart is also called a _____.

   (b) A column of the periodic chart is also called a _____.

   (c) Elements in the same column have the same number of _____.

   (d) Elements in the same (row, column)* tend to have similar properties.

2. Consulting a periodic chart which is *not* labeled with these categories, tell whether each of the following is (A) an active metal; (B) a representative metal; (C) a non-metal; (D) a transition metal; (E) a semi-metal; or (F) an inert gas.

   _____(a) fluorine

   _____(b) cesium

   _____(c) ytterbium

   _____(d) antimony

   _____(e) radon

   _____(f) molybdenum

   _____(g) gold

   _____(h) lead

   _____(j) strontium

   _____(k) cobalt

2.2 The active metals

1. Write a balanced equation for the reaction of strontium with water.

---

* For this sort of question, pick which of the choices in parenthesis best completes the statement.

Chapter 2   The Periodic Table and Inorganic Nomenclature

2. Cesium ion has a charge of _____.

3. Write a balanced equation for

    (a) The formation of sodium peroxide from the elements.

    (b) The formation of rubidium superoxide from the elements.

4. Give the formula for calcium nitride.

2.3 The rare gases

1. Which rare gases form chemical compounds, and with which element or elements?

2. Write chemical formulas for the following gases

    (a) nitrogen

    (b) neon

    (c) fluorine

    (d) argon

    (e) oxygen

    (f) krypton

2.4 The nonmetals

1. Pure carbon can exist as graphite or as diamond. These two are _____ of carbon.

2. When nonmetals form ions, they take on a (positive, negative) charge.

3. Name two elements which occur naturally in liquid form, and give their formulas.

Chapter 2   The Periodic Table and Inorganic Nomenclature

4. (a) Name two halogens

   (b) Name two chalcogens

5. Write balanced chemical equations for

   (a) the formation of magnesium chloride from the elements

   (b) the formation of hydrogen sulfide from the elements

   (c) the formation of ozone from oxygen

   (d) the formation of sulfuric acid from SO_2 in the atmosphere

2.5 The unique position of hydrogen

1. Give the formula for

   (a) hydrogen bromide

   (b) lithium hydride

   (c) the hydronium ion

2.6 The transition metals

1. A given transition metal will tend to form (more, fewer) kinds of ions than a given active metal.

2. A transition metal is (more, less) likely to be found in its pure state in nature than an active metal.

Chapter 2 — The Periodic Table and Inorganic Nomenclature

3. the transition elements exhibit (greater, less) chemical diversity than the representative elements.

2.7 Ionic crystalline solids
and 2.8 Molecular compounds

1. If an exothermic reaction is taking place in a beaker, the beaker will get (warmer, cooler).

2. Which of the following are ionic?

   _____(a) HI

   _____(b) SrF_2

   _____(c) SO_3

   _____(d) Fe_2O_3

   _____(e) NaH

3. Ionic solids tend to melt (lower, higher) than molecular solids, to be (harder, softer) and to form solutions that conduct electricity (better, more poorly).

2.9 Some simple inorganic nomenclature

1. Name the following substances

   (a) PCl_3

   (b) NI_3

   (c) Na_3P

   (d) $AgNO_3$

   (e) SO_2

   (f) NO_2

   (g) $(NH_4)_2SO_4$

2. Provide formulas for the following.

   (a) cobalt(II) chloride

   (b) ferric oxide

   (c) strontium phosphate

Chapter 2   The Periodic Table and Inorganic Nomenclature

    (d) mercurous chloride

    (e) cupric sulfate

    (f) chromium(III) oxide

    (g) potassium permanganate

    (h) sodium hydrogen phosphate

    (j) nitrogen tetroxide

    (k) dinitrogen pentoxide

2.10 Non-stoichiometric compounds

1. Suppose a sample of "FeS" has the exact composition $Fe_{.9}S$.

    (a) What percent of iron ions are present as Fe^{3+}?

    (b) What percent of iron sites in the crystal will be vacant?

Chapter 2 — The Periodic Table and Inorganic Nomenclature

Answers

2.1 The sections of the periodic table

1. (a) period
   (b) family or group
   (c) valence electrons
   (d) column

2. (a) C
   (b) A
   (c) D
   (d) E
   (e) F
   (f) D
   (g) D
   (h) B
   (j) A
   (k) D

2.2 The active metals

1. $Sr + 2H_2O \rightarrow Sr^{2+} + 2OH^- + H_2$

2. $+1$

3. (a) $2Na + O_2 \rightarrow Na_2O_2$
   (b) $Rb + O_2 \rightarrow RbO_2$

4. Ca_3N_2

2.3 The rare gases

1. Xe and Kr, with F and O.

2. (a) N_2
   (b) Ne
   (c) F_2
   (d) Ar
   (e) O_2
   (f) Kr

2.4 The nonmetals

1. allotropes

2. negative

3. Hg (mercury) and Br_2 (bromine)

4. (a) Fluorine and iodine are possibilities
   (b) Oxygen and tellurium are possibilities

5. (a) $Mg + Cl_2 \rightarrow MgCl_2$
   (b) $H_2 + S \rightarrow H_2S$
   (c) $3O_2 \rightarrow 2O_3$
   (d) $2SO_2 + 2H_2O + O_2 \rightarrow 2H_2SO_4$

2.5 The unique position of hydrogen

1. (a) HBr
   (b) LiH
   (c) H_3O^+

Chapter 2   The Periodic Table and Inorganic Nomenclature

2.6 The transition metals

1. more
2. more
3. less

2.7 Ionic crystalline solids
and 2.8 Molecular compounds

1. warmer
2. (b), (d) and (e)
3. higher; harder; better

2.9 Some simple inorganic nomenclature

1. (a) phosphorus trichloride
   (b) nitrogen triiodide
   (c) sodium phosphide
   (d) silver nitrate
   (e) sulfur dioxide
   (f) nitrogen dioxide
   (g) ammonium sulfate

2. (a) $CoCl_2$
   (b) Fe_2O_3
   (c) $Sr_3(PO_4)_2$
   (d) Hg_2Cl_2
   (e) $CuSO_4$
   (f) Cr_2O_3
   (g) $KMnO_4$
   (h) Na_2HPO_4
   (j) N_2O_4
   (k) N_2O_5

2.10 Nonstoichiometric compounds

1. (a) Suppose we start with 90 Fe ions and 100 S^{2-}; all the Fe ions contribute at least +2 to the charge, and if 20 contribute +3 instead then all their charges add up to 200 (200=70×2 + 20×3), so 20 Fe^{3+} is (20)(100)/90≈22%.
   (b) Obviously, we must have started with as many Fe as S sites; directly from the formula, 10% of the Fe sites must be vacant.

CHAPTER 3: THE GAS LAWS AND STOICHIOMETRY INVOLVING GASES

Scope

This chapter further develops the ideas of Chapter 1; there will be much emphasis on the quantitative aspects of chemical combination, and little on why the particular reactions used as examples in fact occur. For example, in this chapter you will see that when we write

$$2H_2(g) + O_2(g) \rightarrow 2H_2O(g)$$

it means not only that two hydrogen molecules and one oxygen molecule react to form two water molecules, and not only that two moles of H_2 react with one of O_2 to form two moles of H_2O, but also that two *volumes* of H_2 react with one *volume* of O_2 to form two *volumes* of H_2O, provided that the pressure and temperature are constant throughout, and provided, in this case, that the pressure and temperature are such that water is a vapor, not a liquid. (If *liquid* water is formed, the assertion about the volumes of H_2 and O_2 remains true, however.)

Avogadro

Gay-Lussac found in the early nineteenth century that when gases react chemically, the combining volumes are in ratios of small whole numbers; thus, in the above example, $V_{H_2}/V_{O_2} = 2$ for the formation of water. Note that if hydrogen peroxide were being formed

$$H_2 + O_2 \rightarrow H_2O_2$$

V_{H_2}/V_{O_2} would be equal to unity; thus the combining ratio of gases depends on the chemical reaction taking place. Also, if hydrogen and oxygen were reacting to form H_2O and H_2O_2 simultaneously, V_{H_2}/V_{O_2} could be anything between one and two; ratios of combining volumes are guaranteed to be ratios of small whole numbers only if a single chemical reaction is taking place by itself.

Amadeo Avogadro proposed in 1811 that Gay-Lussac's *law of combining volumes* was due to the fact that equal volumes of any two gases at the same temperature (T) and pressure (P) contain the same number of molecules. In order for this to be true, certain geseous elements, such as hydrogen and oxygen, must exist as polyatomic molecules, not isolated atoms. To see this, suppose hydrogen and oxygen were monatomic. Then the balanced chemical equation for the formation of water would read

$$2H + O \rightarrow H_2O$$

which, if Avogadro was correct, would imply that three volumes of H and O would form one volume of H_2O. This is still in accord with Gay-Lussac, but the problem is that in fact two, not one, volumes of product are formed. Dalton, the founder of the modern atomic theory, was Avogadro's greatest opponent. He resisted the idea that elements could be polyatomic. Dalton argued that if two hydrogen *atoms* and one oxygen *atom* combine to form two water *molecules*, then the atoms of oxygen must be split in the process; each water molecule must contain one atom of H and a half an atom of O. Since Dalton correctly believed that atoms remain intact in chemical interactions, he argued that equal volumes of gases *cannot* contain equal numbers of atoms or molecules.

Of course, if elemental gases can be diatomic, the objection disappears. When we write

$$2H_2 + O_2 \rightarrow 2H_2O$$

we allow the oxygen *molecule* to be split. The idea that elements can be polyatomic was Avogadro's great contribution, but, as the text mentions, it was not accepted until almost fifty years after he proposed it. One possible reason for this was that even in Avogadro's time chemists were thinking about what kinds of forces must be responsible for chemical combination. Early experiments had been performed with electricity, and the prevailing view (which turned out to be correct) was that the attraction between atoms in molecules had something to do with the attraction between positive and negative charges. It was felt that different atoms must possess different charges, accounting for molecule formation. This turned out to be incorrect, but from this point of view it did not make sense that two hydrogen atoms should combine to form a hydrogen molecule.

This objection eventually yielded to Stanislao Cannizzaro's demonstration (1858) that if Avogadro's hypothesis is accepted, then a self-consistent scale of atomic weights can be determined. For example, it was known, as described above, that two volumes of hydrogen combine with one volume of oxygen to form two volumes of water. It was also known that one gram of hydrogen combines with eight grams of oxygen to form nine grams of water. Cannizzaro pointed out that these two *independent* results make sense if an H atom has a mass of 1 and an O atom has a mass of 16, and if hydrogen exists as H_2, oxygen as O_2 and water as H_2O. Furthemore, using the result for O_2, the data on the reaction

$$N_2 + 2O_2 \rightarrow 2NO_2$$

can be understood by assuming that nitrogen has a relative mass of 14, and, in fact, exists as N_2. All the elements that react to form gases can have atomic weights assigned in this manner. By Cannizzaro's time enough experimental data had been collected on enough reactions to convince chemists of the truth of Avogadro's hypothesis, regardless of the fact that the theorists of the day could not explain it. The mystery of why it is that two hydrogen atoms combine to form H_2 was not resolved until the 1920's; the solution is discussed in Chapter 14.

If we accept that equal volumes of gases contain equal numbers of atoms or molecules, then a mole of any gas must exhibit the same volume, provided that P and T are specified. The question then arises, "How many particles does a mole of gas contain?" We call this number Avogadro's number (N_A), but you should be aware that accurate determinations of N_A became available only in the early twentieth century. Recall that $N_A = 6.02 \times 10^{23}$. It is also useful to know what the volume is that a mole of gas takes up. This turns out to be 22.4L at *standard temperature and pressure (STP)*, defined as 0°C and 1atm pressure. This value is very useful in solving numerical problems; for example, recall that in Chapter 1 we determined from elemental composition that the empirical formula of ethane was CH_3, but that it would require an independent determination of molecular weight to obtain the molecular formula. Avogadro's law supplies one possible method of determining molecular weight. Suppose I find that 1L of ethane at STP weighs 1.3g; then 22.4L must weigh (1.3g/L)(22.4L) = 29g; therefore, the molecular weight is about 29g. Within experimental error, this is equal to 30g, the weight of two moles of CH_3 fragments, so that ethane must have the formula C_2H_6.

As you will see in greater detail in Chapter 4, Avogadro's law is not 100% correct; it is one of a set of laws that pertain exactly only to *ideal gases*. There are no ideal gases in the real world, but most gases come pretty close, at pressures and temperatures not too far removed from STP. The higher the temperature and the lower the pressure, the more ideal a gas becomes. The rest of the laws discussed in

Chapter 3 — The Gas Laws and Stoichiometry Involving Gases

this chapter pertain exactly only to ideal gases, and are good approximations for all common gases, even at STP. In Chapter 4 you will see what makes real gases nonideal, and how the gas laws can be modified to take the nonideality into account.

Boyle

Boyle's law states that, at constant temperature, the pressure and the volume of a gas sample are inversely related.

$$PV = k$$

where k is a constant. (k is constant only for a given temperature and a given number of moles of gas.) To understand Boyle's law, you have to understand what P and V are. I assume you know what volume is; this leaves pressure, which is defined as force divided by area, and force is mass times acceleration.

$$P = F/A$$

$$F = ma$$

Atmospheric pressure is the force generated by the action of the acceleration due to gravity on the mass of the air in the atmosphere. A gas sample experiences the same pressure througout. It is a characteristic of gases and liquids, but not solids, that the molecules move so as to equalize the pressure in all directions; thus a steel cylinder standing on a desktop experiences atmospheric pressure on its top and sides, but on its bottom experiences an additional pressure equal to its mass times the acceleration due to gravity, divided by its cross-sectional area. Without information about the internal mechanics of the rod, it is not possible to say anything about the pressure at an arbitrary point within it; however, within the immediately surrounding atmosphere, the pressure is the same everywhere.

The SI unit of pressure is the *pascal* (Pa)

$$1\text{Pa} = 1\frac{\text{N}}{\text{m}^2} = 1\frac{\text{kg·m/s}^2}{\text{m}^2} = 1\frac{\text{kg}}{\text{m·s}^2}$$

where N is the *newton*, the SI unit of force. Other common units are the *bar*, equal to 10^5 Pa, the *atmosphere* (atm) which is about 1% bigger than a bar, and the *mmHg* or *torr*, equal to (1/760)atm. This last unit is directly related to the use of mercury barometers to measure pressure, as illustrated in

Chapter 3   **The Gas Laws and Stoichiometry Involving Gases**

Figure 3.1.

The Boyle's law equation, PV=k, is the equation of a hyperbola, if P is plotted against V (Figure 3.1); different values of T give different values of k, and each gives its own hyperbola. Each curve is called an *isotherm* (meaning "same temperature"). It is unusually easiest to see whether an equation adequately describes experimental behavior if a method can be found to plot the data so that the predicted data points lie on a straight line. One way to do this for Boyle's law is to plot PV against P (Figure 3.5 and 3.6); since PV is predicted to not change with P, the points should all lie on the horizontal straight line whose y-intercept is k. For one mole of an ideal gas at STP, V=22.414L, so that PV=(1atm)(22.414L)=22.414L·atm. As you can see from Figure 3.6, Boyle's law can be a few percent off at P=1atm, but at lower pressures real gases converge toward ideal behavior.

Charles and Gay-Lussac

So far we have presented two ideal gas laws:

$$\text{Avogadro: } V = k_1 n \text{ (constant T,P)}$$

$$\text{Boyle: } V = \frac{k_2}{P} \text{ (constant n,T)}$$

(Here n stands for the number of moles.) The third ideal gas law is due to the French chemists Jacques Alexandre Charles and Joseph Louis Gay-Lussac, and can be expressed

$$V = k_3 T \text{ (constant P,n)}$$

where T is the temperature on a new kind of scale, called an *absolute temperature scale*. If we let t represent temperature in °C, then what is actually found experimentally, at constant P and n, is that

$$V = mt + b$$

This says that if the volume is plotted against the temperature in °C, a straight line results (Figure 3.7). b represents the y-intercept of the data (where the x-coordinate, namely t, equals zero), and can be symbolized V_o; this can be seen by setting t=0; we then have V=b=V_o. What Charles and Gay-Lussac found was that when the pressure was changed, or when the quantity of gas was changed, both V_o and m changed; however, these changed together in such a way that the *x-intercept* of the straight line

Chapter 3 — The Gas Laws and Stoichiometry Involving Gases

remained constant. Starting with Equation 3-17

$$V = V_o(1+\alpha t)$$

where $\alpha = m/V_o$, we can calculate this common value. The equation can be re-expanded to give

$$V = \alpha V_o t + V_o$$

which is identical to $V=mt+b$, with $m=\alpha V_o$ and $b=V_o$. To get the x-intercept — the value of t where the curve crosses the x-axis — we set y=0, giving $0=mt_o+b$, so that $t_o=-b/m=-(V_o)/(\alpha V_o)=-1/\alpha$. α is found to be equal to $1/273.15°C$ for all gases. As shown in Equation 3-20, this allows the ratios of gas volumes at two temperatures to be given as

$$\frac{V_1}{V_2} = \frac{1+\alpha t_1}{1+\alpha t_2} = \frac{1/\alpha + t_1}{1/\alpha + t_2} = \frac{273.15+t_1}{273.15+t_2} = \frac{T_1}{T_2}$$

where the capital T's now represent the temperature on new scale, the *Kelvin scale*, in which K=273.15 + °C; thus 0°C=273.15K. (Note that we do not use the "°" sign when denoting Kelvin temperatures.)

The temperature 0K=-273.15°C is called the *absolute zero*. In terms of our gas experiments, it is the temperature at which any initial sample of gas would shrink to zero! In fact, real substances liquify and solidify before 0K is achieved, and in addition you should recall that gas behavior becomes less and less ideal at lower temperatures, so that Charles' law probably ceases to hold even before liquifaction sets in. However, absolute zero has a unique role in chemistry and physics, one not confined to the study of the volumes of gases. As you will see in later chapters, temperature is directly related to the average kinetic energy of the atoms or molecules comprising a substance. Kinetic energy decreases, and temperature decreases, when molecules move more slowly, and at absolute zero molecular motion essentially ceases; for this reason, 0K is the coldest temperature theoretically approachable.

PV=nRT

The three ideal gas equations given at the beginning of the last section can be combined into a single equation. If V is directly proportional to T (Charles) and to n (Avogadro), and inversely proportional to P (Boyle), then we have

$$V \propto \frac{nt}{P}$$

Chapter 3 — The Gas Laws and Stoichiometry Involving Gases

$$PV \propto nT$$

$$PV = nRT$$

where R is a universal constant of proportionality known as the *gas constant*. In one common set of units, $R = 0.0821 \, L \cdot atm/mol \cdot K$. Like the other ideal gas laws, $PV=nRT$ holds exactly only at low pressures and high temperature.

This equation has many uses; among the more important ones is the writing of freshman chemistry problems; the text gives several illustrative examples (3.6-3.8), which you should study carefully. Here are some "helpful hints" for solving ideal gas problems.

1. Sometimes you do not need all the information contained in the formula $PV=nRT$, and in these cases you can sometimes save work by using Boyle's, Charles' or Avogadro's law, often in one of the following forms:

$$\text{Boyle:} \quad P_1 V_1 = P_2 V_2 \quad \text{(constant T, n)}$$

$$\text{Charles:} \quad \frac{V_1}{T_1} = \frac{V_2}{T_2} \quad \text{(constant P, n)}$$

$$\frac{P_1}{T_1} = \frac{P_2}{T_2} \quad \text{(constant V, n)}$$

$$\text{Avogadro:} \quad \frac{V_1}{n_1} = \frac{V_2}{n_2} \quad \text{(constant P, T)}$$

$$\frac{P_1}{n_1} = \frac{P_2}{n_2} \quad \text{(constant V, T)}$$

For example, if a gas starts out at 0°C, and the temperature is raised until the volume is doubled, at constant pressure, what will be the temperature?

Since P and n are constant, it is convenient to use Charles' law. We are told that $P_2 = 2P_1$; $T_1 = 273K$, and we are asked to find T_2. This is easily done.

$$\frac{P_1}{T_1} = \frac{P_2}{T_2}$$

$$\frac{P_1}{273K} = \frac{2P_1}{T_2}$$

Chapter 3   The Gas Laws and Stoichiometry Involving Gases

$$T_2 = (2)(273K) = 546K$$

2. Generally, problems that ask a question about a *given* state of an ideal gas — rather than about a *change* of state — require the use of PV=nRT. For example, 2.0g of nitrous oxide is placed in a sealed 1.0L flask. What will the pressure be at 20°C?

$$P = \frac{nRT}{V}$$

$$= \left(\frac{2.0g}{44g/mole}\right)\frac{(0.00821 L\cdot atm/mol\cdot K)(293K)}{1.0L}$$

$$= 1.1 atm$$

Note the explicit inclusion of units for all quantities; very often the correct cancellation of units (to give pressure in atmospheres, in this case) is the best way to check that the problem is set up correctly.

3. Very often gas law problems are combined with stoichiometry problems. This usually involves at least implicit use of Avogadro's law, since the number of moles of gas can change. In complicated cases these problems usually lead to sets of simultaneous equations, but here is a simpler example.

2.0 moles O_2, 2.0 moles H_2 and 2.0 moles Ar are pumped into a tank, and the total pressure is 10atm. Then a spark is passed through the mixture, forming liquid water. What is the new pressure, assuming constant temperature?

H_2 is the limiting reagent. The reaction

$$2H_2 + O_2 \rightarrow 2H_2O$$

goes to completion, so that all the H_2 and one mole of O_2 are used up. Nothing happens to the Ar. The following table pertains. (This sort of table can be very useful in solving gas law problems.)

	n_{H_2}	n_{O_2}	n_{Ar}	n_{total}
before	2	2	2	6
after	0	1	2	3

The volume taken up by the liquid water is negligible, so we can write

Chapter 3 — The Gas Laws and Stoichiometry Involving Gases

$$\frac{P_1}{n_1} = \frac{P_2}{n_2}$$

$$\frac{10\,\text{atm}}{6\,\text{mol}} = \frac{P_2}{3\,\text{mol}}$$

$$P_2 = 5\,\text{atm}$$

Gas law problems have infinite variety, and can be very tricky, so you should work many of them.

The equation PV=nRT is an example of an *equation of state*. The *state* of a substance is specified by all the variables which serve to uniquely identify a sample of the substance; for example, if n, P, V and T are identical for two samples of the same gas, then all the other physical properties of the gas — such as the electrical conductivity, the viscosity, the color, the mass, the energy, and so on — will also be identical. An equation of state is an equation which relates all the parameters that specify the state. Note that if any three of n, P, V and T are specified, the fourth can be calculated from the equation of state. Another useful form of the ideal gas equation of state (besides PV=nRT) involves the density, δ, and molecular weight, M, instead of V and n. If m is the mass, then $\delta = m/V = nM/V$, so that $n/V = \delta/M$. This gives

$$P = \frac{nRT}{V} = \frac{\delta RT}{M}$$

This form of the ideal gas law is useful in certain kinds of problems (Example 3.11).

Dalton

We have implictly been assuming still an additional gas law: namely, that a mixture of one mole of gas A with two moles of gas B acts just like three moles of a single ideal gas, provided that A and B do not react. The view that is taken (and which will be easier to understand after you read Chapter 4) is that in such a mixture, both gases occupy the full volume of the container, but have their own *partial pressures*. For our gases A and B, if the total pressure is given by P_T = 6atm, then P_A = 2atm and P_B = 4atm. This behvior is described by *Dalton's law of partial pressures*.

$$P_A = \frac{n_A}{n_A + n_B} P_T = X_A P_T$$

If there had been more than two gases present, then all of their n's would have been included in the denominator of this expression. X_A denotes the *mole fraction* of A, the fraction of all the gas molecules that are A molecules. Note that

$$P_A + P_B = \frac{n_A}{n_A+n_B} P_T + \frac{n_B}{n_A+n_B} P_T = P_T$$

so that the total pressure is the sum of the partial pressures. Thus we can write

$$P_T = \sum_i P_i = \sum_i \frac{n_i}{\sum_j n_j} P_T = \sum_i X_i P_i$$

$$1 = \sum_i X_i$$

Vapor Pressure

If a liquid is placed in a sealed, initially evacuated container and maintained at some temperature, T, it will evaporate until either the whole sample evaporates or until the pressure in the vessel reaches the *vapor pressure* of the pure substance, sometimes denoted P_o. Evaporation ceases as soon as either of these two conditions is fulfilled. P_o depends on the temperature, and can be obtained from tables, such as Table 3.3. Let us illustrate this behavior using Example 3.12 in the text. The products of the combustion of 1.5388g of benzoic acid (M=122.12g/mole) are contained in a 6.00L vessel. Since 1.5388g/(122.12g/mole)=0.012601 moles of benzoic acid are burned, we know from the balanced equation that three times this number of moles of water are produced, or 0.038 moles. (For brevity, I will confine my calculations to two significant figures.) Now the vapor pressure of water at 298K (25°C) is 23.8 mmHg = 23.8 mmHg/(760 mmHg/atm)=0.031atm. At this partial pressure, the volume that 0.038 mole of any ideal gas would take up is given by

$$V = \frac{nRT}{P}$$

$$= \frac{(0.038 \text{mol})(0.0821 \text{L·atm/mol·K})(298\text{K})}{0.031 \text{atm}}$$

$$= 30\text{L}$$

Since only 6.00L are available, some of the water must exist as liquid, and $P_{H_2O} = P_o = 23.8$ mmHg. But now suppose the sample of benzoic acid burned had been ten times smaller, namely 0.15388g. Then the volume taken up by the complete evaporation of the sample, at its equilibrium vapor pressure, would have been only 3.0L, rather than 30L, and there would have been no liquid water in the 6L container. The partial pressure of this 0.0038 mole of water vapor would then have been given not by P_o, but by

$$P = \frac{nRT}{V}$$

$$= \frac{(0.0038 \text{mole})(0.0821 \text{L·atm/mol·K})(298\text{K})}{6.00\text{L}}$$

$$= 0.015 \text{atm} \left(\frac{760 \text{mmtg}}{1 \text{atm}} \right)$$

$$= 12 \text{mmHg}$$

The moral of this story is that before you can assume that the partial pressure of a substance in a sealed vessel is equal to its equilibrium vapor pressure, you have to make sure that some liquid, as well as vapor, is present. If not, then the partial pressure of the substance is given by the ideal gas law, and is less than the vapor pressure; in no case can the partial pressure of a substance at equilibrium exceed its vapor pressure. But whether the partial pressure is obtained from the vapor pressure or from the ideal gas law, the vapor behaves just like any other gas in the application of Dalton's law of partial pressures and the other ideal gas laws.

Chapter 3 — The Gas Laws and Stoichiometry Involving Gases

Questions

Note: Leave your answers in terms of constants, such as N_A and R, where applicable.

3.1 Avogadro's law and its use in determining molecular formulas

1. How many molecules are in 11.2L of gas at STP?

2. How many moles are in 11.2L of gas at STP?

3. If 11.2L of a gas weighs 46g at STP, what is its molecular weight?

4. If the gas in Question 3 has the empirical formula NO_2, what is its molecular formula?

3.2 The pressure of gases: definition, measurement and units

1. No matter how good a vacuum I create, I cannot use it to pump mercury up a height greater than about 760mm. (True, False).

2. The deepest point in the ocean is the Marianas Trench, which is about 30,000 feet deep. Which of the following is closest to the pressure at the bottom of the Marianas Trench?

    _____ 10atm

    _____ 100atm

    _____ 1000atm

    _____ 10000atm

3.3 Boyle's law

Chapter 3 — The Gas Laws and Stoichiometry Involving Gases

1. If a balloon containing 1L of air at STP is weighed and allowed to fall to the bottom of the Marianas Trench, what will be the new volume? (Ignore any possible temperature change.)

2. A gas in a 1L vessel at a pressure of 3 atm is allowed to escape through a stopcock into a second 1L vessel. What is the final pressure?

3. Real gases tend to behave more ideally as the pressure *increases* and the temperature *increases*.

4. Which of these plots will be a straight line at constant T and n, for an ideal gas?

   _____ (a) P against V

   _____ (b) V against P

   _____ (c) PV against P

   _____ (d) PV against V

   _____ (e) P against 1/V

   _____ (f) V against 1/P

3.4 The law of Charles and Gay-Lussac

1. If a straight line is represented by the equation y=mx+b, then

   (a) the y-intercept is the value of _____ when _____ = 0.

   (b) the x-intercept is the value of _____ when _____ = 0.

   (c) m is equal to the _____.

   (d) b is equal to the _____.

   (e) the x-intercept is equal to _____ in terms of m and b.

2. If in the linear relation given in Question 1 y is V, the volume of an ideal gas, and x is t, the temperature in °C, then m must be (positive, negative) and b must be (positive, negative).

Chapter 3　　　　　　　　　　　　　　　The Gas Laws and Stoichiometry Involving Gases

3. If in the linear relation given in Question 1 y is V, the volume of an ideal gas and x is T, the temperature in K, then b must be equal to _____.

4. A gas sample initially at 0°C has a volume of 100mL and a pressure of 100atm. The temperature is raised to 273°C.

   (a) If the pressure is still 100atm, what is the volume?

   (b) If the volume is still 100mL, what is the pressure?

3.5 The ideal gas law

1. A gas sample is originally at a pressure of 1atm, a temperature of 0°C and a volume of 22.4L. The density is 2g/L.

   (a) Under these conditions,

   (i) How many moles are present?

   (ii) What is the molecular weight?

   (b) P doubles, while n and V are constant.

   (i) What is the new value of T?

   (ii) What is the new value of δ?

   (c) Instead, V doubles while n and T remain constant.

   (i) What is the new value of P?

(ii) What is the new value of δ?

(d) Instead, n doubles *and* V doubles, at constant T.

(i) What is the new value of P?

(ii) What is the new value of δ?

(e) Instead, T and P both double while n remains constant.

(i) What is the new value of V?

(ii) What is the new value of δ?

3.6 Dalton's law of partial pressures

1. Three moles of gas A and two moles of gas B are in a container. The total pressure is 30 atm.

    (a) What is X_A?

    (b) What is X_B?

    (c) What is P_A?

    (d) What is P_B?

Chapter 3 The Gas Laws and Stoichiometry Involving Gases

2. A mixed gas sample has a temperature of 0°C, a volume of 22.4L and a total pressure of 2atm. The partial pressure of gas A. in the mixture is 0.5atm

   (a) How many moles of A are present?

   (b) If instead the total pressure had been only 0.5 atm, how many moles of A would have been present?

3. The vapor pressure of water at 21°C is about 0.025 atm. One liter of a gas is collected and stored above water, and the total pressure of the sample is 1atm.

   (a) What is the partial pressure of the gas?

   (b) If the gas had been collected above mercury, rather than above water, what volume would it have taken up, at the same temperature and total pressure?

   (c) In the original gas sample, one molecule out of every _____ molecules is a water molecule.

4. A container has a volume of 22.4L, and is kept at a temperature of 0°C. A volatile liquid is added to the container drop by drop, and the partial pressure of its vapor measured after each drop. This value is found to increase until 10g of the liquid is added, and then it remains constant at 0.1 atm. What is the molecular weight of the liquid?

Chapter 3 — The Gas Laws and Stoichiometry Involving Gases

Answers

3.1 Avogadro's law and its use in determining molecular formulas

1. $N_A/2$
2. $1/2$
3. 92
4. empirical formula NO_2 has M=46, so gas must be N_2O_4

3.2 The pressure of gases: definition, measurement, and units

1. True (at the earth surface)
2. 1000 atm

3.3 Boyle's law

1. 1 mL
2. 1.5 atm (new V=2L)
3. decreases; (true)
4. (c), (d), (e), (f).

3.4 The Law of Charles and Gay-Lussac

1. (a) y; x
   (b) x; y
   (c) slope
   (d) y-intercept
   (e) -b/m
2. positive; negative
3. zero
4. (a) 200 mL
   (b) 200 atm

3.5 The ideal gas law

1. (a)
     (i) 1
     (ii) ≈40 (2g/L × 22.4L)
   (b)
     (i) 273°C
     (ii) 2 g/L
   (c)
     (i) 0.5 atm
     (ii) 1 g/L
   (d)
     (i) 1 atm
     (ii) 2 g/L
   (e)
     (i) 22.4 L
     (ii) 2 g/L

3.6 Dalton's law of partial pressures

1. (a) 0.6
   (b) 0.4
   (c) 18 atm

Chapter 3   The Gas Laws and Stoichiometry Involving Gases

        (d) 12 atm

2.   (a) 0.5
     (b) 0.5

3.   (a) 0.975 atm
     (b) 0.975L
     (c) 40

4.   The first 10g completely vaporizes, then liquid starts to be present. 10g filling 22.4L at 0.1 atm at 0°C must be 0.1 moles, so that M=100.

CHAPTER 4: THE PROPERTIES OF GASES AND THE KINETIC-MOLECULAR THEORY

Scope

In Chapter 3 you became acquainted with many of the properties of gases. One of the great accomplishments of early nineteenth century chemistry was the elucidation of the stoichiometry of gas reactions, based on Avogadro's law and Cannizaro's scale of atomic weights. As mentioned, scientists were also thinking about what atoms and molecules must be like on a microscopic level; for example, it was a cause of concern that no one could explain why two hydrogen atoms should attract each other to form a hydrogen molecule. On the other hand, attempts to explain the gas laws themselves on the basis of reasonable assumption about atoms and molecules were more successful. This explanation is embodied in the *kinetic theory gases*, whose development reached its culmination in the late nineteenth century.

The text introduces a number of *postulates* about gas molecules and shows how these are consistent with the observed behavior of gases. Boyle's law is then derived by applying the known laws of physics to a system of molecules satisfying the postulates. All this really shows is that the particular *model*, or microscopic picture, embodied in the postulates is in accord with this particular aspect of reality (Boyle's law). It does not rule out the possibility that there may be other, very different, microscopic postulates which explain the same reality equally well; however, our postulates not only explain Boyle's law but with extensions (such as the explicit incorporation of molecular diameter) lead to correct predictions about other gas properties than P and V; for example, the viscosity and the thermal conductivity.

In the last chapter we tried to indicate that even though you probably learned before you were ten years old that water is H_2O, and that H_2O is a molecule made up of hydrogen and oxygen atoms, these notions were not obvious one hundred and fifty years ago. Similarly, the postulates introduced in Section 4.1 may (or may not!) seem self-evident to you, but their acceptance was gradual and was based upon the fact that from them experimentally demonstrable propositions (e.g., Boyle's law) could be derived. As a further illustration of how much we take for granted in the late twentieth century, consider that even the remarkable success of the kinetic theory of gases in explaining the properties of gases did not convince all physicists that atoms and molecules were real; for many, the real proof came with the demonstration of x-ray diffraction from crystals in 1912.

Postulates and Observations

Postulate 1. A pure gas is composed of a huge number of identical molecules.

> Now, of course, we know that a mole of gas contains 6.02×10^{23} molecules, but even in the early 19th century it must have been clear that if a gas consists of molecules at all, they must be very small. The microscope had been around since the seventeenth century, and even at the magnifications available in Dalton's time gas molecules were not visible. That gases, as well as other substances are, in fact, made up of molecules was one of the tenets of Dalton's atomic theory.

Postulate 2. The diameter of a gas molecule is negligible compared with the average distance between gas molecules.

> Unlike Postulate 1, this one distinguishes gases from liquids. It accounts for the low density and high compressibility of a gas, compared to a liquid or solid.

Postulate 3. Gas molecules are in rapid random motion.

> This postulate may not be so obvious. It turns out (see below) that Boyle's law can be derived by assuming that the pressure on the walls of a vessel containing a gas is due to collisions of the gas molecules with the walls. Once this explanation of the origin of gas pressure is accepted, the fact that a gas exerts equal pressures on all the walls of any vessel shows that the motion must be random: the molecules collide with all the walls at equal frequency per unit area, regadless of direction.

> Also, opening a stopcock between two vessels, each containing a different gas at the same pressure, results in the slow *diffusion*, or mixing, of the gases. If gas molecules were not in motion, this would not occur.

Chapter 4   The Properties of Gases and the Kinetic-Molecular Theory

Postulate 4. Gas molecules do not interact with each other.

In the derivation of Boyle's law we will assume that molecules are "point masses" — that is, they have finite mass but zero diameter — and therefore that they do not even collide with each other. This assumption also embodies Postulate 2 (diameters negligible compared to intermolecular distance). We will find that Boyle's law — and, by extension, the other ideal gas laws — can be derived with essentially no further assumptions. In a real gas, however, molecules do interact. When they are very far apart, which corresponds to low pressure, this interaction can be ignored, but as they come closer they may experience an attractive force, and as they come even closer they begin to "bump," which is an effect of finite volume, and which acts like a repulsive force ("bumping" keeps the molecules at least a molecular diameter apart). Thus, Postulates 2 and 4 pertain to ideal, not real gases, and deviations from these postulates lead to deviations from ideal gas behavior. One such deviation is liquifaction, which occurs to any real gas at sufficiently low T and high P. This must be due to attractive intermolecular forces; nevertheless, for any real gas the intermolecular forces must be weak, or else the gas wouldn't expand to fill any size vessel — that is, it would not be a gas!

Postulate 5. Pressure is due to collisions of gas molecules with the vessel walls.

Strictly speaking, this is a consequence of Postulate 3; if molecules are in rapid, random motion, then they *will* collide with the walls, and this *will* produce a pressure, provided only that the molecules have mass, which they must have, since gas samples have mass. As the derivation of Boyle's law given below shows, these assumptions are sufficient to explain the observed pressure behavior of gases.

Postulate 6. All collisions of gas molecules are elastic.

This means that the kinetic energy before and after a collision is the same. Of course, in our model of an ideal gas, the only possible collisions are between gas molecules and the wall, since molecules with zero diameter cannot collide with each other. The idea of elastic collisions with the walls is consistent with Boyle's Law, as we shall see, and if no kinetic energy is lost to the walls then the molecules never lose speed. Since we said that wall collisions are the source of pressure, this implies that the pressure of a gas in a sealed vessel at constant temperature does not diminish with time, as it would if the molecules instead slowed down gradually.

When kinetic theory is extended beyond our simple model to allow molecules to have diameters — or even shapes more complicated than spheres — the postulate of elastic collisions can be left in, and this allows properties such as viscosity and thermal conductivity to be explained mathematically. When kinetic theory is further extended to allow attractive and repulsive interactions, however, this postulate breaks down, since, to give an example, two molecules initially at rest will accelerate toward each other, if they experience an attractive force, increasing the kinetic energy; then, after they collide and fly apart, the kinetic energy will decrease again as they slow down. Furthermore, real molecules have *internal degrees of freedom*, such as the ability to rotate and vibrate, and energy can be transfered back and forth from *translational degrees of freedom* (the overall motion of the molecule in space) to these internal modes on collision. Even when all these possibilities are added to kinetic theory, however, the *average* translational kinetic energy remains constant. Thus the pressure remains constant indefinitely in a sealed vessel of a real gas at constant temperature, despite the fact that the collisions are not elastic.

Boyle's Law from Kinetic Theory

Before deriving Boyle's law, we will discuss a preliminary result, based on the Pythagorean theorem, which will be useful in the derivation. Recall that the Pythagorean theorem states that for a right triangle, the square of the length of the hypoteneuse is equal to the sum of the squares of the two sides. Thus, if a square has sides of length X and Y, the diagonal of the square, D, is related to X and Y by

$$D^2 = X^2 + Y^2$$

In three dimensions, the long diagonal of a cube is related to the length of its sides by

$$D^2 = X^2 + Y^2 + Z^2$$

If X, Y and Z now represent components of the velocity in the X, Y and Z directions, then D^2 represents the square of the speed.

$$u^2 = u_x^2 + u_y^2 + u_z^2$$

Now, on average, there is no reason to believe that, on average, the velocity of a gas particle favors the x, y or z direction, so that, all three terms on the right are equal, for a system of gas molecules.

$$<u_x^2> = <u_y^2> = <u_z^2>$$

The symbol $<x>$ means the average value of x; thus $<u_x^2>$ has the meaning that if you knew u_x for each molecule, squared each value separately, added the squares and divided the sum by the number of molecules, this would equal $<u_x^2>$. Of course we cannot know u_x for each molecule, but you will soon see that we can know $<u_x^2>$, which is called the *mean square velocity component* in the x-direction; $<u^2>$ is known as the *mean square speed*. The statement that the three mean square velocity components of a gas are equal is simply a mathematical way of stating Postulate 3: the motion of gas molecules is random; no particular direction in space is preferred, on average. Furthermore, since the three velocity components are equal, we can write

$$<u^2> = 3<u_x^2> = 3<u_y^2> = 3<u_z^2>$$

This result will be used later on.

Now, suppose N molecules of gas are confined to a cubic box whose edge is L. How do collisions with the wall produce a pressure? Here is a simple analogy that should make this clear. Suppose you are standing holding a large board which someone else is bombarding with raquetballs. Every time a ball hits, you feel a force due to its impact, but if the balls are coming fast enough you will experience a constant force. This force divided by the area of the board is the pressure experienced by the board. In the same way, the force experienced by the wall of a vessel being bombarded by gas molecules, divided by the area of the wall, is the pressure of the gas.

Figure 4.6 represents our hypothetical box. The wall perpendicular to a given axis undergoes collisions only with molecules with a velocity component along that axis; thus, collisions with the walls perpendicular to the y-axis (Figure 4.7) experience a pressure due only to the y-velocity component of the particles. In an elastic collision with a wall perpendicular to its direction of travel, the particle

Chapter 4 — The Properties of Gases and the Kinetic-Molecular Theory

rebounds with a velocity equal and opposite to its incoming velocity. Thus, if the velocity before impact was $-u_y$, the velocity after impact will be u_y, and the total change in velocity is $u_{after} - u_{before} = 2u_y$. If the distance between the two parallel walls perpendicular to the y-axis is L, then the particle must travel the distance 2L between successive collisions with the same wall; the y-velocity component changes sign, but not magnitude, every time it undergoes a collision with one of the walls, so that its y-motion is a shuttling back and forth. The time it takes the particle to go the distance 2L is

$$t = \frac{2L}{u_y}$$

Recall that the velocity change per collision is $2u_y$; this divided by t is the acceleration experienced by the particle

$$a = \frac{2u_y}{2L/u_y} = \frac{u_y^2}{L}$$

(In our model the particle really travels at constant velocity between the walls, then experiences an infinite acceleration at the moment of impact; this formula really gives the average acceleration of the particle over several collisions.)

Force is defined by F=ma; therefore, if the particle has mass m, we have

$$F = \frac{mu_y^2}{L}$$

Pressure is force/area, and the area of a wall of the cube is L^2, so that

$$P = \frac{F}{L^2} = \frac{mu_y^2}{L^3} = \frac{mu_y^2}{V}$$

where V is the volume of the box. This is the pressure on a wall due to a single particle; recall, however, that we said we had N particles, and each can have its own value of u_y. The total pressure is given by

$$P = \frac{m \sum_{i=1}^{N} u_{yi}^2}{V}$$

where u_{yi} is u_y for the i'th particle. But if $<u_y^2>$ represents the average of all the u_{yi}'s, then the sum $\sum_{i=1}^{N} u_{yi}^2$ is equal to $N<u_y^2>$, so that

$$P = \frac{mN\langle u_y^2\rangle}{V}.$$

(As a simpler example of this averaging procedure, suppose I have three apples and you have five; how many apples do we have total? I can either simply add the two numbers together, to give eight, or first take the average — on average, we have four apples each — and multiply by the number of people, two, to give the same answer.)

Now, from our previous result using the Pythagorean theorem,

$$\langle u_y^2\rangle = \frac{1}{3}\langle u^2\rangle$$

This gives

$$P = \frac{mN\langle u^2\rangle}{3V}$$

Furthermore, if I multiply the top and bottom of this equation by N_A, and note that $N/N_A = n$, the number of moles of material present, I have

$$P = \frac{nN_A m\langle u^2\rangle}{3V}$$

κ, the average kinetic energy of a gas molecule, is defined by

$$\langle \kappa\rangle = \frac{1}{2}m\langle u^2\rangle$$

This gives

$$PV = \frac{nN_A m\langle u^2\rangle}{3} = nN_A\left(\frac{2}{3}\langle \kappa\rangle\right)$$

Note that this looks very much like the ideal gas law: $PV=nRT$. This allows the hypothesis that

$$RT = \frac{2}{3}N_A\langle \kappa\rangle$$

which in turn implies that the average kinetic energy of a molecule is constant when the temperature is constant, and that, in fact, $\langle \kappa\rangle$ is directly proportional to the temperature! When the kinetic theory is extended to accommodate internal as well as translational degrees of freedom, it turns out that only the translational kinetic energy comes into the relationship; the kinetic energy associated with internal vibrations and rotations do not. $N_A\langle \kappa\rangle$ is thus E_{trans}, the average *molar* translational kinetic energy, so

Chapter 4   The Properties of Gases and the Kinetic-Molecular Theory

that

$$PV = n\frac{2}{3}\bar{E}_{trans} = n\frac{2}{3}N_A \langle \epsilon_K \rangle = nRT$$

Several other useful forms of this equation are sometimes used. *Boltzmann's constant*, k, is defined as R/N_A; it is the gas constant per molecule, rather than per mole, so that

$$PV = n(N_A k)T = \left(\frac{N}{N_A}\right)N_A kT = NkT$$

$$= \frac{2}{3}N\langle \epsilon_K \rangle$$

$$\bar{E}_{trans} = \frac{3}{2}RT$$

$$\langle \epsilon_K \rangle = \frac{3}{2}kT$$

where the last two equations give the mean translational kinetic energy of a mole of ideal gas and a single ideal gas molecule, respectively.

Note that these equations imply that RT must have units of energy per mole, since $\bar{E}_{trans}$ has units of energy per mole. Common units of RT are L·atm/mol, so that a liter-atmosphere must be a unit of energy. Recall that pressure has units of force/area, or force/length². Multiplying this by a volume (length³) gives (force)(length); but work is a form of energy and is defined as a force applied through a distance, so that (force)(length), and (pressure)(volume) have units of energy. One liter-atmosphere is numerically equal to about 101 joules. The physical interpretation of the liter-atmosphere as a unit of energy is as follows. Suppose a sample of compressed gas is confined to a cylinder with a movable piston, and is allowed to expand against an external pressure of one atmosphere until its volume increases by one liter. Then it has done 1L·atm of work on the external world. This notion will be discussed in detail when thermodynamics is taken up, in Chapter 17.

$\langle u^2 \rangle^{1/2}$, the *root-mean-square speed*, is defined simply as the square root of $\langle u^2 \rangle$, and is sometimes denoted u_{rms}. Since $\langle u^2 \rangle$ has units of velocity², $\langle u^2 \rangle^{1/2}$ has units of velocity. $\langle u \rangle$, the *mean velocity*, has units of velocity also, but differs numerically from $\langle u^2 \rangle^{1/2}$. Recall the above example concerning apples; if I have three and you have five, then

$$\langle \text{apples} \rangle = \frac{3+5}{2} = 4$$

$$\langle \text{apples}^2 \rangle = \frac{3^2+5^2}{2} = 17$$

$$\langle \text{apples}^2 \rangle^{1/2} = \text{apples}_{rms} = 17^{1/2} \approx 4.1$$

Thus $\langle \text{apples} \rangle < \langle \text{apples}^2 \rangle^{1/2}$. We had that

$$\langle \epsilon_K \rangle = \frac{3}{2}kT = \frac{1}{2}m\langle u^2 \rangle$$

From this we can write

$$\langle u^2 \rangle = \frac{3kT}{m}$$

$$u_{rms} = \left(\frac{3kT}{m}\right)^{1/2} = \left(\frac{3RT}{M}\right)^{1/2}$$

where M is the molecular weight; the last relationship comes from multiplying the top and bottom of the previous expression in parentheses by N_A. Since the translational kinetic energy of a gas depends only on temperature ($\overline{E}_{trans} = (3/2)RT$), any two gases at the same temperature have the same translational kinetic energy; however, in order for this to occur, the lighter molecules must be moving more slowly; they have a lower value of u_{rms}, as shown in Example 4.6.

Before leaving the general topic of the Boyle's law deviation, it is worthwhile to recap what we have done. We showed that if gases are infinitesimal particles which, however, have mass, and if they are in random motion, then $PV = n(2/3)\overline{E}_{trans}$. If we assume that $(2/3)\overline{E}_{trans} = RT$, (which the text gives as *Postulate* 7), then we have derived not only Boyle's law, but all the ideal gas laws, as embodied in PV=nRT. This also implies an interpretation of temperature as being fundamentally related to the kinetic energy of random translational motion of the particles, and this even allows us to determine the root-mean-square velocity of the molecules of a gas, provided only that we know the temperature and the molecular weight! Although this is not identical with the average velocity, it is close (see below). From reasonable postulates, plus the known laws of physics, we have calculated a microscopic variable — u_{rms} of the gas molecules in a gas sample at rest — which, at least at the time the derivation was originally done, there was no way of measuring directly. (Such methods exist now.) This kind of interpretation of macroscopic observations — such as P and V — in terms of microscopic variables — such as u_{rms} — goes to the heart of what much of chemistry and physics are about.

Graham's Law

Recall from our example with the apples that u_{rms} was greater than $<u>$. It turns out that for any gas sample,

$$<u> = \left(\frac{8}{\pi}\frac{RT}{M}\right)^{1/2}$$

We had

$$u_{rms} = \left(3\frac{RT}{M}\right)^{1/2}$$

so that

$$\frac{u_{rms}}{<u>} = \left(\frac{3\pi}{8}\right)^{1/2} \approx 1.085$$

Thus u_{rms} is about 8.5% bigger than $<u>$ for any gas. (The derivation of $<u>$ is hard, and we will not carry it out.) Note that $<u>$ and u_{rms} both are proportional to $(RT/M)^{1/2}$, but the proportionality constant is different.

Thomas Graham found in the mid-nineteenth century that the rate of *effusion* of a gas through a small hole also is directly proportional to $(RT/M)^{1/2}$; thus if one gas, call it A, is allowed to effuse at two temperatures, T_1 and T_2, then

$$\frac{\text{rate of effusion of A at } T_1}{\text{rate of effusion of A at } T_2} = \left(\frac{T_1}{T_2}\right)^{1/2}$$

If gas B is then allowed to effuse through the same hole at T_1, we have

$$\frac{\text{rate of effusion of A at } T_1}{\text{rate of effusion of B at } T_2} = \left(\frac{M_B}{M_A}\right)^{1/2}$$

Thus heavy molecules effuse more slowly than slow ones. The fact that effusion obeys the same dependence on T and M as u_{rms} and $<u>$ is a strong indication that the phenomenon depends on gas velocities; in fact, the following argument should make this obvious.

Chapter 4 — The Properties of Gases and the Kinetic-Molecular Theory

Consider light and heavy molecules at the same P and T. They have the same average kinetic energies, but the lighter molecules move faster, and therefore hit the walls more often. The heavier molecules hit the walls less often, but exert a greater force per collision; these opposing effects cancel out exactly when the pressures are identical. Now imagine making a little hole in the wall, to allow effusion. The light molecules will "hit" the hole more often than the heavy molecules, and therefore will effuse more rapidly. Relative effusion rates of a known and unknown gas provide a method of molecular weight determination, as shown in Example 4.7. Effusion also affords a method of separating gases based on their molecular weight.

Maxwell-Boltzmann Distribution

So far we have shown how to calculate u_{rms} for a gas, and we gave the result for $<u>$. We also mentioned in passing that we do not know the instantaneous velocity of every molecule; we do, however, know more than just u_{rms} and $<u>$. The sum total of our knowledge about the velocities of molecules in a pure gas sample is embodied in a relationship called the *Maxwell-Boltzmann distribution*, which we will discuss, without deriving it. The Maxwell-Boltzmann distribution is a mathematical function that is constructed in such a way that the fraction of molecules having velocities between u_1 and u_2 is equal to the area under the distribution curve between u_1 and u_2. It is the function that describes how the molecules are distributed among "bins", each bin consisting of a range of velocities, such as the bin "$u_1\ u_2$."

Let us, in fact consider a bin of constant width Δu, and talk about what happens as we slide this along the x-axis in Figure 4.8. The area under the curve in any such bin will have flat sides and bottom, but a curved top, formed by the distribution, $D(u)$. As we slide Δu along, starting at the left, this area increases, goes through a maximum and decreases; thus, few molecules have very small or very large velocities. The maximum of the curve is the *most probable velocity*; more molecules have velocities in the bin which includes it than in any other. The total area under the curve, from $u=0$ to $u=\infty$, must equal unity, since the fraction of molecules having velocities between 0 and ∞ must be one; every molecule has *some* velocity.

The Maxwell-Boltzmann distribution pertains to an ensemble of molecules all with the same mass; that is, it pertains to a sample of a pure gas at some temperature T. Note that a slow molecule has less kinetic energy than a fast molecule with the same mass, and thus contributes less to the pressure than a fast molecule; the total pressure is due to the sum of the contributions of all the molecules, which, as we saw, involved $<u^2>$: $P=(1/3V)Nm<u^2>$. The most probable speed, that corresponding to the peak of the distribution, turns out to be $(2RT/M)^{1/2}$; thus u_{rms} is $(3/2)^{1/2} \approx 1.22$ times the most probable speed.

Note (Figure 4.9) that at high temperature the maximum of the curve shifts to higher velocities; therefore u_{rms} does as well. This is expected, because we know that as T goes up, E_{trans} goes up, so that more molecules must be moving faster. Note also that the distribution broadens. Applying our bin of constant Δu to the most probable velocity, the area under the curve is less for the higher temperature than for the lower. This means that a smaller fraction of the molecules have velocities within, say, 1m/s of the most probable velocity at a high than at a low temperature.

Van der Waals' Equation

Recall that an equation of state is an equation that describes the relationship among the variables specifying the state of a system. The equation of state for an ideal gas, $PV=nRT$, contains only a single constant, R, but describes real gases only approximately. It is desirable to have an equation of state that describes real gases more precisely, for two reasons. First of all, it is important to have precise data on real gases for many technological purposes, and an equation of state is handier than tables and tables of data. Second of all, an equation of state is a challenge to predict on the basis of a microscopic model; the fact that we were able to predict the ideal gas law based on a very simple physical model led to a much deeper understanding of what pressure is and what an ideal gas is. Predicting a more complicated equation of state from a microscopic model should give further insight into what makes gases non-ideal.

One of the most useful equations of state for real gases is the van der Waals' equation, which contains two constants, called a and b, in addition to R. a and b vary from gas to gas, and are

Chapter 4 — The Properties of Gases and the Kinetic-Molecular Theory

determined empirically. The van der Waals equation reads

$$(P - \frac{n^2}{V^2}a)(V - nb) = nRT$$

Note that if a and b are zero, this becomes the ideal gas equation. Bigger values of a imply stronger attractive forces, and bigger values of b imply larger molecular diameters, so it is not surprising that when a and b disappear the gas is ideal, since an ideal gas molecule has no attractive forces and a zero diameter.

You can view the term $(P + \frac{n^2}{V^2}a)$ as the pressure experienced by a gas molecule, whereas P is the pressure measured by an outside observer. The gas molecule experiences a greater pressure than that recorded by a pressure gauge because it experiences the attractive forces of all the surrounding molecules. As P goes up, V goes down, and (n^2a/V^2) goes up, so the deviation between what is measured and what a gas molecule "feels" gets greater. This is in accord with the fact that gases become less ideal at higher pressures.

The term (V−nb) may be viewed as the volume available to a gas molecule, whereas V is the volume of the container. The gas molecules can only move in the space not occupied by the other molecules, so that b is related to how much volume a mole of gas molecules takes up. Note that as V gets bigger, the term (nb) stays the same size, so that at big enough V (low enough P) nb becomes negligible, and ideality is again approached, as we expect. Note (Figure 4.10) that when PV is plotted against P, the (nb) term causes PV to be greater than the Boyle's law value at high P; on the other hand, the (an^2/V^2) term causes PV to be less than the Boyle's law value at high P.

Chapter 4								The Properties of Gases and the Kinetic-Molecular Theory

Questions

4.1 The postulates of the kinetic theory of gases
4.2 Correlations between the observed properties of gases and the postulates of the kinetic-molecular theory
and 4.3 Criticism of two of the postulates of the kinetic theory

1. Gases are (more, less) compressible than liquids.

2. The walls of a container containing a gas sample experience the (same, different) pressures.

3. If the diameter of a gas molecule is large compared to the average distance between molecules, then the *substance would be not a gas, but rather a liquid or solid.*

4. If gas molecules were not in motion, but stationary, and repelled each other, this would also cause a gas to exhibit pressure. (True, False)

5. The fact that two gases will diffuse into each other after opening a stopcock connecting them proves that the two gas move preferentially toward the stopcock. (True, False)

6. The fact that an ideal gas expands to fill the entire container proves that the molecules repel each other. (True, False)

7. Since a real gas expands to fill a container, the molecules must have *zero* attractive force.

8. The fact that gases have lower density than liquids or solids means that gas molecules are *smaller* than the molecules of liquids or solids.

9. For a real gas, collisions are not elastic. For this reason the pressure of a sealed sample at constant T will slowly decrease with time. (True, False)

10. If the molecules of a real gas attract each other, then

    (a) P will be (greater, smaller) than nRT/V.

Chapter 4   The Properties of Gases and the Kinetic-Molecular Theory

    (b) and PV will be (greater, smaller) than the Boyle's law prediction.

11. If a real gas is compressed to the point that the volume taken up by the molecules is a significant fraction of the volume of the vessel, then

    (a) P will be (greater, smaller) than nRT/V

    (b) and PV will be (greater, smaller) than the Boyle's law prediction.

12. The fact that liquifaction occurs at high P and low T proves that the molecules of a real gas must attract each other. (True, False)

13. The ideal gas law is approached for real gases at _____ temperatures and _____ pressures.

4.4 An outline of the derivation of Boyle's law from the postulates of kinetic theory

1. (a) What is the diagonal of the unit square equal to?

    (b) What is the diagonal of the unit cube equal to?

    (c) What would you guess the diagonal of a unit four-dimensional hypercube is equal to?

2. If a molecule hits a wall perpendicular to the y-axis,

    (a) the only velocity component that changes is _____.

    (b) the magnitude of the change in this component as a result of collision is _____.

3. (a) If I am travelling 60mph, how long will it take me to go 90 miles?

    (b) If I am travelling at speed u, how long will it take me to go the distance L?

Chapter 4   The Properties of Gases and the Kinetic-Molecular Theory

(c) If my y-velocity component is u_y, how long will it take me to go the distance 2L in the y-direction?

(d) If my y-velocity component is u_y, how many times per unit time will I go the distance 2L?

4. If my velocity changes by $2u_y$ every $2L/u_y$ seconds, what is my average acceleration?

5. If my acceleration due to collisions is u_y^2/L and my mass is m what is the pressure on the wall against which I am colliding, if its area is L^2?

6. If my motion is random, restate the pressure you calculated in Question 5 in terms of my speed, u.

7. Now restate the pressure in terms of my kinetic energy, E_K.

8. Suppose now that I am not alone, and there exists a mole of others like me, each with a different value of E_K. Replace E_K in the above expression so that the total pressure is still correct.

9. In the derivation of Boyle's law

   (a) We assumed that the diameter of a gas molecule is equal to _____.

   (b) We built in *no* attractive or repulsive forces.

   (c) We proved that $PV = (2/3)nE_{trans}$ for n moles of ideal gas. (True, False)

   (d) We proved that $E_{trans} = (3/2)RT$; thus we proved that $PV = nRT$. (True, False)

Chapter 4   The Properties of Gases and the Kinetic-Molecular Theory

10. (a) Units of Boltzmann's constant (k) are

(b) Express R in terms of k.

4.6 Energy units for PV and RT

1. One joule equals _____ ergs.

2. When using the SI system of units, mass is in units of _____, time is in _____, length is in _____, area is in _____, volume is in _____, force is in _____ and energy is in _____.

3. A joule is *bigger* than a calorie.

4. According to Table 4.2 in the text, a 10mg fly doing a 1mm push-up expends one erg of energy.

    (a) In SI units, how much energy does a 50kg human being doing a 10cm push-up expend?

    (b) Restate this answer in kcal.

    (c) If this same person's food intake for a day is 2500 kcal, how many push-ups would he or she have to do to completely use up this energy?

Chapter 4 — The Properties of Gases and the Kinetic-Molecular Theory

(d) Suggest an animal which would expend about one joule doing a push-up.

4.7 The mean-square speed, the root-mean-square speed, and the mean speed

1. Given the three numbers -3, -2 and 8,

   (a) What is the mean?

   (b) What is the mean square?

   (c) What is the root-mean-square?

2. A sample containing O_2 and O_3 is enclosed in a sealed container.

   (a) The O_2 molecules have *lower* kinetic energy, on average, than the O_3 molecules.

   (b) The O_2 molecules have *greater* speed, on average, than the O_3 molecules.

3. For any gas sample, if the conditions are changed so that u_{rms} increases, then

   (a) $<u^2>$ must also increase (True, False)

   (b) $<u>$ must also increase (True, False)

4.8 Graham's law of effusion

1. A gas sample contains equimolar quantities of H_2 and N_2 at the same T and P. The vessel contains a small hole. After some time, the vessel will contain *more* moles of H_2 than N_2.

Chapter 4   The Properties of Gases and the Kinetic-Molecular Theory

2. A gas sample contains equimolar amounts of oxygen and of hydrogen gas. The hydrogen will effuse through a small hole _____ times (faster, slower) than the oxygen.

3. If the absolute temperature doubles, the effusion rate (increases, decreases) by a factor of _____.

4. A gas sample contains hydrogen and oxygen gas only, and $X_{H_2} = 0.1$. The mixture is allowed to effuse through a small hole. What will X_{H_2} be in the first gas sample effusing through the hole?

4.9 The distribution of molecular velocities

1. When the temperature is increased, u_{rms} of an ideal gas (increases, decreases, remains the same).

2. When the pressure is increased, u_{rms} of an ideal gas (increases, decreases, remains the same).

3. Over time, the velocity distribution of a gas sample at constant P and T (changes, remains the same).

4. Over time, the velocity of an individual gas molecule (changes, remains the same).

5. The velocity at the peak of the Maxwell-Boltzmann distribution is called the _____ velocity.

6. At a high temperature a (greater, smaller) fraction of molecules in a gas sample have velocities within the same neighborhood of the most probable velocity than at a lower temperature.

4.10 Van der Waals' equation

1. Larger values of the van der Waals a have the physical interpretation of larger _____.

2. Larger values of the van der Waals b have the physical interpretation of larger _____.

3. For a given real gas, as the pressure decreases and the temperature increases, a and b (increase, decrease, do not change).

4. For an ideal gas, a= _____ and b= _____.

5. (a) Bigger values of a cause a gas molecule to in effect experience a (greater, smaller, equal) pressure than/to the measured pressure.

(b) Bigger values of b cause a gas molecule to in effect experience a (greater, smaller, equal) volume than/to the measured volume.

Chapter 4 — The Properties of Gases and the Kinetic-Molecular Theory

Answers

4.1 The postulates of the kinetic theory of gases
4.2 Correlations betweeen the observed properties of gases and the postulates of the kinetic molecular theory and
4.3 Criticism of two of the postulates of the kinetic theory

1. more
2. the same
3. substance could be a non-ideal ("real") gas, a liquid or a solid
4. True; however, this model breaks down for the following reason (among others). Suppose the gas were confined to a long, thin cylinder. The molecules would be further apart, on average, than if they were confined to a sphere of the same volume, so the pressure would be lower in the cylinder. But in fact the pressure of a gas does not depend on the shape of the container.
5. False; random motion will do.
6. False; random motion explains this.
7. a low (if any)
8. False; the molecules are just farther apart
9. False; see **Scope**
10. (a) smaller
    (b) smaller
11. (a) greater
    (b) greater
12. True
13. high; low

4.4 An outline of the derivation of Boyle's law from the postulates of kinetic theory

1.
   (a) $2^{1/2}$
   (b) $3^{1/2}$
   (c) $4^{1/2} = 2$
2. (a) u_y
   (b) $2u_y$
3. (a) 1.5 hr.
   (b) L/u
   (c) $2L/u_y$
   (d) $u_y/2L$
4. u_y^2/L
5. mu_y^2/L^3
6. $(1/3)mu^2/L^3$
7. $(2/3)E_K/L^3$
8. $(2/3)N_A <E_K>/L^3$

Chapter 4 The Properties of Gases and the Kinetic-Molecular Theory

9. (a) 0
   (b) True
   (c) True
   (d) False; we *assumed* or *postulated* that $E_{trans} = (3/2)RT$

10. (a) L·atm/molecule·K
    (b) $R = N_A k$

4.6 Energy units for PV and RT

1. 10^7

2. kilograms
   seconds
   meters
   meters^2
   meters^3
   newtons (1N=1kg m/s^2)
   joules (1J=1N·s)

3. smaller

4. (a) about 50 J
   (b) about 12 cal, or 0.012 kcal
   (c) about 200,000 push-ups
   (d) a small cat, weighing 5lb (2.5 kg) doing a 4cm (.04m) push-up.
   $$E = (2.5 kg)(.04m)(\frac{10m}{s^2}) = 1J$$

4.7 The mean-square speed, the root-mean-square speed, and the mean speed

1. (a) 1
   (b) about 26
   (c) $26^{1/2} \approx 5$

2. (a) the same
   (b) True

3. (a) True
   (b) True

4.8 Graham's law of effusion

1. fewer

2. $(\frac{32}{2})^{1/2} = 4$ times faster

3. increases by $2^{1/2}$

4. $X_{H_2,out} = X_{H_2,in}(\frac{32}{2})^{1/2} = (.1)(4) = .4$

4.9 The distribution of molecular velocities

1. increases

2. remains the same

3. remains the same

4. changes

5. most probable

6. smaller

Chapter 4

4.10 Van der Waals' equation

1. intermolecular attractions
2. molecular volumes (or diameters)
3. do not change
4. 0,0
5. (a) greater
   (b) smaller

Chapter 5 — Intermolecular Forces, Condensed Phases, and Changes of State

CHAPTER 5: INTERMOLECULAR FORCES, CONDENSED PHASES, AND CHANGES OF STATE

Scope

You saw in the last chapter that ideal gas molecules experience no attractive or repulsive forces; an ideal gas exhibits smaller and smaller volumes as T is lowered, with V going to zero as T goes to absolute zero. Real substances, however, liquify as T is lowered. The fact that the molecules then cease to fill an arbitrarily large container proves that there must be attractive forces that overcome the strictly random molecular motion characteristic of a gas. Recall that E_{trans} increases with T. At low enough temperature E_{trans} becomes small enough so that the attractive forces dominate, forming a cohesive liquid. Furthermore, the fact that the volume of any real substance never goes to zero no matter how low the temperature gets means that there must be repulsive forces as well, that prevent the molecules from getting too close together. The easiest such repulsive force to envision is one based on molecules taking up space; for example, being spheres of finite size which cannot penetrate each other.

Van der Waals Forces

We distinguish between forces between bonded atoms (the forces that hold atoms together in a molecule) and forces between non-bonded atoms. The main basis of this distinction is the strength of the interaction: non-bonded forces are much weaker than bonding forces. It is the non-bonded forces that are responsible for the attractions and repulsions that cause real gases to deviate from ideality, and that hold liquids and molecular crystals together. These non-bonded forces are called *van der Waals* forces, since it is they that are responsible for the attractive and repulsive interactions embodied in the constants a and b in the van der Waals equation. It is of interest that van der Waals interactions also exist between atoms within the same molecule that are not bonded to each other; for example, a long chain molecule will curl up because the distant atoms in the molecule attract each other in much the same way as do atoms in two separate molecules; however, our discussion will be confined to *intermolecular* interactions.

The various sorts of van der Waals attractive forces are all *shorter-range* forces than bonding forces, and the repulsive force is shorter in range than the attractive forces. This is why in a plot of PV

Chapter 5 **Intermolecular Forces, Condensed Phases, and Changes of State**

against P for a typical real gas (Figure 4.12) the attractive forces tend to dominate at the lower pressures, where the molecules are still fairly far apart, whereas repulsive forces always dominate at very high pressure, where the molcules tend to "bump".

Dipole moments and polar covalent bonds were briefly discussed in Chapter 2. Recall that when an atom from the right-hand side of the periodic chart bonds to an atom from further left, the former tends to attract electron density from the latter, so that in HCl, for example, the Cl carries a partial negative charge, and the H carries a partial positive charge, so that HCl has a *dipole moment*. (We say that Cl is more *electronegative* than H; Chapter 13 develops this concept in further detail.) Two HCl molecules sufficiently close together will therefore orient in space so that the H of one points toward the Cl of the other, and their mutual attraction will tend to pull the molecules closer. This is called a *dipole-dipole* force.

Another polar molecule is H_2O, which is V-shaped; the H's are more positive than the O, so H_2O has a dipole moment. On the other hand, CO_2 is a linear molecule, with C in the middle. Each C-O bond is polar, with the O the more negative atom, but because the molecule is linear, the dipole moments of the two bonds cancel out and the molecule has no net dipole moment.

In general, a *dipole* is defined in physics as two charges of the same magnitude, Q, but opposite signs, separated by a distance d. The magnitude of the dipole moment, μ, is defined as Qd, so that large charges close together have the same dipole moment as small charges far apart. A convenient unit for expressing the dipole moment of molecules is the *debye*. 4.80 debye is the dipole moment if Q is the charge on an electron and d is one angstrom. As Figure 5.5 shows, each bond can be considered to have a dipole moment, and their vector sum is the dipole moment of the molecule. The reason CO_2 has no dipole moment is that the two C-O bonds have equal and opposite dipole moments, leading to a zero resultant.

Even non-polar molecules exhibit attractive forces; for example, nitrogen, N_2, is non-polar, since two nitrogen atoms have the same electronegativity. Nevertheless, N_2 has a van der Waals a equal to 1.39, showing that the molecules attract each other. This is not due to dipole-dipole interactions, but rather to another kind of attractive force, the *London dispersion force*. The electrons in any atom or molecule are in rapid motion, and when any two atoms approach each other, the electrons from one experience a repulsion from the electrons in the other, and therefore the electrons move in a *correlated* manner.

Consider two identical non-polar atoms, A and B, interacting in this manner. Suppose that an electron on A is temporarily on the side of the A nucleus opposite to the direction of B. There is then a momentary dipole moment within the A atom, with its positive end toward B. This will tend to make the electrons on B move toward the side of B closer to A, creating a dipole within B whose negative end is closer to the positive end of the A dipole. This results in a net attraction. At this instant, the electron cloud on A is *polarized* away from B, and that on B is polarized toward A. Of course, a moment later the situation may be reversed, and the momentary dipoles on A and B may be pointing the other way; however, the net effect is always an attraction.

The attraction caused by this effect is also called an *induced-dipole-induced-dipole attraction*, because the proximity of two neutral atoms *induces* momentary dipoles in both. This force is purely quantum-mechanical in origin, since it is quantum mechanics that dictates that electrons must always be in motion. The London force is the weakest attractive force that exists between non-bonded molecules; however, it is still the strongest force attracting two non-polar molecules. As you can imagine from the above explanation, the magnitude of this force tends to rise as the number of electrons in a molecule goes up; thus for non-polar compounds, as the molecular weight goes up, so does the van der Waals a, usually (Table 4.4, except H_2O and SO_2, which have dipole moments). Since the attractive forces are responsible for liquifaction, higher molecular weights usually lead to higher boiling points for non-polar compounds (Table 5.1). When dipole-dipole interactions exist, they add a further attraction, so that among substances with similar molecular weights, those with greater dipole moments tend to boil higher (Table 5.2). These relationships are only approximate, however.

Although not mentioned at this point in the text, there are several other intermolecular attractive forces that are important in chemistry. For example, a molecule with a permanent dipole moment will cause polarization of the electron clouds of a non-polar molecule, leading to a *dipole-induced-dipole* attractive force, which is intermediate in magnitude between a dipole-dipole force and an induced-dipole-induced-dipole force; obviously, this is important only in mixtures, such as HCl in Ar. Similarly, a charge will attract a dipole, and this is why liquids with strong dipole moments dissolve ionic solids: the energy it takes to pull apart the ions is recouped in the orientation of the dipolar solvent molecules around the separated ions. Obviously, this *charge-dipole force* also only becomes important in mixtures.

There is another type of interaction, *hydrogen bonding*, which is usually not considered a van der Waals force, since it is considerably stronger than even typical dipole-dipole forces. When hydrogen is bonded to a very electronegative atom — N, O, or F — the bond is highly polarized, and has a large dipole moment, with the H positive. The H of one molecule is therefore strongly attracted to N, O or F of another molecule, as in the text's illustration of the zig-zag chains formed by liquid HF. There are several things to note. In the structure H-F···H-F the unit F···H-F forms a straight line; however, the unit H-F···H is bent.

If hydrogen bonding were purely based on molecular dipoles, we would expect the entire ···H-F···H-F··· chain to be linear, not zig-zag. The reason it is bent instead is that the F in HF has lone pairs of electrons, as you learned in Chapter 2. The lone pairs stick out from the F at angles from the H, and the H from another molecule coordinates to the lone pair of an F. Thus hydrogen bonding is largely a dipolar effect, but the dipoles are the H-F bond dipole from the *donor* molecule (the one supplying the H) and the F-lone-pair dipole from the *acceptor*, and not the overall molecular dipoles. (The terms "donor" and "acceptor" are not used in the text.)

Some of the best evidence for hydrogen bonding is shown in Figure 5.9, for the hydrides of the group IVA, VIA and VIIA elements. (Although the hydride ion refers to H^-, the term hydride is also used for binary compounds of hydrogen; thus H_2O and H_2S are sometimes called hydrides of oxygen and sulfur, respectively.) You can see at a glance that something funny is happening with HF and H_2O: their boiling points are much higher than the trends in their columns of the periodic chart predict; therefore they experience unusually high attractive forces. This is due to hydrogen bonding.

Chapter 5 Intermolecular Forces, Condensed Phases, and Changes of State

HF has three lone pairs which can accept hydrogen bonds, but only one H to donate. For this reason, HF can only form hydrogen-bonded chains. Similarly, ammonia, NH_3, with three donor H's but only one lone pair, can only form chains. But H_2O has two donor H's plus two lone pairs, and therefore can form extended three-dimensional hydrogen-bonded structures in space. This is what happens in ice: as the text mentions, each water molecule in ice is surrounded by four neighbors; the molecule is acting as a hydrogen bonding donor to two neighbors and as an acceptor to two others. The result is a very "open" structure, with low density. This is why ice, unlike most solids, floats on its own melt. As the text mentions, the unique properties of ice and of liquid water, which are due to hydrogen bonding, are intrinsically tied up with the chemistry of life processes, both on the molecular level and on the ecological level.

Potential Energy

Potential energy is something we invent so that we can say that energy is conserved. It represents stored energy, which can be released later. For example, if I lift a book from the floor to the desktop, I have to do work against gravity. The book can do this same amount of work by falling to the floor. When the book is on the desk, it has higher (more positive) potential energy than when it is on the floor. When the book falls, it gains kinetic energy. Just before it hits the floor it has a kinetic energy equal to the potential energy it gained from being lifted onto the desk in the first place. After it hits the floor, its kinetic energy is transformed into random kinetic energy of the molecules in itself and the surroundings, which corresponds to an increase in temperature, though a very small one.

There are other sorts of potential energy besides gravitational. If two atoms which attract each other are pulled apart, their potential energy increases, much like lifting a book. If the molecules are released, they will fly together and accelerate, and the potential energy will decrease as the kinetic energy increases. This is true no matter what the nature of the attractive force — whether the attraction is between an Na^+ and Cl^- ion, between two H atoms which "wish" to form an H_2 molecule, between two HCl molecules in the gas phase experiencing a dipole-dipole attraction or two Ar molecules experiencing the London dispersive force. For all of them, as well, at big enough separation (r), the attractive force goes to zero, and it is convenient to define the potential energy as zero as well at r=∞*. If two bodies that attract have zero potential energy at large separations, and if the potential energy gets lower as they come together, then when they get closer they must have negative potential energy. As shown in Figure 5.12, however, as they come even closer the potential energy curve turns upward, and eventually becomes positive. Recall that we said earlier that the repulsive force experienced by two molecules is shorter in range than any of the attractive forces. This turning upward of the potential curve at short r is the result of the repulsive force. Recall that the more negative the potential energy, the more stable the system (potential energy gets more negative when a book falls, or when two molecules that attract come together). The fact that potential energy goes positive at very low r means that the two molecules are in an unstable configuration when they are this close. In fact, two molecules will tend to adopt the distance where their mutual potential energy is lowest, which is the value of r where ϵ, the maximum *depth of the potential well*, obtains.

We have said much about the nature of the attractive forces between atoms. What is the nature of the repulsive force? Naturally, as atoms come together their electron clouds repel, just because two negative charges repel. There is more to it than that, however. In Chapter 13 you will learn about the Pauli exclusion principle, which states in effect that certain kinds of particles, including electrons, cannot occupy the same region of space. The repulsive force experienced by two atoms coming together is due to this principle, as well as to the Coulombic interaction of the electron clouds. How does the curve in Figure 5.12 differ for the different types of attractive force? First of all, the depth, ϵ, will be deeper for the stronger forces. Second, starting at the minimum of the curve and going to longer r, the curve will approach the x-axis faster for the shorter range forces. Recall that London forces are shorter in range than dipole-dipole forces, and that these in turn are shorter in range than charge-charge forces.

---

*The zero of potential energy is arbitrary. I can define gravitational potential energy to be zero at sea level, or at the floor of my fourth storey apartment or anywhere else I choose; no matter which choice I make, however, the same book lifted to the top of the same desk at any location gains the same potential energy, assuming only that gravity is constant.

Chapter 5   Intermolecular Forces, Condensed Phases, and Changes of State

Solids, Liquids and Gases

In solids, the molecules or ions reside at fixed points in a regular arrangement or *lattice*. As the temperature rises, the molecules or atoms can vibrate about their lattice positions, and molecules can sometimes rotate or vibrate internally, but the lattice positions are fixed. Gas molecules, on the other hand, are in random motion, as we have seen. Liquids are the hardest of the three to describe precisely, because they exhibit a degree of order intermediate between the strict order of a solid and the chaos of a gas. Individual molecules still move randomly, but at any moment the immediate environment of a molecule in a liquid is somewhat ordered. Solids and gases have been well understood for some time, but achieving a correct understanding of the liquid state on a molecular level is still an active area of research.

You know, of course, that as a gas is cooled it becomes a liquid, and when the liquid is further cooled it becomes a solid. Both of these phenomena are examples of what are called *phase transitions*. A phase is a region of matter that is uniform throughout, physically as well as chemically, so the gas, liquid and solid states of a substance represent three different phases. Sometimes a substance exists in two different crystalline forms, such as diamond and graphite, allotropes of carbon. These are then two different solid phases of the same substance.

In general, energy is absorbed or released in a phase transition. The liquid-gas phase transition is called *vaporization*. It is an *endothermic* process, meaning that heat must be added as the process takes place. The reverse process is called *condensation* and is *exothermic*, meaning that it gives off heat. The heat content of a substance is called the *enthalpy*, denoted H, and represents the amount of heat it would take to raise the temperature of the substance from absolute zero to whatever temperature it is at, with the entire process taking place at constant pressure. The rationale for this definition will not be clear until Chapter 17, but until then the main thing to remember is that *enthalpy changes*, ΔH, involve the transfer of heat at constant pressure. When heat is added to a substance its heat content goes up, and its ΔH is positive; this we call endothermic — an endothermic process is one that absorbs heat, so the enthalpy of the substance undergoing the process must be increasing. Vaporization requires heat (is

Chapter 5 — Intermolecular Forces, Condensed Phases, and Changes of State

endothermic), and thus for the process

$$H_2O(l) \rightarrow H_2O(g)$$

$\Delta H > 0$. For vaporization, ΔH is denoted ΔH_{vap}, sometimes with a superscript denoting the temperature. For water, $\Delta H_{vap} = 44.10 kJ/mol$ or $2.45 kJ/g$.

The reason vaporization is endothermic should be clear from the previous section. When gases form liquids attractive forces come into play, and the molecules get closer together. In fact, near neighbors may approach the minimum of the potential energy-distance curve. To overcome this attraction and form the gas, energy must be added, in the form of heat. Note (Table 5.4) that the van der Waals a, the depth of the potential well, ϵ, ΔH_{vap} and the boiling point all increase together, supporting this explanation.

The solid-liquid transition is called *fusion*, which is just a fancy word for melting. Fusion is endothermic, but not as endothermic as vaporization: ΔH_{fus} for H_2O is $6.01 kJ/mol$, compared to $40.7 kJ/mol$ for ΔH_{vap}. Most solids expand on melting; for these solids the fact that $\Delta H_{fus} < \Delta H_{vap}$ can be explained by saying that in the solid the separation of minimum potential energy is more closely achieved than in the liquid, and to melt the solid requires that the molecules move farther apart, so that the potential energy increase on separation must be overcome by the addition of heat; of course, the substance expands less on fusion than on vaporization, so that not as much heat is needed for fusion as for vaporization. This cannot be the whole story, however, since, unlike most solids, water molecules get closer together when ice melts, and still ΔH_{fus} is positive. When ice melts the water molecules twist out of their ideal configurations, the hydrogen bonds become bent, and the whole structure partially collapses inward. The liquid configuration is the one of higher energy because the bent hydrogen bonds are not as stable (i.e., are not as low in energy) as the linear ones which are found in ice. Thus Figure 5.12 does not contain all the information necessary to explain the solid-liquid phase transition. The orientation of the molecules, as well as the intermolecular separation, is often important.

The solid-gas transition called *sublimation*. Most solids, as they are heated at 1 atm pressure, become liquids, then gases. Some, however, go directly to the gas without an intervening liquid stage.

Chapter 5   Intermolecular Forces, Condensed Phases, and Changes of State

For all solids, however, the liquid will be formed on heating if the pressure is high enough. Because of this, ΔH_{fus} and ΔH_{vap} can be measured for all substances, and, as shown in Figure 5.14, $\Delta H_{subl} = \Delta H_{fus} + \Delta H_{vap}$

Vapor Pressure

In Chapter 3 you learned about vapor pressure: that in a sealed vessel at constant T a liquid will evaporate until its partial pressure in the container is equal to its vapor pressure, $P°$, at this temperature. Recall also (Table 3.3) that $P°$ increases with T for any substance. We are now in a position to understand why this is. Recall that the velocities of a gas are given by the Maxwell-Boltzmann distribution (Figure 4.9); this is true for a liquid as well. At a given temperature, some fraction of the molecules in the liquid will have enough velocity (enough kinetic energy) to overcome the attractive forces and evaporate. At higher temperatures, a larger number of molecules have this much kinetic energy, and therefore evaporation is facilitated by higher temperatures. (Incidentally, the molecules left behind will have lower velocities, on average, than the molecules which have evaporated, and therefore will have lower kinetic energies and a lower temperature, unless heat is added from an outside source. This is why evaporating a dab of alcohol on your skin cools your skin. It is also in accord with evaporation's being endothermic: the cooling liquid absorbs heat from your skin.)

If the volume above the evaporating liquid is enclosed, vapor begins to build up in the vessel, and the velocities of the molecules in the gas also exhibit a Maxwell-Boltzmann distribution. As the pressure builds up on evaporation, however, the gas molecules get closer and closer together; the closer they get, the more the attractive forces are felt. Gas molecules with low velocities (toward the left of the Maxwell-Boltzmann curve in Figure 4.9) have low kinetic energies, and the closer the molecules are the greater the fraction of them that will experience an attractive force sufficient to lead to condensation. Thus, as the vapor pressure rises, the faster condensation takes place, until at some point it just equals the rate of evaporation. At this point, the pressure does not change further, and is equal to $P°$, the *equilibrium vapor pressure* of the substance at this temperature.

Chapter 5   Intermolecular Forces, Condensed Phases, and Changes of State

This is an excellent example of the chemical notion of equilibrium. By definition, equilibrium is a state where nothing macroscopic is changing: at P°, the amounts of gas and liquid in the vessel are remaining constant. Nevertheless, on the microscopic level, molecules are evaporating and condensing all the time. It is just that the *rate of evaporation* and the *rate of condensation* are equal. We call this a *dynamic equilibrium*; all equilibrium states in chemistry are dynamic in this same sense. Note that this picture, in which incessant motion on a microscopic scale gives rise to a static macroscopic state, has already been encountered in the explanation of the ideal gas laws as resulting from chaotic molecular motion.

Entropy

Note that in a sealed vessel the liquid and its vapor at partial pressure P° are in equilibrium, despite the fact that the liquid has lower potential energy, and despite the fact that condensing all the vapor would release heat and lower the energy of the system. This would seem to contradict the general observation that systems tend to move toward situations of lower potential energy; for example, a book moved off the edge of my desk will tend to fall. It turns out that the tendency toward lower energy is often counterbalanced by another intrinsic tendency of the universe: a tendency toward greater "freedom," "disorder," or *entropy*. This will be discussed further in Chapter 18, but note now that molecules in the gas phase are much freer to move about than molecules in the liquid phase; they have higher entropy, as well as higher energy. Higher energy is unfavorable, but higher entropy is favorable. Gas molecules evaporate just until these two factors balance. At this point, the pressure is equal to P° and the system is in equilibrium. Entropy is an important factor in many chemical and physical processes.

Boiling

We noted that P° rises as T rises. This is shown for water and for chloroform ($CHCl_3$) in Table 5.6. What happens when P° = 760mmHg (1atm) in an open beaker? At this point, the entire mass of liquid tries to vaporize, and produces bubbles throughout its mass in order to do so. P° is plotted against T for water and for chloroform in Figure 5.20. The normal boiling point for each substance is the temperature where its curve intersects the horizontal line in the figure corresponding to $P^\circ = 760$ mmHg. Note that if you were at a high altitude, such as Pike's Peak, where the atmospheric pressure was less than 760mmHg, the boiling point would be lower than the normal boiling point.

Several other phenomena are discussed at this place in the text. First, that of *volatility*. One substance is more volatile than another if it has a higher vapor pressure; thus chloroform is more volatile than water. This is just a point of vocabulary. Second is the phenomenon of *superheating*. Sometimes liquids can be heated above their boiling points before boiling actually begins. In such a case, boiling is likely to be violent when it does set in, and what is known as *bumping* can occur, and is especially common with viscous liquids.

Le Chatelier

Another new concept is Le Chatelier's principle. This is the idea that if a dynamic equilibrium is subjected to a stress, the equilibrium will shift so as to accommodate the stress. For example, you already know that if a liquid is placed in a cylinder with a piston, with space above the liquid, evaporation takes place until the space is filled with vapor at partial pressure P°. If the piston is now raised, making the space bigger, more liquid will evaporate; from the point of view of Le Chatelier's principle, the disturbance has been to make the volume bigger. The system must expand to accommodate this, and the way it does so is to turn some liquid into gas, since gases have bigger volumes than liquid. This process proceeds until the partial pressure of the vapor is once again equal to P°.

Another important example of Le Chatelier's principle involves heat effects. If a liquid and its vapor are at equilibrium, and the temperature is raised, then heat is being added to the system. Since a gas has a greater heat content than a liquid ($\Delta H_{vap} > 0$), the system accommodates this extra heat by turning some liquid into gas — so that P° increases with increasing T. Le Chatelier's principle will be discussed further in Chapter 8.

Chapter 5 Intermolecular Forces, Condensed Phases, and Changes of State

Phase Diagrams

According to the last paragraph, *any* endothermic process should take place to a greater extent at high than at low temperatures. Recall that $\Delta H_{subl} > 0$ for a solid; therefore the vapor pressure of a solid should increase with temperature. This along with the liquid-vapor curve, is shown for a typical substance in Figure 5.22. To interpret this diagram, consider the following. Suppose that we have a *one-component system* (i.e., a single pure substance). At any temperature along the x-axis of Figure 5.22, if the total pressure (which is equal to P° for a one component system) is less than the vapor pressure given by the curve, the substance will be entirely in the vapor form. If the total pressure is greater than the vapor pressure given by the curve, then the substance will be entirely in the solid or liquid form, depending on the temperature. Only if the total pressure is equal to the vapor pressure given by the curve can the vapor exist in equilibrium with the solid or liquid. At the *triple point*, where the curves intersect, all three phases can coexist at equilibrium.

By adding an additional curve to the diagram (Figure 5.23, 5.24) we obtain a *one-component phase diagram*. Here the y-axis really does represent the total pressure. The new line (OA) represents the liquid-solid equilibrium curve. Note that at low enough pressures there is no liquid phase (recall that solids sublime at low pressures). At pressures greater than the triple-point pressure, the solid melts, rather than sublimes, on heating. In Figure 5.23 higher pressures cause the melting point to increase. At any temperature above the triple-point temperature, applying pressure to the liquid will cause it to solidify. By Le Chatelier's principle, this must mean that the solid takes up less volume (is denser) than the liquid. This is the typical behavior. In Figure 5.24, on the other hand, applying pressure to the solid causes it to liquify. This must mean that the liquid takes up less space than the solid, and is denser. Figure 5.24 is, in fact, the phase diagram for water.

Critical Behavior

There is one more aspect of one-component phase diagrams to consider, a rather surprising one. It turns out that beyond a certain temperature, called the *critical temperature*, a gas cannot be liquified, no matter how high the pressure. This is due to the fact that the attractive forces that form the liquid are always competing against kinetic energy. As the temperature rises E_{trans} keeps increasing, until, beyond the critical temperature, the attractive force can no longer dominate over the combination of E_{trans} and the repulsive interactions, no matter how high the pressure.

For this reason, the liquid-vapor equilibrium curve (Figure 5.23, 5.24) ends at B, the *critical point*. The pressure at the critical point is called the *critical pressure*. Substances at temperatures above the critical temperature are called *supercritical fluids*. They are gases, not liquids, even though they can be pressurized to achieve the densities of liquids. Because they are gases they exhibit no surface tension, and very low viscosity. On the other hand, they exhibit good solvent properties at high density. This has made them useful for a variety of industrial processes, such as the extraction of caffeine from coffee beans, and of petroleum from partially depleted reservoirs.

Chapter 5 Intermolecular Forces, Condensed Phases, and Changes of State

Questions

5.1 Van der Waals forces

1. Van der Waals forces are the forces between (bonded, non-bonded) atoms and molecules.

2. An electron and a proton one angstrom apart have a dipole moment of _____ debye.

3. Charges of +1 and -1 one angstrom apart have the same dipole moment as charges of +2 and -2 _____ angstrom apart.

4. In a molecule AB, if B is the more electronegative atom, the positive end of the dipole will usually be on atom _____.

5. The molecule ABA, with B more electronegative than A, will have a dipole moment only if the molecule is (bent, linear).

6. Two dipoles tend to *repel* each other.

7. Name two types of van der Waals forces.

8. If electrons were not in continual motion, there would be no London dispersive force (True, False).

9. The London force is *a repulsive* interaction.

10. The London force tends to increase with increasing _____.

11. Which member of each of the following pairs will exhibit stronger London forces?

    (a) He, Xe

    (b) C_2H_6, C_4H_{10}

    (c) HI, HBr

Chapter 5   Intermolecular Forces, Condensed Phases, and Changes of State

5.2 The intermolecular potential energy function

1. List the electronegative atoms which exhibit hydrogen bonding.

2. Which is the strongest of the following interactions?

    _____ dipole-dipole

    _____ London dispersion

    _____ hydrogen bonding

3. Which of the following forces tend to increase the boiling point?

    _____ dipole-dipole

    _____ London dispersion

    _____ hydrogen bonding

4. (a) Name a substance that sinks in its own melt.

    (b) Name a substance that floats on its own melt.

5.3 The hydrogen bond

1. If two particles attract, bringing them closer together *raises* the potential energy.

2. If two particles repel, bringing them further apart *raises* the potential energy.

3. (a) Two molecules at infinite separation have potential energy of _____.

    (b) As they are brought together, the potential energy _____, and they begin to experience an (attractive, repulsive, zero) force.

Chapter 5 — Intermolecular Forces, Condensed Phases, and Changes of State

(c) When they are brought closer than the minimum depth of the potential well, the potential energy _____, and they experience an (attractive, repulsive, zero) force.

(d) At potential energy ϵ, the particles experience an (attractive, repulsive, zero) force.

4. As ϵ increases, the van der Waals a tends to _____ and the boiling point tends to _____.

5. (a) For two molecules of similar molecular weight, ϵ _____ as dipole moment increases.

   (b) For two molecules of similar polarity, ϵ _____ as molecular weight increases.

5.4 Molecular arrangements in the solid, liquid and gaseous states

1. (a) A gas (generally, rarely, never) has lower density than the corresponding liquid.

   (b) A liquid (generally, rarely, never) has higher density than the corresponding solid.

2. The most orderly arrangement of molecules is found in a (gas, liquid, solid).

5.5 Vaporization, fusion, and sublimation

1. When a process absorbs heat, it is (endothermic, exothermic), tends to (heat, cool) its environment and exhibits a (negative, positive) ΔH.

2. For vaporization, $\Delta H (<, >, =) 0$.

3. Vaporization is (more, less) (exothermic, endothermic) than fusion, and (more, less) (exothermic, endothermic) than sublimation.

4. Hydrogen bonding tends to (increase, decrease, have no effect on) ΔH_{vap}.

5. ΔH_{vap} and ΔH_{fus} are (exactly, approximately) constant with temperature.

6. Consider the two processes

   $A(l) \rightarrow A(s)$ 1.

   $A(s) \rightarrow A(g)$ 2.

   (a) For the first process, ΔH is (positive, negative, zero).

Chapter 5   Intermolecular Forces, Condensed Phases, and Changes of State

(b) For the *overall* process (1+2), ΔH is (positive, negative, zero).

5.6 The equilibrium between a liquid and its vapor

1. (a) For a liquid in a sealed vessel in equilibrium with its vapor, evaporation and condensation *are not taking place*.

    (b) If the vessel is suddenly expanded at constant T, the evaporation rate (increases, decreases, remains constant), and the condensation rate (increases, decreases, remains constant).

    (c) If instead the temperature is raised at constant volume, the evaporation rate (increases, decreases, remains the same), and the condensation rate (increases, decreases, remains the same).

    (d) If instead more liquid is added to the vessel at constant P and T, the evaporation rate (increases, decreases, remains constant) and the condensation rate (increases, decreases, remains constant).

2. The *relative humidity* of air is defined as $P/P°$, where P is the actual partial pressure of H_2O in air, and $P°$ is the equilibrium vapor pressure.

    (a) On a hot day, air of 60% relative humidity will contain (more, less, the same amount) of water vapor than on a cool day.

    (b) At the same temperature and total pressure, air at 60% relative humidity will contain (more, less, the same amount) of water vapor than/as N_2 at 60% relative humidity.

3. A liquid and its vapor are in equilibrium in a cylinder fitted with a piston and maintained at constant temperature.

    (a) The piston is forcibly depressed until the pressure rises. The cylinder now contains (pure liquid, pure vapor, liquid plus vapor).

    (b) Instead, the piston is forcibly raised until the pressure lowers. The cylinder now contains (pure liquid, pure vapor, liquid plus vapor).

Chapter 5 — Intermolecular Forces, Condensed Phases, and Changes of State

5.7 The nature of the equilibrium state: the tendency to decrease the potential energy and to increase the molecular disorder

1. For each of the following pairs, which has the lower energy and which the lower entropy?

    (a) gas, liquid

    (b) liquid, solid

    (c) solid, gas

2. (a) For vaporization, ΔH is (positive, negative, zero) and ΔS is (positive, negative, zero).

    (b) For freezing, ΔH is (positive, negative, zero) and ΔS is (positive, negative, zero).

    (c) A process tends more to go if its ΔH is more (positive, negative) and its ΔS is more (positive, negative).

5.9 The temperature dependence of the equilibrium vapor pressure and the normal boiling point

1. The normal boiling point is the temperature where the vapor pressure is equal to _____.

2. (a) In a pressure cooker, water will boil (at, above, below) the normal boiling point.

    (b) On the moon, water will boil (at, above, below) the normal boiling point.

5.10 Phase diagrams for pure substances

1. (a) At a pressure and temperature along a line on a phase diagram, how many phases can coexist in equilibrium?

    (b) At a pressure and temperature between lines on a phase diagram, how many phases can coexist in equilibrium?

    (c) At a pressure and temperature where three lines meet on a phase diagram, how many phases can coexist in equilibrium?

    (d) What is the point described in part (c) called?

2. If the liquid-solid line has a positive slope, then

    (a) At a constant T (smaller than, greater than) the triple point the liquid can be made by (increasing, decreasing) the pressure on the solid.

Chapter 5 — Intermolecular Forces, Condensed Phases, and Changes of State

(b) The solid has a (lower, higher) density than the liquid.

3. If the liquid-solid line has a negative slope, then

   (a) At a constant T (smaller than, greater than) the triple point the liquid can be made by (increasing, decreasing) the pressure on the solid.

   (b) The solid has a (lower, higher) density than the liquid.

4. Under each of the following circumstances, tell whether a substance will be a gas or a liquid.

   _____(a) Just below the critical temperature and just below the critical pressure.

   _____(b) Just below the critical temperature and just above the critical pressure.

   _____(c) Just above the critical temperature and just below the critical pressure.

   _____(d) Just above the critical temperature and just above the critical pressure.

5. Hydrogen bonding tends to (raise, lower) the critical temperature.

Chapter 5   Intermolecular Forces, Condensed Phases, and Changes of State

Answers

5.1 Van der Waals forces

1. non-bonded
2. 4.80
3. 1/2
4. A
5. bent
6. attract
7. dipole-dipole, London dispersive
8. True
9. an attractive
10. molecular weight, number of electrons
11. (a) Xe
    (b) C_4H_{10}
    (c) HI

5.2 The intermolecular potential energy function

1. F, O, N
2. hydrogen bonding
3. (all three)
4. (a) (almost anything)
   (b) water

5.3 The hydrogen bond

1. lowers
2. lowers
3. (a) zero
   (b) decreases, attractive
   (c) increases, repulsive
   (d) zero
4. increase, increase
5. (a) increases
   (b) increases

5.4 Molecular arrangements in the solid, liquid and gaseous states

1. (a) generally
   (b) rarely
2. solid

5.5 Vaporization, fusion, and sublimation

1. endothermic, cool, positive
2. $\Delta H > 0$
3. more, endothermic, less, endothermic
4. increase

5. approximately
6. (a) negative
   (b) positive

5.6 The equilibrium between a liquid and its vapor

1. (a) are taking place at equal rates
   (b) remains constant, decreases
   (c) increases, decreases
   (d) remains constant, remains constant

2. (a) more
   (b) the same amount

3. (a) pure liquid
   (b) pure vapor

5.7 The nature of the equilibrium state: the tendency to decrease the potential energy and to increase molecular disorder

1.

	Lower E	Lower S
(a)	liquid	liquid
(b)	solid	solid
(c)	solid	solid

2. (a) positive, positive
   (b) negative, negative
   (c) negative, positive

5.9 The temperature dependence of the equilibrium vapor pressure and the normal boiling point

1. 1atm (or 760mmHg)
2. (a) above
   (b) below

5.10 Phase diagrams for pure substances

1. (a) two
   (b) one
   (c) three
   (d) the triple point

2. (a) greater than, decreasing
   (b) higher

3. (a) less than, increasing
   (b) lower

4. (a) gas
   (b) liquid
   (c) gas
   (d) gas

5. raise

CHAPTER 6: PROPERTIES OF DILUTE SOLUTIONS

In the last chapter you learned about the dynamic equilibrium that exists between a liquid and its vapor in a sealed vessel; recall that in such a situation molecules are evaporating and condensing all the time, but that the rates of evaporation and condensation are equal. In this chapter you will see that the same sort of dynamic equilibrium exists between a *solute* and its *saturated solution* in some *solvent*. In fact, there are certain analogies which we will point out between a solute dissolving and a liquid vaporizing. These can be used as an aid in remembering what happens in solutions, but do not take them too seriously, because the comparison is not a rigorous one.

Saturation

First, some terminology. When two or more components mix to form a homogenous *solution*, if one is present in large excess, it is called the *solvent*, and the other is called the *solute*. The ideas of solute and solvent are especially useful when liquids dissolve solids, because usually in such cases only a small amount of solid will dissolve in a given quantity of liquid. Extra solid added beyond this sinks or floats, as the case may be, but once the *solubility* is reached the amount of solute per gram or milliliter solvent or solution stays constant. The solution is then termed *saturated*.

This is somewhat analogous to the vaporization of a liquid in a sealed container, where the volume of solution is like the volume of the sealed container. The process only occurs until a certain concentration (vapor pressure, P°, for evaporation) is reached, and then extra liquid (for vaporization) or solute (for solution) is left behind. Also, in the same way that the vaporized liquid molecules experience the entire space above the liquid, the dissolved solute molecules experience the entire volume of the solution. For many solvent-solute pairs, the solubility increases with temperature, but for many others it decreases, so that here the analogy between vaporization and solubility breaks down: for all substances, P° increases with increasing temperature.

Many pairs of liquids are only partially soluble in each other; for example, only 8.11mL of ether will dissolve in 100mL of water at 22°C, and for them the terms solvent and solute are still useful. Many others pairs are *miscible*, however, meaning that they form solutions in all proportions; one example is

Chapter 6   **Properties of Dilute Solutions**

ethyl alcohol and water. In a 50:50 mixture of the two it does not make sense to consider one the solute and the other the solvent. The vaporization/solution analogy breaks down here, too, since saturation does not occur.

The equilibrium between a pure solute and its saturated solution is analogous to that between a liquid and its vapor, in that both are dynamic. In the saturated solution, molecules are continually dissolving and precipitating out, but the rates of dissolution and precipitation are equal.

Mole Fraction, Molality and Molarity

These are all commonly used concentration units in chemistry, and are all based on the number of moles of one or more component. There are also concentration units based on weight, such as grams solute per 100mL solvent, and it is important to be able to convert back and forth between any two concentration units. Something to remember when you are asked to do this is that if you have to convert between a weight-based and a mole-based unit you need to know a molecular weight, and if you are converting between one unit that involves the volume of solvent or solution and another that involves the weight, you need to know a density.

The following table sets out the definitions of the commonly used concentration units.

Concentration Units

Name	Abbreviation	Definition
mole fraction	X	moles solute/mole solution
molality	m	moles solute/kilogram solvent
molarity	M	moles solute/liter solution
weight percent	%	grams solute × 100%/gram solution
"concentration"	c	grams solute/100mL solvent

The last unit listed is named rather arbitrarily; there is no standard name for this unit, which is used in many tabulations of solubility data.

The definition of mole fraction has already been given in Chapter 3. As an example, consider a saturated solution of ether in water at 22°C, which we already said comprises 8.11mL of ether in 100mL of water. In order to calculate X_{ether} in this mixture we need to know the number of moles of both ether

Chapter 6 — Properties of Dilute Solutions

and water in the solution. In order to do this we have to know the molecular weights and the actual weights used of both ether and water. The density of water is 1.00g/mL and that of ether is 0.714g/mL. Thus

$$\text{weight ether} = (8.11\text{mL})(0.714\text{g/mL}) = 5.79\text{g}$$

$$\text{weight water} = (100\text{mL})(1.00\text{g/mL}) = 100\text{g}$$

The molecular weight of water is 18.0, and that of ether is 74.1, so that

$$\text{moles ether} = \frac{5.79\text{g}}{74.1\text{g/mole}} = 0.0781$$

$$\text{moles water} = \frac{100\text{g}}{18.0\text{g/mole}} = 5.56 \text{ mole}$$

$$X_{\text{ether}} = \frac{\text{moles ether}}{\text{total moles}}$$

$$= \frac{0.0781}{5.56 + 0.0781} = 0.0139$$

Of course, we must have $X_{\text{ether}} + X_{\text{water}} = 1$, so that $X_{\text{water}} = 0.9861$.

For the same solution, what is the molality, m? Since 100mL of water weighs 100g, 1000mL must weigh 1kg, and must dissolve 10(5.79)=57.9g, or 0.781 mole, of ether, so that

$$m_{\text{ether}} = \frac{0.781 \text{mole ether}}{1\text{kg water}} = 0.781 m$$

From what we have already done, it is easy to calculate the weight percent ether in the saturated solution.

$$\%\text{ether} = \frac{\text{grams ether} \times 100\%}{\text{grams ether} + \text{grams water}} = \frac{5.79\text{g} \times 100\%}{100\text{g} + 5.79\text{g}} = 5.47\%$$

How about the molarity, M? Note that the molarity requires the number of moles of solute per *liter* of *solution*. It is tempting to say that when 8.11mL of ether is added to 100mL, the resulting solution has volume 108mL, and proceed from there; unfortunately, total volumes often change by a few percent or more on mixing, and it is not even possible to know in advance whether this volume change will be positive (an expansion) or negative (a contraction). It is found experimentally that a saturated solution of ether in water has a density of 0.985g/mL, and this fact can be used to calculate the molarity. A liter of such a solution must have a mass of

Chapter 6 — Properties of Dilute Solutions

$$(1000\text{mL})(0.985\text{g/mL}) = 985\text{g}$$

Since 5.47% of this weight is ether, as determined above, a liter of solution must contain

$$(9.85\text{g})(0.0547) = 53.9\text{g ether}$$

Finally, since the molecular weight of ether is 74.1, a liter of solution contains

$$\frac{53.9\text{g}}{74.1\text{g/mole}} = 0.727\text{mol ether}$$

Thus the molarity is 0.727M.

Note that as the temperature changes, the volume of a solution changes, but the weight and number of moles do not. Therefore the molarity of a given solution will change with temperature, whereas the mole fraction, molality and weight percent will not.

Calculations Involving Molarity

Solutions whose concentrations are expressed in molarity are especially easy to prepare and to use in certain types of chemical calculations. The reason it is easy to prepare solutions of specified molarity is that it is easier (though not as precise) to measure volume than mass. One still has to weigh out the appropriate number of moles of solute, but then this can be added to a *volumetric flask* and diluted, with mixing, to the mark, which guarantees that the final solution has the volume specified on the flask, to within experimental error. Suppose one makes up 1.00L of 0.100M solution of NaCl this way. By using a calibrated pipet to transfer, say 10.0mL of the solution to a 100mL volumetric flask, and then diluting to the mark, one is taking (0.100 moles/L)(0.010 L)=0.001mol of NaCl and diluting this to a final volume of 0.100L, for a final molarity of 0.001mol/0.100L=0.01M. Thus it is easy to use concentrated solutions to make up dilute solutions of known concentration by volumetric techniques, if all concentrations are specified in molarity.

You should know that some chemists use the term *formality* to describe the concentration of ionic substances. Thus one gram-formula weight of NaCl in 1L solution is a 1F solution; most chemists are content to use the term "molarity" even in such cases, however.

Raoult

Solutions, as well as pure substances, have vapor pressures. We consider first a solution of a *non-volatile* solute in a *volatile* solvent; thus all the vapor pressure can be attributed to the solvent. In all such solutions, the vapor pressure goes down as solute is added. In discussing dilute two-component solutions, we usually use subscript 1 for the solvent and 2 for the solute. Thus, P°_1 is the vapor pressure of pure solvent, and P_1 is the actual partial pressure of the solvent above the solution. For a non-volatile solvent, P_1 is also the total vapor pressure of the solution, since P_2 can be ignored. The assertion that addition of a solute depresses the vapor pressure of the solvent can be stated $P_1 < P_1^\circ$.

In the pure solvent $X_1 = 1$, of course, and as solute is added X_1 decreases. Francois Raoult discovered in the late nineteenth century that for certain solutions

$$P_1 = X_1 P_1^\circ$$

that is, the vapor pressure of component 1 is directly proportional to its mole fraction. This is known as *Raoult's law*, and even though only a few solutions exhibit it over the entire concentration range, all solutions exhibit it in the dilute regime; that is, where X_1 is greater than about 0.9 or so. Solutions exhibiting Raoult's law are called *ideal solutions*; all solutions act like ideal solutions if they are sufficiently dilute.

This is somewhat analogous to the fact that all gases act like ideal gases at sufficiently low pressures. You saw in Chapter 4 that the idea behind an ideal gas is that the molecules do not interact. The liquid state owes its very existence to attractive forces; what is behind the idea of an ideal solution is that the solute and solvent molecules have the same degree of attraction to each other as to themselves; in an ideal solution of component 2 in component 1, molecular pairs 1-1, 2-2 and 1-2 exhibit equal attractive (and repulsive) forces. The vapor pressure of a substance can be interpreted as its tendency to escape into the gas phase. In an ideal solution, the addition of component 2 does not cause a molecule of component 1 to be any more tightly or loosely held into the solution than it is held in the pure liquid; however, since the addition of 2 causes the concentration of 1 (X_1) to decrease, the net escaping tendency of 1 is decreased.

Chapter 6 — Properties of Dilute Solutions

The *vapor pressure lowering* of the solution, ΔP, is given by

$$\Delta P \equiv P_1^\circ - P_1 = P_1^\circ - X_1 P_1^\circ$$

$$= P_1^\circ(1-X_1) = P_1^\circ X_2$$

Here we first applied Raoult's law to express P_1 in terms of P_1°, then used the fact that for a two-component system $X_1 + X_2 = 1$. This equation holds for ideal solutions only, but recall that this category includes all dilute solutions.

Vapor pressure lowering is an example of a *colligative property*: one which depends not on what kinds of particles are dissolved, but only on how many. The phenomenon of vapor pressure lowering and its expression for ideal solutions in Raoult's law is important for understanding other properties as well, as we shall see shortly. The direct measurement of vapor pressure can be used for molecular weight determination, as in text Example 6.7, but this is not common experimental practice.

Boiling Point Elevation and Freezing Point Depression

Recall that the vapor pressure of a solution of a non-volatile solute in a volatile solvent will be lower than that of the pure solvent at the same temperature. Recall as well that the vapor pressure of a pure liquid rises with temperature; this is true for a solution as well. Now suppose that at one atmosphere a pure liquid is boiling. Its temperature is then the normal boiling point, so that $P^\circ = 760$ mmHg. Suppose that we add some non-volatile solute to the liquid. This lowers the vapor pressure. In order for the solution to boil, the temperature must be raised until the vapor pressure in once again 760mmHg. Thus addition of non-volatile solute raises the normal boiling point; this can be seen graphically on a phase diagram in Figures 6.8 and 6.9.

For dilute solutions, it is found that the *boiling point elevation*, ΔT_b, is proportional to the molality.

$$\Delta T_b = K_b m$$

K_b is called the *molal boiling point elevation constant*, and differs from solvent to solvent. Boiling point elevation can be used for molecular weight determination; for example, suppose a solution of 10.0g of a non-volatile substance is dissolved in 1kg water, and the resulting solution boils at 100.05°C. Then $\Delta T_b =$ (100.05–100.00)°C = 0.05°C. Since K_b for water is 0.51°C, the solution must be 0.1m, so the gram-molecular weight must be (10.0g/0.1) = 100g.

Several things are worth noting here. First, since K_b's are small, ΔT_b's are also small. Since it is hard to measure temperature to great accuracy, the values of MW, molecular weight, obtained using this technique can be associated with large errors. Often, however, when one has the empirical formula and is trying to determine the molecular formula, one does not need great accuracy. Second, accuracy is improved if one can use as a solvent a compound with a relatively large K_b; however, one's choices are limited by what the unknown will dissolve in. Third, boiling point elevation is another colligative property, so that when we determined that m=0.1, this was really the number of moles of particles, not of "formula units." If the unknown is molecular, then our molecular weight is correct, within experimental error. Suppose, however, that our unknown turned out to be a substance, AB, that breaks up into A^+ and B^- ions in water. Our experiment says that the solution is 0.1m in total ions; thus it is only 0.05m in AB. The formula weight of AB is then 200, not 100. Without other data — for example, knowing whether the unknown is in fact molecular — colligative properties do not allow one to distinguish among such possibilities.

Freezing poing depression is also a consequence of vapor-pressure lowering. It should be clear from Figure 6.9 that the triple point of a liquid is lowered by the addition of a non-volatile solute. In order to deduce this we look at the intersection of the solution line on the phase diagram with that of the pure solid. Note that this lies to the left of (that is, at a lower temperature than) the intersection of the lines for the pure liquid and solid. This means that the triple point of the solution — the unique pressure and temperature where the liquid, solid and gaseous phases can coexist at equilibrium — lies at a lower temperature for the solution than for the pure liquid. The fact that we look at the intersection of the pure solid line with the solution line means that we are assuming that the solution is in equilibrium with the pure solid; in other words, when a solution freezes the solute is excluded from the solid phase; only the solvent freezes out. Because of this, as a solution freezes it gets more concentrated. The remaining solution has a liquid-gas equilibrium curve below the old one on the phase diagram, so that its triple point is a bit below the old one, and as freezing takes places the triple point gets lower and lower. (Something analogous happens in boiling, only there the solvent forms vapor, so that the remaining solution gets more concentrated, successively raising T_b as boiling proceeds.)

The fact is that T_f, the freezing point, is very close to the triple point (the triple point of water is about 0.01°C), and what we have described for the depression of the triple point also occurs for T_f. ΔT_f, the freezing point depression, obeys the following relationship, for dilute solutions

$$\Delta T_f = K_f m$$

K_f, like K_b, depends only on the solvent, and K_f's tend to be larger than K_b's, so that molecular weights tend to be associated with less error when determined using T_f instead of T_b. Note from Table 6.2 that for water $K_f \approx 3.6 K_b$, and that camphor has a very large K_b, equal to 40 kg·K/mol. Incidentally, both the T_b and T_f methods of determining molecular weight are plagued with the problem that the linearity of ΔT with m breaks down at high concentrations (non-dilute solutions), but one would like to use high concentrations to obtain easily measurable ΔT's.

As mentioned earlier, the molecular weights determined by freezing- and boiling- point methods are good only for *non-electrolytes*; an *electrolyte* is a substance that breaks up into ions in solution. I will deviate slightly from the text in my discussion of freezing point depression and boiling point elevation for electrolytes. If a compound breaks up into ν ions (for example, $AB \rightarrow A^+ + B^-$ exhibits $\nu=2$), then, theoretically, we should have

$$\Delta T_f^{theor} = \nu K_f m$$

$$\Delta T_b^{theor} = \nu K_b m$$

Experimentally, however, not all electrolytes dissociate completely, so that in a solution of AB, for example, some of the A^+ and B^- ions exist as A^+B^- ion pairs, lowering the number of moles of particles per kilogram of solvent from (νm) to something less, so that the *experimental* T_b, T_b^{exp}, is less than T_b^{theor}. We now use the symbol i for what the text calls the *van't Hoff mole number* and write

$$\Delta T_f^{exp} = i K_b m$$

$$\Delta T_b^{exp} = iK_b m$$

Thus, for freezing, $i = \Delta T_f^{exp}/K_f m$. In general, as a solution becomes more dilute, ΔT_f^{exp} approaches ΔT_f^{theor}, and i approaches ν. For NaCl (Table 6.4), i increases from 1.87 to 1.96 as m decreases from 0.1m to 0.005m. Of course, $\nu_{NaCl}=2$. In Chapter 7 we will discuss how to use freezing-point and boiling-point results to calculate the percent dissociation of an electrolyte.

The original idea that an acid is a substance which ionizes in solution to form H^+ ions came from measurements of T_f^{exp}. For example, HCl is a gas, and it was known from gas density measurements that the molecular weight is about 36. A 0.100m aqueous solution should then exhibit a ΔT_f of about 0.186K; however, it exhibits close to twice this, leading to the hypothesis that two particles are formed in solution for each HCl gas molecule that dissolves. Arrhenius proposed that this release of H^+ ions occurs for all acids.

Osmosis

Certain membranes, called *semipermeable*, have pores big enough to allow small molecules (such as water) to pass, but not big enough to let big ones (such as proteins) through. If a solution of, say, a protein in water is separated from the pure solvent by such a membrane in an arrangement like the one shown in Figure 6.10, water will pass through the membrane, diluting the solution, until the diluted solution reaches some height, h, and then stop. First let us discuss what causes this phenomenon.

If the protein solution had been added to the beaker of water with no membrane present, gradually the solution and the surrounding water would have diffused into each other, forming a homogeneous, dilute solution. This is the same process as the diffusion of gases discussed in Chapter 4. Recall that this process is driven by entropy: in the homogeneous mixture both components have the freedom to experience the total solution volume, whereas before diffusion takes place the protein only experiences a small volume. Furthermore, whenever a solution becomes more dilute the entropy increases, since the solute can then experience a greater volume. It is this increase in entropy which causes the water to diffuse through the membrane to dilute the protein solution.

How far does this dilution process go? As the height of the column of solution increases, its gravitational potential energy increases, until the (unfavorable) gain in energy just balances the (favorable) gain in entropy; then the process stops. The pressure equivalent to the height of the column of solution is then just equal to the osmotic pressure of the diluted solution.

It is more convenient, however, to know the osmotic pressure of the solution we weighed out and prepared initially. This can be obtained by seeing how much pressure one has to add to the vertical tube to *suppress* the column rise. This is π, the *osmotic pressure*, of the original solution. Osmotic pressure is a colligative property, and obeys the equation

$$\pi = cRT$$

where c is the concentration in molarity and R is the gas constant. Osmosis is a very sensitive technique; low c's give large π's. In Example 6.11 an osmotic pressure of 26.1mmHg, or 0.0343atm, is obtained for a particular aqueous solution at 30°C. This corresponds to a concentration of

$$c = \frac{\pi}{RT} = \frac{0.0343 \text{ atm}}{(0.0821 \text{L·atm/K·mol})(303\text{K})}$$

$$= 0.00138\text{M}$$

For dilute aqueous solutions, molarity is numerically equal to molality; this solution would thus exhibit a ΔT_f of $(0.00138)(1.86)\text{K} = 0.00256\text{K}$, which would be very difficult to measure accurately, whereas a pressure of 26.1 mmHg is easy to measure. From the molarity, the weight of solvent and the volume of solution, all given in the example, it is trivial to show that the molecular weight comes out to 5800 for insulin. Because of its sensitivity, osmotic pressure used to be widely used to determine the molecular weights of large biological molecule; it has now been widely superseded by other techniques, however.

Two Volatile Liquids

In an ideal solution of two volatile components, both obey Raoult's law. Recall that in an ideal solution the molecules of the two components have the same degree of interaction with each other as with themselves; thus the only solutions that really act ideal over their entire concentration range are those where the molecules of the two components are very similar chemically; the text uses the example

Chapter 6 — Properties of Dilute Solutions

of benzene and toluene. Note that if both components are volatile they both have significant partial pressures in the vapor above the solution. Assuming that only the two components are present, we can write

$$P_{total} = P_B + P_T$$

where B and T stand for benzene and toluene respectively. We also use the notation X^{liq} and X^{gas} to refer to mole fractions in the liquid and vapor phase. Then we have

$$X_B^{liq} + X_T^{liq} = 1$$

$$X_B^{gas} + X_T^{gas} = 1$$

The solution is ideal, so that Raoult's law applies to both components

$$P_B = P_B^\circ X_B^{liq}$$

$$P_T = P_T^\circ X_T^{liq}$$

These relationships are illustrated in Figure 6.11: each Raoult's law relationship is a linear one, and the sum of the two linear relationships, giving P_{total}, is another straight line.

These equations for Raoult's law give the partial pressure, P, of each component in terms of X^{liq}, its mole fraction in the liquid. P can also be related to X^{gas}, using Dalton's law of partial pressures

$$P_B = P_{total} X_B^{gas}$$

$$P_T = P_{total} X_T^{gas}$$

Presumably we know the values of X^{liq}, since we measured out our solution compounds carefully, but we do not know the values of X^{gas}. By setting the Raoult's law and Dalton's law expressions for P_B and P_T equal to each other (not in text) we can write

$$P_B^\circ X_B^{liq} = P_{total} X_B^{gas}$$

$$P_T^\circ X_T^{liq} = P_{total} X_T^{gas}$$

By dividing these expressions we get

$$\frac{X_B^{gas}}{X_T^{gas}} = \frac{P_B^\circ}{P_T^\circ} \frac{X_B^{liq}}{X_T^{liq}}$$

This equation is the basis of *distillation*. In order to see what the equation means, suppose we start with a solution of one mole each of benzene and toluene. Then $X_B^{liq}=X_T^{liq}=0.5$, so that $(X_B^{liq}/X_T^{liq})=1$. From the data given in the text, $(P_B^\circ/P_T^\circ) = $ 75mmHg/22mmHg=3.4. The equation then gives

$$\frac{75\text{mmHg}}{22\text{mmHg}} \frac{0.5}{0.5} = \frac{X_B^{gas}}{X_T^{gas}} = 3.4$$

Thus, even though benzene and toluene have equal mole fractions in the liquid phase, the gas phase is enriched in the more volatile component, namely benzene, by a factor of 3.4 to 1. In fact, no matter what the ratio of mole fractions in the liquid phase, the gas phase would be enriched, relative to it, in benzene, the more volatile component, by the factor P_B°/P_T°.

It is important to note that P° for any liquid is a function of temperature, so that a ratio like P_B°/P_T° will change with temperature. Let us now discuss a *constant temperature* process which is analogous to distillation. Suppose we were to start out with the solution we have described, equimolar in benzene and toluene, and let some of it vaporize into some volume. As we have already calculated, $(X_B^{gas}/X_T^{gas}) = 3.4$. If we take this vapor, and compress it, to form a liquid, then, in this new liquid, $(X_B^{liq}/X_T^{liq}) = 3.4$. If we now let some of this new liquid vaporize, we will have, for the new gas phase, $(X_B^{gas}/X_T^{gas}) = (3.4)(3.4) = 11.6$. This was obtained from the same equation as before, starting with $(X_B^{liq}/X_T^{liq}) = 3.4$, instead of 1.0. In fact, after n "distillations," each time compressing the vapor and then allowing the new liquid to vaporize, we will have

$$\frac{X_B^{gas}}{X_T^{gas}} = \left(\frac{X_B^{liq,initial}}{X_T^{liq,initial}}\right) \left(\frac{P_o^B}{P_o^T}\right)^n$$

Since $(P_o^B/P_o^T)>1$, by repeating the process often enough (thereby making n big enough) we can make (X_B^{gas}/X_T^{gas}) as big as we want; that is, we can enrich the final product as much as we desire in the volatile component, benzene. This repeated process is analogous to *fractional distillation*; however, these processes are not identical to the way distillation is really done.

Chapter 6   Properties of Dilute Solutions

Real Distillation

In a *simple*, or one stage *distillation*, the mixture is heated to boiling. The boiling point of the mixture is where P_{total}=760mmHg. As before, the vapor will be enriched in the more volatile component. This vapor is condensed and collected, and as the process proceeds, the undistilled residue in the pot becomes enriched in the *less* volatile component, so that P_{total} tends to decrease; however, heat is constantly supplied to keep the mixture boiling, so that as the composition of the residue changes the temperature of the pot increases. The vapor will continue to be enriched in the more volatile component, but by the factor (P_B°/P_T°) that obtains at the new, higher temperature, and starting with a ratio (X_B^{liq}/X_T^{liq}) that is successively smaller than the initial one. Subsequent drops of distillate will be successively less enriched in the more volatile component than the initial drop, since the residue becomes successively less enriched in the less volatile component. (The ratio (P_B°/P_T°) doesn't change much.) Note that the difference between this process and the analogous process we described earlier is that in our earlier process we compressed and expanded vapors at constant T; we performed our separations by varying P. In real distillations, P_{total} is held constant at 760mmHg, and we allow T to increase.

A real *fractional distillation* can be thought of in the following manner. The liquid in the pot is held at its boiling point, and the vapor and liquid compositions are in equilibrium by the equation we have been using, so that the vapor is enriched in the more volatile component. This vapor is condensed into a new liquid as it ascends the column, and is held at *its* boiling point, which is lower than the pot temperature. The vapor from this distilled then condensed liquid is further enriched in the more volatile component, and is re-condensed into a new liquid, which is maintained at its still lower boiling point, and so on. All these stages take place up a *distillation column*, which, in an industrial facility (such as a petroleum refinery) really is sometimes organized in discrete stages; in the laboratory it is more often a column filled with *packing*, which provides surface area onto which can the liquid condense and make contact with the vapors. As one goes up the column, the temperature decreases, since higher up corresponds to later stages, which have lower boiling points. Vapor is collected from the top of the column and condensed. The more stages of distillation, the greater the enrichment of the final distillate in the more volatile component.

What we have just described constitutes a "snapshot" of a fractional distillation. If no new liquid is added to the pot, then, as the process proceeds, the residue gets more and more enriched in the less volatile component, its temperature rises, and so does the temperature of every stage. The process is stopped when enough of the less volatile component starts appearing in the distillate to violate some predetermined criterion of purity. In most industrial processes, fractional distillations are run not batch by batch, but in a continuous process. Fresh initial liquid is constantly supplied to the pot to maintain a constant volume. Then temperatures and compositions remain constant throughout. This is the stuff of which chemical engineering courses are made.

Non-Ideal Solutions

If the energetic attractions between the two solution components (1-2 interactions) are stronger than the 1-1 and 2-2 interactions, this will decrease the tendency of both materials to escape from the solution, relative to the pure liquids, so that for both components the partial pressure will be less than that expected from Raoult's law. This is called a *negative deviation* from Raoult's law: $P < X^{liq} P^o$, for P_1 and P_2. (Figure 6.13). Since the energetic interaction on forming such a solution must be favorable, energy must be released in the form of heat when the components are mixed, so that negative deviations from Raoult's law are associated with exothermic mixing processes, $\Delta H_{mix} < 0$. An example is ethyl alcohol and water.

Where the 1-1 and 2-2 interactions are energetically favored over the 1-2 interactions, mixing is endothermic, and $\Delta H_{mix} > 0$. Here, the escaping tendency is greater than that expected from the pure liquids, so that such solutions exhibit vapor pressures greater than the Raoult's law value (Figure 6.14). Naturally, solutions exhibiting positive deviations from Raoult's law boil at lower temperature than the corresponding ideal solutions would, since P_{total} is greater, and those exhibiting negative deviations boil higher.

Chapter 6 — Properties of Dilute Solutions

Solutions exhibiting positive deviations from Raoult's law often expand on mixing, and those exhibiting negative deviations often contract. This is because if the molecules in solution exhibit stronger attractive forces than they do in the pure state (negative deviations) the molecules will tend to approach each other more closely when they are mixed, and vice versa for positive deviations.

Henry's Law

Recall that if Raoult's law is obeyed (ideal solutions), then, for each component, the partial pressure is proportional to the mole fraction, with the proportionality constant equal to the partial pressure of the pure liquid

$$P_1 = X_1^{liq} P_1^o$$

$$P_2 = X_2^{liq} P_2^o$$

We also said that if even non-ideal solutions are dilute enough, they act ideal. They really act ideal only in the sense that Raoult's law holds for the solvent (the component in excess)

$$P_1 = X_1^{liq} P_1^o$$

For the solute, the vapor pressure is found to be proportional to the concentration, but with a different proportionality constant, which we will call k_1'

$$P_2 = X_2^{liq} k_1'$$

$$X_2^{liq} = \left(\frac{1}{k_1'}\right) P_2$$

this is called *Henry's law*. At this point in the text, the solute is called A, and its concentration is expressed in moles per liter, so that k_H, the Henry's law constant, is defined by

$$[A] = k_H P_A$$

This equation is most useful to describe how the solubility of sparingly soluble gases depends on their pressure. k_H depends on both the gas and the solvent, and the numerical value of k_H depends on which concentration untis are being used in the liquid phase (molarity, mole fraction or whatever).

Chapter 6   Properties of Dilute Solutions

Questions

6.1 Dynamic equilibrium in saturated solutions

1. The solubility of sodium chloride is 38g per 100mL water.

   (a) 50g NaCl is added to 100mL water. The mixture is shaken once and filtered, and the dried precipitate weighs 25g. Just before the solution was filtered, the rate of precipitation was *greater than* the rate of dissolution.

   (b) 50g NaCl is added to 100mL water. The mixture is shaken repeatedly and filtered, and the dried precipitate weighs 12g. Just before the solution was filtered, the rate of precipitation was *greater than* the rate of dissolution.

   (c) 20g NaCl is added to water, shaken repeatedly and filtered, and it is found that 1g of NaCl has not dissolved. How much water was originally added?

   (d) Would you expect a crystal of NaCl immersed in a saturated NaCl solution to change in weight over time? Why or why not?

   (e) Would you expect the crystal in part (d) to change in shape over time? Why or why not?

Chapter 6   Properties of Dilute Solutions

6.2 Concentration units for solutions

1. (a) How many grams of water must I add to 29g of sodium chloride (molecular weight 58) to make a solution whose mole fraction NaCl is 0.1?

    (b) What is the mole percent water in this solution?

2. What additional information, if any, is necessary to calculate

    (a) the molality

    (b) the weight percent NaCl

    (c) the molarity

3. How many grams of sulfuric acid (molecular weight 98) will I need to make 4.0L of a 0.5M solution?

4. If I take 100mL of the solution in Question 3 and dilute it to 250mL, what will be the concentration of the final solution?

Chapter 6   Properties of Dilute Solutions

5. How many milliliters of the solution in Question 3 contains 0.01 moles of sulfuric acid?

6. (a) The molarity of pure water is 55.6. What is the molality?

    (b) What is the molality of a 0.001M aqueous NaCl solution?

7. (a) Acetone has a molecular weight of 58 and a density of 0.79g/mL. The pure liquid is 14M. What is the molality?

    (b) Is the molality of a 0.001M solution of NaCl in acetone greater, less than or the same as the molarity?

8. How many moles of ions are in one liter of a 0.1M $CaCl_2$ solution assuming complete dissociation?

6.3 The vapor pressure of a solution of a non-volatile solute in a volatile solvent: Raoult's law

1. (a) If I add some non-volatile solute to a solvent, the vapor pressure *decreases*.

    (b) If I then add more solvent, the vapor pressure *decreases*

Chapter 6 Properties of Dilute Solutions

2. A saturated solution of NaCl in water at 25°C will always have the same vapor pressure. (True, False)

3. Suppose pure liquids A and B have the same vapor pressure, but B has a greater molecular weight. If I make up 0.1m sucrose solutions in A and B, which solvent will exhibit the greater vapor-pressure lowering?

4. An ideal solution exhibits Raoult's law *over its entire composition range*.

5. The vapor pressure of water at 29°C is 30mmHg. Assuming ideality, what would be the vapor pressure of an aqueous solution

   (a) of sucrose at mole fraction 0.1?

   (b) of sodium chloride at mole fraction 0.1?

6.4 The elevation of the boiling point of dilute solutions of a non-volatile solute and
6.5 The depression of the freezing point of dilute solutions of a non-volatile solute

1. (a) 10g of a substance (molecular weight 100) is dissolved in 1kg solvent, and its boiling point increases from the value of 89.0°C in the pure liquid to 89.5°C. What is K_b for the solvent?

   (b) 5.0g of a solute dissolved in 1kg this solvent gives a solution which boils at 89.2°C. What is the molecular weight of the solute?

Chapter 6   Properties of Dilute Solutions

2. In ebullioscopic or cryoscopic molecular-weight determinations, molecular weight values are usually *more* accurate the larger K_b or K_f.

3. The equation $\Delta T_f = K_f m$ is most accurate at (low, high) molalities.

6.6 Freezing-point depressions and boiling-point elevations for solutions of electrolytes

1. Suppose you use the freezing-point method to determine the molecular weight of a substance. Later you discover that the substance was ionic. Your originally calculated molecular weight is (too low, too high, correct).

2. Give, in terms of K_f, the expected freezing point depressions of the following:

   (a) a 0.1m solution of NaCl in water

   (b) a 0.1m solution of $CaCl_2$ in water

   (c) a 0.1m solution of NaCl in a solvent in which all the ions are in ion pairs

   (d) a 0.1m solution of HCl in water

   (e) a 0.1m solution of NaCl in a solvent in which it is 50% dissociated

3. $K_b = 0.51°C/m$ and $K_f = 1.86°C/m$ for water. For which of the following solutions is the solute an electrolyte? If an electrolyte, what is ν equal to?

Chapter 6   Properties of Dilute Solutions

_____(a) a 0.2m solution which boils at 100.1°C

_____(b) a 0.1m solution which freezes at -0.4°C

_____(c) a 0.15m solution which boils at 100.2°C

4. Ion-pair formation will tend to (increase, decrease) apparent values of ν from freezing-point data on electrolyte solutions.

5. As electrolyte solutions get more concentrated, ν from freezing-point data appears to (increase, decrease).

6.7 Osmotic pressure

1. Solutions of the same solute with the same molarity, but in two different solvents, will exhibit (the same, different) osmotic pressures at a given temperature.

2. Increasing the concentration of the solute *increases* the osmotic pressure of the solution.

3. Increasing the molecular weight of the solute *increases* the osmotic pressure of the solution for a given weight-percent solution.

4. Increasing the temperature *increases* the osmotic pressure of the solution

5. An experimental set-up such as the one in Figure 6.10 is built. The osmotic pressure of the solution above the membrane in Figure 6.10a is (greater than, less than, equal to) that in Figure 6.10b.

6.8 Solutions of two volatile liquids

1. Suppose liquid A has a vapor pressure of 25mmHg and liquid B has a vapor pressure of 75mmHg at 20°C. Assume they form an ideal solution.

    (a) What will be the vapor pressure of a solution with $X_A = 0.2$?

Chapter 6   Properties of Dilute Solutions

(b) What will be the vapor pressure of a solution with $X_A = 0.5$?

(c) What will be the vapor pressure of a solution with $X_A = 0.8$?

2. Given the same pair of liquids, but no longer assuming ideality, suppose a solution with $X_A = 0.4$ has a vapor pressure of 60mmHg

   (a) Is this a positive or a negative deviation from Raoult's law?

   (b) Would you expect ΔH_{mix} to be positive, negative or zero for this solution?

   (c) Would you expect the components to expand or shrink on mixing?

3. Assume the same A and B as in Question 1, and that they form an ideal solution.

   (a) Suppose $X_A^{liq} = 0.5$. What is X_A^{gas}?

   (b) In a distillation, will the distillate be enriched in A or in B?

6.9 Henry's law

1. If the pressure of carbon dioxide in a bottle of Coke doubles, the quantity of the gas dissolved approximately *doubles*.

Chapter 6   Properties of Dilute Solutions

2. Ideal solutions automatically obey Henry's law (True, False).

3. The Henry's law constant depends *only on the solvent*.

4. The reason that an O_2/He mixture, rather than an O_2/N_2 mixture, is breathed by divers is because

   (a) He is (more, less) soluble in the blood than N_2

   (b) He obeys Henry's law more closely than N_2 (True, False)

Chapter 6   Properties of Dilute Solutions

Answers

6.1 Dynamic equilibrium in saturated solutions

1. (a) less than (since more solute was in the process of dissolving)
   (b) equal to (saturated solution)
   (c) 50mL (since 19g, or half the solubility in 100mL, dissolved)
   (d) no
   (e) yes (dissolution and precipitation are continually taking place)

6.2 Concentration units for solutions

1. (a) 81g (I need 9 times as many moles H_2O as NaCl; I had 1/2 mole NaCl, so I need 9/2 mole H_2O, MW 18, and (9/2)(18g) = 81g.)
   (b) $X_{H_2O} = 0.9$, so mole percent H_2O = 90%.
2. (a) no additional info
   (b) no additional info
   (c) density of solution
3. 196g
4. 0.2M (Dilution by a factor 2.5 gives 0.5m/2.5 = 0.2M.)
5. 20mL (1L or 1000mL contains 0.5M, so need (0.01/0.5)(1000mL) = 20mL
6. (a) 55.6 (since 1L H_2O contains 1kg H_2O)
   (b) 0.001m (since the solution is so dilute that a volume of 1L weighs 1 kg to much better than one significant figure.)
7. (a) (14/0.79)m = 18m (since to get 1kg we need more than a liter)
   (b) greater than (since it takes more solution, containing more moles solute, to make up 1kg than 1L)
8. 0.3 moles

6.3 The vapor pressure of a solution of a non-volatile solute in a volatile solvent: Raoult's law

1. (a) True
   (b) increases
2. True
3. B (Since fewer moles of B than A are in 1kg, adding the same number of moles solute to both lowers the mole fraction B more than it does A.)
4. True
5. (a) 27mmHg
   (b) 24mmHg

6.4 The elevation of the boiling point of dilute solutions of a non-volatile solute and
6.5 The depression of the freezing point of dilute solutions of a non-volatile solute

1. (a) 5°C/m
   (b) 125
2. True
3. low

Chapter 6 — Properties of Dilute Solutions

6.6 Freezing-point depressions and boiling-point elevations for solutions of electrolytes

1. too low
2. (a) $0.2K_f$
   (b) $0.3K_f$
   (c) $0.1K_f$
   (d) $0.2K_f$
   (e) $0.15K_f$
3. (a) non-electrolyte
   (b) $\nu = 2$
   (c) $\nu = 3$
4. (a) decrease
   (b) decrease

6.7 Osmotic pressure

1. The same
2. True
3. Decreases
4. True
5. Greater than (because solution is more concentrated)

6.8 Solutions of two volatile liquids

1. (a) 65mmHg
   (b) 50mmHg
   (c) 35mmHg
2. (a) positive
   (b) positive
   (c) expand
3. (a) 0.25
   (b) B

6.9 Henry's law

1. True
2. True
3. on the solvent and the solute
4. (a) less
   (b) False

Chapter 7   Aqueous Solutions and Ionic Reactions

CHAPTER 7: AQUEOUS SOLUTIONS AND IONIC REACTIONS

Scope

Toward the end of the last chapter you saw that the measurement of colligative properties indicates that certain compounds ionize when added to water. The ionic solid NaCl gives $2N_A$ particles for each mole dissolved, and, more surprisingly, so does the molecular gas, HCl. This chapter begins by discussing the unique properties of water that give rise to this behavior, and proceeds to a discussion of typical reactions of ionic substances in aqueous solution.

Solvation

In Chapter 5 you learned about several kinds of attractive forces which may exist between unbonded molecules. Dipole-dipole and London forces were discussed in detail, but in the Study Guide we mentioned that there exists a charge-dipole interaction as well. Consider a positive charge in the vicinity of a dipole. The dipole will orient so that its negative end points toward the positive charge, and the result is a net attractive force, and a lowering of the potential energy. Now, if the charge is a sodium ion, Na^+, and it is surrounded by H_2O dipoles, the ion will be surrounded by a sheath of H_2O molecules, oriented with their oxygens pointed toward the ion. The process is called *hydration*, and the sheath of water molecule is called the *hydration shell*. The number of water molecules in this shell is called the *hydration number*. You should understand that the ion is not permanently bound to the molecules in its hydration shell, but, in a dynamic equilibrium, the molecules in the shell are continually exchanging with those in the bulk water. The hydration shell is not a rigid geometric array, either, so that hydration numbers are really long-time averages over fluctuating structures. "Hydration" always refers to water; *solvation* is a more general term, which refers to how any solvent interacts with a solute.

Recall that coulombic forces, which hold an ionic solid together, are very strong. The reaction

$$MX(crystal) \rightarrow M^+(g) + X^-(g)$$

is very endothermic ($\Delta H >> 0$), since to carry it out positive and negative charges must be highly separated. For this reason ionic solids like NaCl do not form isolated Na^+ and Cl^- ions in the gas phase, but rather ion pairs, tetramers, and so on. In dilute aqueous solutions of NaCl there is little ion pair

Chapter 7 — Aqueous Solutions and Ionic Reactions

formation, however. The reason is that the water molecules hydrate the ions, as described above, and the ion-dipole interactions stabilize the separated ions. This is equivalent to saying that the reaction

$$M^+(g) + X^-(g) + xH_2O \rightarrow M^+(aq) + X^-(aq)$$

is highly exothermic ($\Delta H << 0$), so that for the total process,

$$MX + xH_2O \rightarrow M^+(aq) + X^-(aq)$$

which is the sum of the previous two reactions, the overall ΔH, the *heat of solution*, is close to zero: $\Delta H_{soln} \approx 0$. (When two reactions are summed to give an overall reaction, their ΔH's sum to give an overall ΔH.) Depending on the precise solute and concentration, ΔH_{soln} can be either somewhat negative (exothermic) or somewhat positive (endothermic). These correspond respectively to the ingredients spontaneously giving off heat (getting warmer) or absorbing heat (getting cooler) on mixing (Figure 7.3).

Recall from the last chapter that entropy virtually always favors the solution process, since when the solute dissolves it increases its freedom by increasing the volume it can experience. For this reason, even if a solution process is endothermic, meaning that the energy goes up on dissolution, the entropy effect can be great enough to overcome this, and dissolution may take place anyway. Water has properties that tend to make ΔH_{soln} more favorable (more exothermic) than it is for other solvents for many, many solutes. Consider first ionic solutes. In discussing this topic, we are going to change our perspective somewhat from that in the preceding paragraphs. We have been taking a molecular view in discussing solvation: the orientation of water dipole around an ion; for example. Now we are going to view the solvent as a *continuum*, as if it did not consist of molecules. We will return to our molecular view again shortly.

According to Equation 7.3 in the text, Coulomb's law is

$$F = \frac{kq_1q_2}{Dr^2}$$

This describes the force between two charges q_1 and q_2 immersed in a medium with *dielectric constant* D. For example, q_1 might be +1, the charge on Na^+, and q_2 might be -1, the charge on Cl^-. We will ignore k for the moment, since it is constant within a given system of units. D is an experimentally determined property of a substance, which can be measured by seeing how much force two charges exert on each other when immersed in the substance. D=1 for a vacuum (and is substantially the same for air), and values for some common substances are given in Table 7.2. The fact is that the greater the value of D, the less force is experienced by the same two charges the same distance apart immersed in the substance. Now clearly, D for any substance must result from the detailed atomic-molecular nature of the substance, but D itself is measured in the bulk.

H_2O has a very high value of D ($D_{H_2O} = 79$). In fact, D_{H_2O} is 79 times bigger than D_{vacuum}; this means that the reaction

$$MX(crystal) \rightarrow M^+(g) + X^-(g)$$

would be 79 times less endothermic in water than in a vacuum. Thus, the continuum picture explains, in general, why ionic substances tend to dissolve in water more readily than in, say, cyclohexane, for which D is 2: it takes less energy to separate charges. Dielectric constant cannot be the whole story, however, since if it were, every ionic compound would have a mildly endothermic ΔH_{soln} in water. An extra exothermic contribution to the total ΔH_{soln} comes from the ion-dipole interactions we were discussing earlier. In certain cases, this contribution is sufficient to make the entire solution process exothermic.

What causes water to have such a large value of D? D is a property of the *bulk liquid* (your text uses the term "liquid as a whole"), and is a result of molecular processes taking place within the bulk liquid. These include dipole-dipole London and hydrogen-bonding interactions, but not the ion-dipole interactions we discussed earlier, since these involve the interaction of the liquid with ions. Consider an electrical field placed across a substance. This could come about by having a positively and a negatively charged ion separated in the substance, or else by placing the substance between a positively and a negatively charged electrode. For the same field strength, the greater the dielectric constant, the lower the energy. The electric field can induce temporary dipole moments in the molecules of the substance, which can then induce dipole moments in adjacent molecules; thus non-polar molecules which exhibit greater London forces tend to exhibit greater values of D. Similarly, the external field can cause molecules with permanent dipole moments to line up "head to tail;" this also contributes to D. When intermolecular aggregates can form, the entire aggregate can have a dipole moment, and this can be larger than the dipole moment of individual molecules, since the charge can be separated over a greater distance. Water forms such aggregates, due to hydrogen bonding, and this contributes to its high value of D as well.

So far our discussion has been limited to aqueous solutions of *electrolytes*, substances that form ions in solution. Water dissolves many *non-electrolytes* as well. Substances that hydrogen bond are often soluble in water, since hydrogen bonding can take place between them and the water molecules. Ammonia (NH_3) and hydrofluoric acid (HF) are examples. These form ions, but only to a small extent. Polar non-electrolytes are rarely soluble in water if they cannot hydrogen bond; for example, chloroform, $CHCl_3$, has a dipole moment of 1.01debye, but is only sparingly soluble in water (about 1g per 100mL at room temperature). Most polar substances do contain oxygen or nitrogen atoms, however, and therefore can act as hydrogen-bonding acceptors, and therefore are somewhat soluble. Acetone (CH_3COCH_3), for example, is *miscible* in water (can form solutions at all compositions). Non-polar substances, such as benzene (C_6H_6) and carbon tetrachloride (CCl_4) are usually *immiscible* in water, meaning that they do not dissolve at all. (Actually, immiscibility is an idealized concept, since anything will dissolve in anything else to *some* extent!). The reason for this insolubility is that in order for a non-polar molecule to enter liquid water, some water-water hydrogen bonds actually have to break to make room, and the only energy gained in return is the relatively weak London energy between the water and the solute molecule.

Electrolytes

An electrolyte is *strong* if it is completely (or nearly completely) ionized in solution, and *weak* otherwise. It is important to distinguish between the strength of an electrolyte and its solubility. Calcium hydroxide, $Ca(OH)_2$, is only slightly soluble in water, but is a strong electrolyte because virtually all of what does dissolve is in ionic form. Acetic acid, on the other hand, is miscible with water, but is only a weak electrolyte, since only a very small fraction of what is in solution is ionized; most is in molecular form.

Electrolytes can be classified as *acids*, *bases* or *salts*. Acids tend to taste sour, to turn certain dyes (called *indicators*) a characteristic color — for example, *litmus* red — and to dissolve many metals with the release of H_2 gas. Bases taste bitter, give different colors with indicators than acids — a base turns litmus blue — and have a slippery feel. Recall from the last chapter that HCl is a gas of molecular weight 36, but that dilute aqueous HCl solutions freeze and boil at temperatures that indicate that each gas molecule gives two ions in solutions. In 1884 Arrhenius proposed that what defines an acid is its dissociation in water into a hydrogen ion and some anion, e.g. $HCl \rightarrow H^+ + Cl^-$. Arrhenius proposed that a base is a substance which, when dissolved in water, gives a hydroxyl ion, OH^-, and some cation, e.g., $NaOH \rightarrow Na^+ + OH^-$. A *strong* acid or base is one for which this process is virtually complete, and a *weak* one is one which is only partly dissociated.

Arrhenius

It was known that acids and bases *neutralize* each other, and in the Arrhenius picture, the product is a salt and water.

$$NaOH + HCl \rightarrow NaCl + H_2O$$

In this context, a salt is an electrolyte that is neither an acid nor a base; in other words, one whose cation is not H^+ and whose anion is not OH^-. Later on, in Chapter 9, you will see that even salts may sometimes exhibit acidic or basic properties, such as the ability to color litmus.

Table 7.3 provides rules for recognizing weak and strong electrolytes, and Example 7.1 shows how the rules are used. This is information that must be memorized. To condense the information, however, the most common strong acids are sulfuric, H_2SO_4, nitric, HNO_3, hydrochloric, HCl, and the other haloacids (HBr, etc.) except for HF. The group IA and IIA hydroxides (except for $Be(OH)_2$) are strong bases. Most salts are strong electrolytes. Two important weak electrolytes are acetic acid, CH_3COOH, and ammonia, NH_3, which is weakly basic by virtue of the reaction

$$NH_3 + H_2O \rightarrow NH_4^+ + OH^-$$

Lowry-Bronsted

Consider the following reaction

$$NH_3(g) + HCl(g) \rightarrow NH_4Cl(s)$$

Since it takes place in the gas phase, and since no H^+ or OH^- is involved, and no water, this cannot be an acid-base reaction in the Arrhenius sense, even though, according to Arrhenius, NH_3 is a base, HCl is an acid, and NH_4Cl is a salt! Acids and bases have been redefined several times in the history of chemistry, and a definition that extends the Arrhenius definition so as to include examples such as this one is the Lowry-Bronsted theory, developed in the 1920's. This theory is probably the most useful one for most of the material in the course, though a further extension (the Lewis theory) will be discussed in Chapter 20.

According to Lowry-Bronsted, an acid is a *proton donor*, and a base is a *proton acceptor*. Thus, in the neutralization reaction we discussed earlier

$$HCl(aq) + NaOH(aq) \rightarrow NaCl(aq) + H_2O$$

the Lowry-Bronsted acid is H^+ and the base is OH^-. Note the subtle distinction from the Arrhenius idea; for Arrhenius, the acid was HCl and the base was NaOH. Note that the reaction

$$NH_3(g) + HCl(g) \rightarrow NH_4Cl(s)$$

is an acid base reaction according to Lowry-Bronsted: HCl is the acid, and NH_3 is the base, since HCl donates a proton to NH_3 to form the ammonium ion, NH_4^+. The Lowry-Bronsted definition is more general than the Arrhenius definition, and includes the Arrhenius definition as a special case.

The Lowry-Bronsted idea of an acid includes all *proton-transfer* reactions. Even the very dissolution of an acid in water constitutes such a reaction.

$$HCl(g) + H_2O \rightarrow Cl^-(aq) + H_3O^+(aq)$$

The H^+ ion does not exist "naked" in water solution. Instead it is incorporated into the water structure. We symbolize this by writing H_3O^+ (the *hydronium ion*) to indicate hydrated H^+, but, strictly speaking, this is not much better than writing H^+, since we could just as easily write $H_5O_2^+$ or $H_7O_3^+$, to signify the hydration of H^+ with two or three water molecules instead of one. In liquid water all the molecules

Chapter 7 Aqueous Solutions and Ionic Reactions

are hydrogen-bonded to each other anyway, so in the formula $H(H_2O)_n^+$, n is arbitrary (Figure 7.7). We often write H^+ to indicate simple ionization processes, and H_3O^+ when we wish to emphasize hydration. In the dissolution of HCl, HCl is a Lowry-Bronsted acid and H_2O is a base, since H_2O accepts a proton from HCl. In the dissolution of ammonia, on the other hand,

$$NH_3(aq) + H_2O \rightarrow NH_4^+(aq) + OH^-(aq)$$

water acts like an acid, since it donates a proton to NH_3. These ideas will be further elaborated in Chapter 9.

Percent Ionization

In Chapter 6 we discussed how measurements of colligative properties can distinguish between electrolytes and non-electrolytes. Accurate measurements can quantify the difference between weak and strong electrolytes. Recall that a strong electrolyte is one which is almost completely dissociated, and a weak one is one which is only partly dissociated. The percent dissociation of an electrolyte of known formula can be calculated in the following manner. Recall from Chapter 6 that experimental values of T_f^{exp} for solutions of known m give experimental values for i, the van't Hoff mole number. Comparing i to ν then allows the fraction or percent dissociation to be determined. For example (Table 6.3) 0.100m Na_2SO_4 exhibits $\Delta T_f^{exp} = 0.434K$ in water. Then

$$i = \Delta T_f^{exp}/K_f m = 0.434K/(1.86 \cdot K/m)(0.100m) = 2.33$$

Na_2SO_4 "should" form $2Na^+ + SO_4^{2-}$ in solution, so that $\nu=3$. Suppose x is the fraction dissociation. Then, in the actual solution, the concentration of species will be as follows:

Na_2SO_4:   0.100 − 0.100x = 0.100 (1−x)

Na^+:   2(0.100)x = 0.200x

SO_4^{2-}:   0.100x

Total: 0.100(1−x) + 0.200x + 0.100x

= 0.100 + 0.200x

= im = (2.33)(0.100) = 0.233

x = 0.665

Chapter 7 — Aqueous Solutions and Ionic Reactions

Thus the Na_2SO_4 is 66.5% dissociated. Note that we used the stoichiometry to determine the concentration of each species, based on x; then we added up all these expressions, and set the total equal to the known value of im, which represents the true (experimental) molality of all particles in solution. This allowed us to solve for x.

Solubility

The solubility of an electrolyte depends on the strength of the lattice forces in the crystal as well as the strength of the ion-dipole interactions in solution. Table 7.4 summarizes general trends in the solubility of electrolytes, and Example 7.3 gives specific examples. Again, these have to be memorized, but here is a summary for the most frequently occurring substances. Salts of group IA metals and of ammonia are usually soluble. So are halides, except fluorides, of most elements (notable exception: silver halides). Except for ammonium and group IA metals, hydroxides, sulfates, carbonates, phosphates and sulfides are generally insoluble. Nitrates, acetates and perchlorates are generally soluble.

Ionic Reactions

We have already looked at a few acid-base neutralization reactions in aqueous solution; for example

$$HCl(aq) + NaOH(aq) \rightarrow NaCl(aq) + H_2O$$

To correctly designate the species that actually exist under these conditions, however, we should write

$$H^+(aq) + Cl^-(aq) + Na^+(aq) + OH^-(aq) \rightarrow Na^+(aq) + Cl^-(aq) + H_2O$$

Note, however, that Na^+ and Cl^- appear unchanged on both sides of the reaction (they are *spectator ions*) so that they can be left out, giving

$$H^+(aq) + OH^-(aq) \rightarrow H_2O$$

If one prefers, H_3O^+ can be used instead of H^+, giving, for the overall process,

$$H_3O^+(aq) + OH^- \rightarrow 2H_2O$$

In general, we leave out spectator ions, write electrolytes in solution in their ionic forms and write the full formulas for everything else. Thus, for the reaction of silver nitrate with sodium chloride to form the insoluble silver chloride, we might write, before applying these rules

Chapter 7

Aqueous Solutions and Ionic Reactions

$$AgNO_3(aq) + NaCl(aq) \rightarrow AgCl(s) + NaNO_3(aq)$$

By applying the rules, this simplifies to

$$Ag^+(aq) + Cl^-(aq) \rightarrow AgCl(s)$$

This is a *precipitation reaction*. Sometimes a precipitation reaction can be reversed. For example, calcium carbonate, $CaCO_3$, is insoluble, but upon adding acid the carbonate reacts to form H_2CO_3, carbonic acid, which decomposes into CO_2 and water. For the overall process,

$$CaCO_3(s) + 2H^+(aq) \rightarrow Ca^{2+}(aq) + H_2O + CO_2(g)$$

Suppose I have 100mL of a solution containing an unknown amount of NaCl in water, and a silver nitrate solution that I know is 0.2M. There are two ways (at least!) that I can use the $AgNO_3$ solution to determine how much NaCl was present originally. These illustrate several important points about ionic reactions.

One method is the basis for the *gravimetric* (based on weighing) determination of chloride. The reaction is

$$Cl^-(aq) + Ag^+(aq) \rightarrow AgCl(s)$$

All I do here is to add enough silver nitrate solution to the unknown to be sure I have precipitated all the chloride; it really doesn't matter how much excess I add. Then I filter, dry and weigh the AgCl. Suppose I find it weighs 1.43g. The molecular weight of AgCl is 143, so I have 0.01 moles. Therefore, the original solution must have contained 0.01 moles of NaCl. The volume was 0.1L, so the concentration was (0.01mol/0.1L) = 0.1M.

How much silver nitrate solution did it take to complete the precipitation? Since one mole of Ag^+ precipitates one mole of Cl^-, and because a total of 0.01 mole of AgCl precipitated, I must have needed 0.01 mole of $AgNO_3$. 0.01 mole was contained in (0.01mol)/(0.2mol/L) = 0.052L, or 50mL, of solution. By the time this much was added, what were the concentrations of all the ions in solution? No silver or chloride ions were in solution, since they were all in the precipitate, so we have only Na^+ and NO_3^- to be concerned with. We started out with 0.01 mole Na^+ and added 0.01 mole NO_3^-, and nothing happened to these, so this number of moles of both must be present, but the total volume is 150mL, or 0.15L, so

Chapter 7   Aqueous Solutions and Ionic Reactions

that each ion has the concentration (0.01mol/0.15L) = 0.067M. (Actually, you will learn in later chapters that the concentrations of Ag^+ and Cl^- are not quite zero, but they are much lower than 0.067M.)

Now, however, suppose in our zeal to make sure we precipitated all the AgCl we had added a total of 75mL of 0.2M $AgNO_3$. This is (0.2mol/L)(0.075L) = 0.015mol $AgNO_3$. Now, we still only have 0.01mol Na^+, as before, and certainly all the chloride is gone, but we have added 0.015mol NO_3^-, which must all still be there. For Ag^+, 0.01 mol of the 0.015mol added must have precipitated with the Cl^-, leaving 0.005mol. The total volume is 175mL=0.175L, so that $[Na^+]=$ (0.01mol)(0.175L)= 0.057M, $[NO_3^-]=$ 0.086M and $[Ag^+]=$ 0.029M. (Square brakets are often used to indicate molar concentration.) Simlar calculations at these and additional stages of adding $AgNO_3$ are summarized in the following table.

Titration of 100mL 0.1M NaCl with 0.2M $AgNO_3$

mL $AgNO_3$ Added	mL Total	$[Na^+]$	$[Cl^-]$	$[Ag^+]$	$[NO_3^-]$	[all ions]
0	100	0.100	0.100	0	0	0.20
25	125	0.080	0.040	0	0.040	0.16
50	150	0.067	0	0	0.067	0.13
75	175	0.057	0	0.029	0.086	0.17

Note that the *total concentration* of ions goes through a minimum when just enough $AgNO_3$ has been added to react with all the NaCl. This is because before that point there are extra Cl^- ions around, and after that point there are extra Ag^+ ions around. The electrical conductance of a solution is proportional to the total ionic concentration, so that the conductance also goes through a minimum at the *equivalence point*, the point where the stoichiometrically correct number of moles of *tritant* (the solution being added) are present to just react with the substance being tritated. If the $AgNO_3$ solution had been added dropwise from a buret, and the conductance jotted down upon each addition, the concentration of NaCl originally present could have been determined by noting that the conductance had reacted its minimum value at 50mL titrant added. This is called a *conductance* (or *conductometric*) *titration.*

We have made several assumptions: first, that all the silver chloride that can possibly precipitate does so. This is sometimes given the expression, "the reaction *goes to completion,*" or "proceeds *quantitatively.*" In fact, no reaction goes to completion, and in Chapter 10 you will learn how to calculate just how much Ag^+ and Cl^- are in solution at the equivalence point. The quantities are so small, however, as to be virtually unweighable on the scale of this experiment. Our second assumption was that everything that doesn't react just stays there, so that changes in $[Na^+]$, for example, result solely from dilution of the original solution.

A more common sort of titration is an acid-base titration; again, the equivalence point is when just the right number of moles of titrant have been added to completely react with the substance originally present. For example, suppose 0.1M NaOH solution is added to a solution of oxalic acid, which has two acidic protons, and which we will symbolize H_2Ox. The reaction can be written

$$H_2Ox + 2OH^- \rightarrow 2H_2O + Ox^{2-}$$

One can add an indicator to the solution, so that the color will change when the solution, originally acidic, just starts to turn basic. This is then the equivalence point. Suppose it takes 50mL NaOH solution to reach the equivalence point; then $(0.1 mol/L)(0.05L) = 0.005$mol NaOH must have been added. From the stoichiometry, it takes two moles NaOH to neutralize one of H_2Ox, so that 0.0025mol H_2Ox must have been present in the original sample. These examples also illustrate how useful the concept of molarity is, since it provides us a mean of knowing how many moles of solute were added if we know how many milliliters of solution were added.

Chapter 7   Aqueous Solutions and Ionic Reactions

Questions

7.1 The role of water as a solvent for ionic crystalline solids

1. Name the force most responsible for the hydration of ions.

2. In the hydration shell of Cl^-, the water molecules have the (H, O) pointed toward the ion.

3. In the hydration shell of Li^+, the water molecules have the (H, O) pointed toward the ion.

4. When the heat of solution is (positive, negative) the ingredients tend to get warmer when mixed.

5. Which would you expect to have the greater crystal lattice energy, MgO or NaCl?

6. (a) The overall entropy of solution is generally (positive, negative).

    (b) This (favors, disfavors) the solution process.

7. The hydration of ions makes ΔH_{soln} more (positive, negative) than it would otherwise be.

8. The greater the dielectric constant, the (larger, smaller) the force between two ions in solution.

9. (a) Dipole moment makes an important contribution to dielectric constant (True, False)

    (b) Dipole moment is entirely responsible for dielectric constant (True, False)

    (c) Formation of molecular aggregates usually tends to (increase, decrease) dielectric constant.

7.2 The role of water as a solvent for molecular compounds

1. (a) Name a substance miscible with water.

    (b) Name a substance immiscible with water.

    (c) No two substances are really 100% miscible. (True, False)

    (d) No two substances are really 100% immiscible. (True, False)

Chapter 7   Aqueous Solutions and Ionic Reactions

2. Tell whether the solubility of each of the following in water is likely to be high or low.

   _____ (a) ionic solid

   _____ (b) polar, hydrogen bonding liquid

   _____ (c) polar, non-hydrogen-bonding liquid

   _____ (d) non-polar liquid

7.3 Strong and weak electrolytes and non-electrolytes

1. If an electrolyte has a low solubility in water, it must be a weak electrolyte (True, False)

2. Tell whether each of the following is an acid, base or salt, and whether it is a weak or strong electrolyte

   (a) HF

   (b) HCl

   (c) NH_3

   (d) Na_2SO_4

   (e) $NaHSO_4$

   (f) Na_2S

   (g) LiOH

3. The higher the concentration of an electrolyte, the *more* completely it tends to be dissociated.

7.4 Proton transfer reactions: the Bronsted-Lowry theory of acids and bases

1. When HCl dissolves in water, the water acts as a Lowry-Bronsted *acid*.

2. When NH_3 dissolves in water, the water acts as a Lowry-Bronsted *acid*.

3. (a) Give the Lowry-Bronsted definition of an acid.

Chapter 7 Aqueous Solutions and Ionic Reactions

(b) Give the Arrhenius definition of an acid.

4. For a substance MX_2, which ionizes according to

$$MX_2 \rightarrow M^{2+} + 2X^-$$

let y be the molality of M^{2+}. Suppose the initial concentration of MX_2 was 0.1m.

(a) Give an expression for M_{X^-} (the molarity of X^-)

(b) Give an expression for M_{X_2}

(c) Give an expression for M_{total}

(d) Suppose, based on freezing-point measurements, it is known that $M_{total} = 0.11$. What is the percent ionization of MX_2?

5. (a) A salt in the pure state consists of *ions*.

(b) An acid in the pure state consists of *ions*.

7.5 A summary of information about the solubilities of electrolytes in water

1. Tell whether the solubility of each of the following substances in water is high or low

_____(a) AgCl

7-14

Chapter 7 Aqueous Solutions and Ionic Reactions

_____(b) NaCl

_____(c) $CaSO_4$

_____(d) Na_2SO_4

_____(e) $Zn(ClO_4)_2$

_____(f) $Hg(NO_3)_2$

_____(g) CdS

_____(h) Li_2S

7.6 Writing correctly balanced net ionic equations

1. Complete and balance the following reactions; only include the species which participate. Assume aqueous solution.

    (a) barium chloride + sodium sulfate

    (b) cadmium nitrate + ammonium sulfide

    (c) sodium carbonate + strontium perchlorate

    (d) strontium carbonate + nitric acid + heat

    (e) ammonia + hydrogen sulfide

2. For each reaction in Question 1, list the spectator ions.

    (a)

    (b)

    (c)

Chapter 7   Aqueous Solutions and Ionic Reactions

    (d)

    (e)

7.7 Stoichiometric of ionic reactions in aqueuous solutions

1. Give the concentration of all major ions in solution for each of the following

    (a) 200mL of 0.1M NaCl is mixed with 50mL of 0.2M HCl

    (b) 200mL of 0.1M NaOH is mixed with 50mL of 0.2M HCl

    (c) 100mL of 0.1M Na_2SO_4 is mixed with 100mL of 0.2M $Ba(NO_3)_2$

2. Define the equivalence point of a titration

3. (a) A conductance titration is most often performed when a _____ reaction takes place.

    (b) In such a tritation, the conductance is at its *maximum* at the equivalence point.

Chapter 7 Aqueous Solutions and Ionic Reactions

Answers

7.1 The role of water as a solvent for ionic crystalline solids

1. ion-dipole (or charge-dipole) force
2. H
3. O
4. negative
5. MgO (ions have greater charge)
6. (a) positive
   (b) favors
7. negative
8. smaller
9. (a) True
   (b) False
   (c) increase

7.2 The role of water as a solvent for molecular compounds

1. (a) ethanol, methanol, acetic acid
   (b) hexane, benzene, cyclohexane, gasoline
   (c) False
   (d) True
2. (a) high
   (b) high
   (c) low
   (d) low

7.3 Strong and weak electrolytes and non-electrolytes

1. False
2. (a) weak acid
   (b) strong acid
   (c) weak base
   (d) strong salt
   (e) strong salt
   (f) strong salt
   (g) strong base
3. less

7.4 Proton transfer reactions: the Bronsted-Lowry theory of acids and bases

1. base
2. True
3. (a) An acid is a proton donor.
   (b) An acid is a substance which, when dissolved in water, produces hydrogen ions.
4. (a) 2y
   (b) 0.1 - y
   (c) 0.1 + y
   (d) 10% (since then y=0.01, and % ionization= $(M_{M^+}/M_{initial}) \times 100\% = (0.01/0.1) \times 100\% = 10\%$)
5. (a) True
   (b) molecules

Chapter 7 — Aqueous Solutions and Ionic Reactions

7.5 A summary of information about the solubilities of electrolytes in water

1. (a) low
   (b) high
   (c) low
   (d) high
   (e) high
   (f) high
   (g) low
   (h) high

7.6 Writing correctly balanced net ionic equations

1. (a) $Ba^{2+}(aq) + SO_4^{2-}(aq) \rightarrow BaSO_4(s)$
   (b) $Cd^{2+}(aq) + S^{2-}(aq) \rightarrow CdS(s)$
   (c) $Sr^{2+}(aq) + CO_3^{2-}(aq) \rightarrow SrCO_3(s)$
   (d) $SrCO_3(s) + 2H^+(aq) \rightarrow H_2O + CO_2(g) + Sr^{2+}(aq)$
   (e) $2OH^-(aq) + H_2S(aq) \rightarrow 2H_2O + S^{2-}(aq)$

2. (a) Cl^-, Na^+
   (b) NO_3^-, NH_4^+
   (c) Na^+, ClO_4^-
   (d) NO_3^-
   (e) NH_4^+

7.7 Stoichiometric of ionic reactions in aqueous solutions

1.
   (a) $[Na^+] = 0.08M$, $[H^+] = 0.04M$, $[Cl^-] = 0.12M$
   (b) $[Na^+] = 0.08M$, $[OH^-] = 0.04M$, $[Cl^-] = 0.04M$
   (c) $[Na^+] = 0.1M$, $[NO_3^-] = 0.2M$, $[Ba^{2+}] = 0.05M$

2. The equivalence point is the point where just the correct number of moles of tritant have been added to completely react with the number of moles of other reactant initially present.

3. (a) precipitation
   (b) minimum

CHAPTER 8: INTRODUCTION TO CHEMICAL EQUILIBRIUM

Scope

In Chapter 5 and 6 you learned about the idea of dynamic equilibrium, as applied to a liquid in equilibrium with its vapor and to a solid in equilibrium with its saturated solution. In Chapter 7 we discussed chemical reactions which go to completion, for example the precipitation reaction

$$Ag^+(aq) + Cl^-(aq) \rightarrow AgCl(s)$$

In fact, however, no reaction goes 100% to completion, and many processes stop far short of completion. One example is the acid base reaction*

$$NH_3 + CH_3COOH = NH_4^+ + CH_3COO^-$$

For this reaction, after the ammonia and acetic acid are mixed, ammonium and acetate ions begin to appear, and quickly rise to a low concentrations, but do not rise further, no matter how long one waits.

The state of chemical equilibrium is the state in which concentrations no longer change with time, but underlying this state is the same sort of dynamic process which characterizes liquid-gas and solid-solution equilibrium; namely, on a molecular level, using this reaction as an example, ammonia and acetic acid are continually reacting to form ammonium ion and acetate ion, but the reverse reaction is also taking place at an equal rate, so that there is no net change. The same goes for the precipitation of silver chloride: at equilibrium, the rate of precipitation is equal to the rate of dissolution, so there is no net change in the amount of precipitate present. For this precipitation reaction, however, in contrast to the ammonia-acetic acid reaction, very little of the material of the left-hand side of the chemical equation is still present by the time equilibrium is achieved.

Chemical Equilibrium

For the reaction

$$\alpha A + \beta B = \gamma C + \delta D$$

---

* Usually a single arrow, "→", is used in a chemical reaction to indicate a process that goes essentially to completion. To indicate a process that goes only part way to completion, it is common to use either a double arrow, which the text uses, or else an equal sign, "=", which we will use in the Study Guide.

taking place in solution, the *equilibrium constant* is defined by the value of the expression

$$K_{eq} = \frac{[C]^\gamma [D]^\delta}{[A]^\alpha [B]^\beta}$$

when the system is at equilibrium; that is, this expression is equal to the equilibrium constant when [A], [B], [C] and [D] are equal to their equilibrium concentrations, in moles per liter. It is found empirically that for a given reaction, at a given temperature, this ratio is always the same, provided that equilibrium has been achieved. Recall that at equilibrium the concentrations are no longer changing with time, so that one may have to measure concentrations repeatedly over time to make sure that equilibrium has been obtained. Only then does this ratio of concentrations equal the equilibrium constant.

The statement that K_{eq} is constant for a given reaction at a given temperature is called the *law of mass action*, or the *ideal law of chemical equilibrium*. By now you should know that when a law is termed "ideal," it means, "it ain't always so." Like most ideal laws having to do with solutions, the constancy of the above expression for K_{eq} is closest to being true at low concentrations. Below, we will give a different expression for K_{eq} that pertains to gas phase reactions. It, of course, will come closest to being perfectly true at low pressures. Throughout this course, we will assume that the law of mass action holds.

K_p and K_c

Note that to write the equilibrium constant expression, one divides the product of the concentrations on the right-hand side of the chemical equation, each raised to the power of its stoichiometric coefficient, by the similar product for the left hand side. When the reactants are gases rather than substances in solution, it is customary to use a different expression for the equilibrium constant. Note that the ideal gas law gives

$$P = \left(\frac{n}{V}\right)RT$$

and that (n/V) is equal to molarity. Since RT is constant at constant temperature, P is proportional to molar concentration, and we can use the partial pressures of gaseous components as measures of concentration in the equilibrium constant expression for gaseous reactions. Recall our reaction

Chapter 8 — Introduction to Chemical Equilibrium

$$\alpha A + \beta B = \gamma C + \delta D$$

Now assume this reaction takes place in the gas phase. Let us use the symbol K_c for our earlier defined equilibrium constant expression based on molar concentrations.

$$K_c = \frac{[C]^\gamma [D]^\delta}{[A]^\alpha [B]^\beta}$$

If we now use the symbol K_p for the equilibrium constant expression based on partial pressures, we have

$$K_p = \frac{P_C^\gamma P_D^\delta}{P_A^\alpha P_B^\beta} = \frac{[C]^\gamma [D]^\delta}{[A]^\alpha [B]^\beta} \frac{(RT)^\gamma (RT)^\delta}{(RT)^\alpha (RT)^\beta}$$

$$= K_c (RT)^{\gamma+\delta-\alpha-\beta} = K_c (RT)^{\Delta n}$$

We have related K_p to K_c by substituting for each P its expression, in terms of molar concentration, taken from the ideal gas law. In the final result, Δn stands for the change in the number of moles of gas between the left and right side of the chemical equation. For example, in the reaction

$$2H_2(g) + O_2(g) = 2H_2O(g)$$

Δn is equal to -1 (2−2−1). If the same reaction had been run under conditions where water was produced in liquid form, we would have had Δn equal to -3 (0−2−1). Note that the numerical value of K_c is different from that of K_p, but that is only because for a gas the numerical value of P is different from that of (n/V); the conversion factor for the concentration units is RT. The net effect is that if gaseous concentrations are given in moles per liter, K_{eq} must be given as K_c; if concentrations are in units of pressure, then K_p must be used.

According to the expressions given, the equilibrium constant should have units; Example 8.2 shows this. You should know, however, that there are good reasons for considering the equilibrium constant to be unitless, and many chemists consider it to be such. The reasons for this will be discussed in the **Perspective** of Chapter 18.

Effect of Constant Concentrations

For the ionization of acetic acid

$$HOAc + H_2O = OAc^- + H_3O^+$$

From our rules for writing down equlibrium constants,

$$K_c' = \frac{[OAc^-][H_3O^+]}{[H_2O][HOAc]}$$

You will see in a moment why we call this K_c', rather than K_c. If the solution is dilute, $[H_2O]$ can be considered constant, and equal to 55.5mol/L, the value for the pure liquid. For a 0.01M solution, even if all the HOAc ionized it would not change the value of $[H_2O]$, to three significant figures. In fact, only a small amount of the HOAc ionizes, so this statement holds even at higher concentrations. We define the *acidity constant* (or *ionization constant*, or *dissociation constant*) of acetic acid by the expression

$$K_a = [H_2O]K_c' = \frac{[OAc^-][H_3O^+]}{[HOAc]}$$

Note that if we had written the ionization of acetic acid as

$$HOAc = H^+ + OAc^-$$

we would have immediately written

$$K_c = K_a = \frac{[H^+][OAc^-]}{[HOAc]}$$

In dilute solution, the solvent acts like a pure liquid, and its concentration may be assumed constant and eliminated from equilibrium constant expressions.

Another application of this rule is in writing the equilibrium constant expression for the vaporization of a liquid.

$$H_2O(l) = H_2O(g)$$

$$K_p = P_{H_2O}$$

There is no term in the denominator for the concentration of the liquid, which remains constant. Likewise, if one component of a reaction is a solid, the concentration of the solid does not appear in K_{eq}, since a pure solid at the same temperature always has the same concentration.

$$Zn(s) + 2Ag^+ = Zn^{2+} + 2Ag(s)$$

$$K_c = \frac{[Zn^{2+}]}{[Ag^+]^2}$$

As another example consider the dissolution of AgCl

$$AgCl(s) = Ag^+(aq) + Cl^-(aq)$$

$$K_c = K_{sp} = [Ag^+][Cl^-]$$

Note that this is the reverse of the precipitation of AgCl, which we said went "to completion." We also said that nothing really goes to completion; in fact, there are small amounts of Ag^+ and Cl^- remaining in solution in equilibrium with solid AgCl. The equilibrium constant expression does not include [AgCl] in the denominator, because it is constant. K_c for the dissolution of a highly insoluble salt is given the name, the *solubility product*, K_{sp}.

Note that for *heterogeneous* reactions (ones involving more than one phase, such as the dissolution of silver chloride, which involved an aqueous and a solid phase) K_{eq} may involve mixed concentration units, as in Example 8.3c.

$$Zn(s) + 2H^+(aq) = Zn^{2+}(aq) + H_2(g)$$

$$K_{eq} = \frac{P_{H_2}[Zn^{2+}]}{[H^+]^2}$$

Reaction Quotient

For an arbitrary reaction

$$\alpha A + \beta B = \gamma C + \delta D$$

we defined the equilibrium constant to be the value of the expression

$$\frac{[C]^\gamma [D]^\delta}{[A]^\alpha [B]^\beta}$$

when all the concentrations are at equilibrium values. One can be certain that a chemical process is at equilibrium only if one has observed concentrations change and then observed the changes to cease — then one knows that the process is fast enough for net changes to occur on the time scale of observation, and that the process has gone as far as it will go. Suppose one measures the concentration of all the components when the system is not at equilibrium. The ratio given above then has the name, the *reaction quotient*, symbolized Q.

$$Q = \frac{[A]^\alpha [B]^\beta}{[C]^\gamma [D]^\delta}$$

Q can always be defined, just as long as the concentrations can be measured, but only when the system is in equilibrium does Q equal K_{eq}.

Now, for illustrative purposes, let us consider the reaction

$$A(aq) = B(aq)$$

$$Q = \frac{[B]}{[A]}$$

Assume $K_{eq}=1$ Suppose this reaction takes some hours to reach equilibrium. Now, I start with pure A, and add it to water, and measure [A] and [B] every few minutes. At my first measurement, [A] will be high and [B] will be low, so that $Q<<1$. As reaction proceeds in the forward direction, [A] will decrease and [B] will increase, and Q will increase, coming closer and closer to K_{eq}, eventually becoming experimentally indistinguishable from it. Now consider a different experiment which is initiated by the addition of pure B to water. Here, $[B]>>[A]$ at first, so that $Q>>1$. The reverse reaction will then predominate, and [B] will decrease, [A] will increase, and Q will simultaneously decrease, eventually falling to the value of K_{eq}.

The moral of the story is embodied in the text's Equation 8-14: if $Q>K_{eq}$, then a net reverse reaction will occur spontaneously, until K_{eq} is achieved. If $Q<K_{eq}$, then a net forward reaction will occur spontaneously, until Keq is achieved. If $Q=K_{eq}$, then no net reaction will occur, since the system is at equilibrium. The equilibrium is dynamic, however, and both forward and reverse reactions will be occurring even here, though at equal rates. Furthermore, recall that even if $Q<<K_{eq}$, indicating a spontaneous forward reaction, there is no guarantee that the reaction will in fact occur quickly enough to be readily observable. Despite this caveat, the concepts of K_{eq}, Q and spontaneity are extremely useful in chemistry.

Note that K_{eq} is always a positive number, since it is a ratio of concentrations, which are positive numbers, raised to various powers. Using our simple example, the reaction A=B, suppose that at equilibrium there is 100 times more B than A present; then $K_{eq}=100$; if there is 100 times more A than B, then $K_{eq} = 1/100 = 0.01$. It is common to find equilibrium constants for chemical processes as great as 10^{20} and as small as 10^{-20}; outside this range, K_{eq} is very difficult to measure.

Recall that we said that K_{eq} is a function of temperature; that is, K_{eq} will differ at different temperatures, and may in general increase or decrease as the temperature rises. This behavior will be discussed later in the chapter and in greater depth in Chapter 18. In addition, the same process taking place in different solvents will, in general, have different values of K_{eq}. K_{eq} does not, however, depend on the total concentrations of the components. In fact, it is in the invariance of K_{eq} with concentration that makes the equilibrium constant so useful. Suppose, again for the simple reaction A=B, that K_{eq}=10. If 1.1mol of A is added to water to make 1L solution, then at equilibrium [A]=0.1M and [B]=1.0M. If instead 0.11mol of A is present initially then at equilibrium [A]=0.01M and [B]=0.10M. Concentrations change, but K_{eq} stays constant.

LeChatelier

Recall from Chapter 5 that LeChatelier's principle states that if a system in equilibrium is subjected to a stress, then the equilibrium will shift so as to relieve the stress. We will discuss stresses induced by concentration changes, pressure changes and temperature changes.

For the dissolution of the precipitate $Zn(OH)_2$

$$Zn(OH)_2(s) = Zn^{2+}(aq) + 2OH^-(aq)$$

$K_{eq} = K_{sp} = [Zn^{2+}][OH^-]^2$. (By now you should be able to write this expression by inspection from the balanced chemical reaction. Note that no concentration term appears for $Zn(OH)_2$, since it is a solid, and its concentration is constant.) It is found that $K_{sp} = 5 \times 10^{-17}$ for this substance at 25°C. We can now ask, what are the concentrations of Zn^{2+} and OH^- in a solution in equilibrium with the solid $Zn(OH)_2$? If we let x represent $[Zn^{2+}]$, then $[OH^-]$ must be 2x, by the reaction stoichiometry. Then we have $(x)(2x)^2 = 4x^3 = 5 \times 10^{-17}$, which is easily solved, in this era of the electronic calculator[*], to give $x=2.3\times 10^{-6}M=[Zn^{2+}]$, and $[OH^-]=4.6\times 10^{-6}M$.

---

[*] If you cannot do this calculation almost instantly on your pocket calculator, then you should devote some time to practicing similar operations. You will need this facility often throughout the next few chapters.

Chapter 8 — Introduction to Chemical Equilibrium

Now let us consider the effect of raising the OH^- concentration, which can be done by making the solution say 0.1M in NaOH. First of all, note the effect on Q. Before any shift in equilibrium takes place, the new value of $[OH^-]$ is $(0.1 + 4.6 \times 10^{-6})M \approx 0.1M$, and that of $[Zn^{2+}]$ is $2.3 \times 10^{-6}M$, as just calculated. Then Q is equal to $(0.1)^2(2.3 \times 10^{-6}) = 2.6 \times 10^{-8}$, which is fifty million times greater than K_{eq} for the reaction! Recall that if $Q > K_{eq}$, then the reverse reaction, i.e., the precipitation of $Zn(OH)_2$, will be spontaneous. This is just a statement of LeChatelier's principle for the special case of concentration: if the stress is the increase in concentration of something on the right-hand side of the reaction, then the shift which relieves this stress is a net reverse reaction. It is easy to calculate what $[Zn^{2+}]$ will be after the shift takes place, using the fact that K_{sp} is still 5×10^{-17}.

First of all, note that $[OH^-] = 0.1M$ after the shift, since even if all the Zn^{2+} in solution were to precipitate, carrying OH^- with it, the new OH^- concentration would be $(0.1 - 4.6 \times 10^{-6})M \approx 0.1M$. So now we can write

$$K_{sp} = 5 \times 10^{-17} = [Zn^{2+}][OH^-]^2$$
$$= [Zn^{2+}](0.1)^2$$

This gives $[Zn^{2+}] = 5 \times 10^{-16}$. Note that the Zn^{2+} concentration has been lowered by a factor of about 2×10^{-10}, or twenty billion-fold, by the addition of 0.1M hydroxyl ion! Note also that despite the addition of extra OH^-, $K_{sp} = 5 \times 10^{-7} = [Zn^{2+}][OH^-]^2$

Incidentally, there is a name for this sort of application of LeChatelier's principle; it is called the *common ion effect*, and refers to "pushing" an equilibrium in the desired direction by adding an excess of one of the reacting ions that appears on the other side of the equation. Here, OH^- appeared on the right of the equation, so to push the reaction to the left, we added excess OH^-. Clearly, if a way could be found to *remove* OH^-, instead of adding it, this would lead to a shift in the equilibrium to the right. As the text shows, the addition of acid will do this.

$$H^+ + OH^- \rightarrow H_2O$$

so that a $Zn(OH)_2$ precipitate may be solubilized by the addition of acid, according to the following reaction

Chapter 8 — Introduction to Chemical Equilibrium

$$Zn(OH)_2 + 2H^+ \rightarrow Zn^{2+} + 2H_2O$$

What really happens is that the H^+ reacts with the OH^- already in solution, forming H_2O. This lowers $[OH^-]$, so that $Q<K_{eq}$. Therefore the reaction goes to the right (more $Zn(OH)_2$ dissolves), but the resulting $OH^-(aq)$ quickly reacts with more H^+ to form more water, Q is still less than K_{eq}, and the forward reaction continues, until either the H^+ is all used up or the $Zn(OH)_2$ is completely dissolved.

Finally, there is one more complication in this particular reaction which we will mention, but not show how to deal with until Chapter 11. It turns out that Zn^{2+} enters into the following reaction

$$Zn^{2+}(aq) + 4OH^-(aq) = Zn(OH)_4^{2-}(aq)$$

Thus, increasing $[OH^-]$ exerts a common ion effect, as we saw, tending to precipitate $Zn(OH)_2$, but also removes Zn^{2+} from solution by formation of the soluble $Zn(OH)_4^{2-}$ complex; thus, depending on the concentration, excess OH^- might dissolve a $Zn(OH)_2$ precipitate, rather than form more of it! Therefore the calculation we did above, showing the common ion effect, is incomplete, because it does not take this complication into account. We will do so later, in Chapter 11. This goes to show, however, that a chemical equilibrium calculation can only be trusted when you are sure that all chemical equilibrium processes have been taken into account. (At this stage of your chemical career, you need to be told about complications such as the one just mentioned.)

Now consider the gaseous reaction

$$A(g) = B(g) + C(g)$$

for which

$$K_p = \frac{P_B P_C}{P_A}$$

Consider these P's now to represent their equilibrium values in the initial volume. If we now halve the volume of the container, all the partial pressures are doubled, so that, before the equilibrium shifts, we have

$$Q = \frac{(2P_B)(2P_C)}{2P_A} = 2K_{eq}$$

Since now $Q>K_{eq}$, a reverse reaction will take place: P_A will increase and P_B and P_C will decrease, until once again $Q=K_{eq}$. The best way to sum up this result is as follows: if the partial pressure of any gaseous component at equilibrium is increased, then the equilibrium will shift so as to reduce the partial pressure of that component; if the partial pressure of all the gaseous components are simultaneously increased, then the reaction will shift toward the side with fewer moles of gaseous components — the left, in this reaction.

Note that this is identical to what happens in the liquid phase, except that in the gas phase the concentration unit is partial pressure, rather than molar concentration. For the analogous reaction

$$A(aq) = B(aq) + C(aq)$$

$$K_{eq} = \frac{[B][C]}{[A]}$$

if we simultaneously doubled the concentrations of all components, then, before an equilibrium shift, we would once again have $Q=2K_{eq}$, and a net reverse reaction would occur, until Q was equal to K_{eq}. So even in solution, simultaneously increasing the concentration of all components shifts the equilibrium to the side with fewer moles of reactants.

Suppose, for the gaseous reaction, the total pressure had been raised, not by decreasing the volume, but by adding an inert gas. What does LeChatelier's principle predict will happen? Suppose I have pure A in a vessel at 1atm, then add an inert gas, such as argon, to make the total pressure 2atm. P_A is still 1atm, by Dalton's law. Thus adding an inert gas does not alter the partial pressures of the components. Since it is these partial pressures that appear in the equilibrium constant, and these are not changed, the equilibrium is not stressed in any way, and no shift takes place. Moral: whenever calculating a shift in equilibrium for a gas-phase reaction, look at changes in *partial* pressures! Note that this lack of a shift has an analog in solution processes. If to the reaction

$$A(aq) = B(aq)+C(aq)$$

I add a non-reacting substance, D, it raises the total concentration of all the components, but not the concentrations of A, B, and C, and therefore does not affect the equilibrium.

The effect of a temperature change is different from the effect of a concentration or pressure change, because K_{eq} changes with temperature, but not with concentration or pressure; nevertheless,

Chapter 8 — Introduction to Chemical Equilibrium

LeChatelier's principle still applies, though the way to handle this mathematically will only become clear in Chapter 18. Recall that exothermic processes are ones which give off heat ($\Delta H < 0$), and that endothermic processes are ones which absorb heat ($\Delta H > 0$). Raising the temperature of a system tends to add heat to the system; therefore this will inhibit a process that gives off heat (an exothermic process). You can think of heat as one of the products of an exothermic reaction; therefore, adding heat to an exothermic process, by raising the temperature, pushes the process to the left (K_{eq} decreases). Similarly, heat can be thought of as one of the reactants of an endothermic process, and here increasing the temperature increases K_{eq} (pushes the process to the right). Note that this is in accord with what you learned about vapor pressure in Chapter 5; vaporization is always endothermic, and for the process

$$A(l) = A(g)$$

the equilibrium constant is $K_p = P_A^\circ$. Since the process is endothermic, increased temperature should lead to increased K_p, and in fact P_A° (the vapor pressure of A) is always observed to increase with increasing T.

K_{eq} for Altered and Combined Reactions

Consider the simple reaction

$$A = B + C$$

$$K_c^f = \frac{[B][C]}{[A]}$$

The superscript "f" refers to the forward direction. Now consider the reverse reaction

$$B + C = A$$

$$K_c^r = \frac{[A]}{[B][C]}$$

From this it should be clear that *when a reaction is reversed, the new equilibrium constant is the reciprocal of the old equilibrium constant, $K^r = 1/K^f$*.

Now consider twice the original reaction

$$2A = 2B + 2C$$

$$K_c = \frac{[B]^2[C]^2}{[A]^2} = \left(\frac{[B][C]}{[A]}\right)^2 = (K_c^f)^2$$

Thus *multiplying a reaction by any number, n, gives a new equilibrium constant that is the n'th power of the old one.* Our first rule, above, is really a special case of this second rule, if we adopt the convention that reversing a chemical equation is equivalent to multiplying it by -1. Then, for such a reversal, we would have $K^r = (K^f)^{-1} = 1/K_f$. Also, n need not be an integer. If an equation is multiplied by 1/2, the new equation will have $K_{new} = (K_{old})^{1/2} = \sqrt{K_{old}}$.

Now suppose we have the following system of reactions

$$A = B + C \tag{I}$$

$$2B = D \tag{II}$$

$$K_I = \frac{[B][C]}{[A]}$$

$$K_{II} = \frac{[D]}{[B]^2}$$

If I and II are taking place in the same system, then multiplying I by 2 and adding it to II gives

$$2A = 2C + D \tag{III}$$

$$K_{III} = \frac{[C]^2[D]}{[A]^2}$$

The rule for adding chemical equations is that anything that appears on the right of one equation and on the left of another cancels, mole for mole. Thus, B is produced in I and used up in II. In order to produce enough B in I for II to proceed, we have to multiply I by 2. Then adding 2I + II gives III, as we showed. Now note that

$$K_{2I} = K_I^2 = \frac{[B]^2[C]^2}{[A]^2}$$

$$K_{II} = \frac{[D]}{[C]^2}$$

$$K_{2I}K_{II} = \frac{[B]^2[D]}{[A]^2} = K_{III}$$

This leads to our third rule: *when equations are added, the K of the overall reaction is the product of*

Chapter 8 — Introduction to Chemical Equilibrium

the K's of the original reactions.

The three rules for obtaining K_{eq} after reversing, multiplying or adding an equation to another enables K_{eq} to be calculated for any overall system of equations, given the K's for the individual equations. One application of this is the calculation of the overall K for the dissociation of polyprotic acids, as illustrated in Equations 8.32 through 8.37 for H_2S.

There are many cases where one is given individual reactions, and asked to combine them into some overall reaction, as in Example 8.13. Here the overall reaction is

$$N_2O + 1/2\ O_2 = 2NO \qquad \text{(I)}$$

and the individual reactions are

$$N_2 + 1/2\ O_2 = N_2O \qquad \text{(II)}$$

and

$$N_2 + O_2 = 2NO \qquad \text{(III)}$$

Note that if II is reversed and added to III, I results. Therefore

$$K_I = \frac{K_{III}}{K_{II}}$$

The reverse of II has $K = 1/K_{II}$, and this multiplied by K_{III} (since III is added to the reverse of I) gives K_I as shown.

K_{eq}'s with Special Names

Certain types of chemical reactions have equilibrium constants which are given special names. One of these is the *solubility product*, which is K_{eq} for the dissolution of an almost insoluble precipitate; for example

$$Ag_2CrO_4(s) = 2Ag^+(aq) + CrO_4^{2-}(aq)$$

$$K_{sp} = [Ag^+]^2\,[CrO_4^{2-}]$$

Recall that $[Ag_2CrO_4]$ does not appear in the denominator of K_{sp}, since it is a solid, and therefore of

Chapter 8

Introduction to Chemical Equilibrium

constant concentration.

For the ionization of an acid, K_{eq} is called an *acidity constant*; the terms *dissociation constant* and *ionization constant* are sometimes used.

$$H_2S(aq) = H^+(aq) + HS^-(aq)$$

$$HS^-(aq) = H^+(aq) + S^{2-}(aq)$$

$$K_{a1} = \frac{[H^+][HS^-]}{[H_2S]}$$

$$K_{a2} = \frac{[H^+][S^{2-}]}{[S^{2-}]}$$

As this example shows, acids with more than one *acidic hydrogen* have several K_a's. K_a is only used for weak acids; strong acids, such as HCl, are considered 100% ionized, which would correspond to $K_a = \infty$!

Likewise, K_b is the *basicity constant* for a weak base.

$$NH_3(aq) + H_2O = NH_4^+(aq) + OH^-(aq)$$

$$K_b = \frac{[OH^-][NH_4^+]}{[NH_3]}$$

Recall that since we assume the solution is dilute, $[H_2O]$ is considered constant and does not appear in the denominator.

Sometimes complex ions are formed in solution. K_{eq} for their formation is called a *formation constant*, and K_{eq} for the reverse of formation is called the *dissociation constant*.

$$Ag^+(aq) + 2NH_3(aq) = Ag(NH_3)_2^+(aq)$$

$$K_f = \frac{[Ag(NH_3)_2^+]}{[Ag^+][NH_3]^2}$$

$$K_d = \frac{[Ag^+][NH_3]^2}{[Ag(NH_3)_2^+]}$$

This particular formation reaction causes ammonia to dissolve AgCl precipitates, since the reaction lowers $[Ag^+]$. Note that we said that the term "dissociation constant" is sometimes used for K_a. "Dissociation constant" is the more general term, and can be used for any reaction where a substance breaks apart into its components.

Chapter 8

Finally, a process like

$$2A = B$$

is called a *dimerization*; an example is

$$2NO_2(g) = N_2O_4(g)$$

The NO_2 is called the *monomer* and the N_2O_4 is called the *dimer*. A dimerization is where two identical subunits join to form a molecule, and K_{eq} for such a reaction can be called an *association constant*. Of course, K for the reverse reaction would be a dissociation constant. In your study of chemistry you will encounter numerous other reactions which are given special names.

Gas-Phase Equilibrium Problems

This section (8.6) of the text consists almost entirely of worked examples. I will simply add some additional comments to these.

What many students find hardest in these sorts of problems is to be able to use the reaction stoichiometry (the balanced chemical equation) to write down expressions for the number of moles of all components present at equilibrium. For the reaction (from Example 8.14)

$$2SO_2(g) + O_2(g) = 2SO_3(g)$$

we are told that initially $P_{SO_2} = 1.00$ atm and $P_{O_2} = 0.50$ atm and that after reaction $P_T = 1.35$ atm. The question then is, "What is K_p?" In order to know K_p one needs the equilibrium values of P_{SO_2}, P_{SO_3} and P_{O_2}. It is very helpful when working such a problem to make a table such as the following. (All data in atm.)

	P_{SO_2}	P_{O_2}	P_{SO_3}	P_T
before	1.00	0.50	0	1.50
after	**(1.00-x)**	**(0.50-x/2)**	x	1.35

What students find hard is to fill in the table entries given in bold face; the other information is essentially given in the problem. If we simply let x be equal to P_{SO_3}, then we note that for each mole of SO_3 formed one mole of SO_2 reacted; therefore the number of moles of SO_2 present after equilibrium is

Chapter 8 — Introduction to Chemical Equilibrium

achieved is (moles SO_3) fewer than the original; since for gases moles are proportional to partial pressure, $P_{SO_3}^{after}=(1.00-x)$ atm. Now, for O_2, note that for each *two* moles SO_3 formed, one mole of O_2 reacts; thus for each mole SO_3 formed, 1/2 mole O_2 has reacted. Therefore $P_{O_2}^{after} = (0.50-x/2)$ atm. From here the problem is solved using Dalton's law: $P_T = P_{SO_2} + P_{SO_3} + P_{O_2}$, which gives x, then the P's are used to calculate K_p, as shown in the text.

Example 8.15 is a bit like 8.14 in reverse; here one is given K=6, and asked to find concentrations at equilibrium. The same kind of table is useful. The reaction is

$$H_2(g) + I_2(g) = 2HI(g)$$

Note that here, $K_c=K_p$, since $\Delta n=0$ (same number of moles on both sides of the equation). Data are given as concentrations. Generally, K for a gas reaction is understood to be K_p. If Δn had not been zero, it would have been necessary either to convert K_p to K_c by multiplying by $(RT)^{\Delta n}$, or else to use K_p, and first convert the concentrations to partial pressures. As given, however, here is the table

	$[H_2]$	$[I_2]$	$[HI]$
before	1.00	1.00	0
after	1.00-x	1.00-x	2x

Here, x has been defined as the number of moles of I_2 reacting; this choice is arbitrary (see below). By this definition we have (1.00-x) for $[I_2]$ at equilibrium. Since 1 mole of H_2 reacts for each mole of I_2 that reacts, and since H_2 also started out at 1.00M concentration, (1.00 -x) is also equal to $[H_2]$ at equilibrium. Finally, since 2 moles of HI are formed for each mole of I_2 reacting, $[HI] = 2x$ at equilibrium. These concentrations are plugged into the expression for K_c, and the result obtained straightforwardly, as shown in the text.

Suppose, however, we had decided to use, say, y to represent the number of moles of HI formed at equilibrium. We would then have

	$[H_2]$	$[I_2]$	$[HI]$
before	1.00	1.00	0
after	1.00 -y/2	1.00 -y/2	y

Chapter 8 — Introduction to Chemical Equilibrium

The stoichiometry tells us that half as many moles of I_2 or H_2 disappear as the number of moles of HI formed; therefore H_2 and I_2 have equilibrium concentration $(1.00 - y/2)$. This leads to

$$K = 6 = \frac{y^2}{(1.00 - y/2)^2}$$

$$2.45 = \frac{y}{1.00 - y/2}$$

$$2.45\,(1.00 - y/2) = y$$

$$4.90 - 2.45y = 2y$$

$$y = \frac{4.90}{4.45} = 1.10 = [HI]$$

$$(1.00 - y/2) = 0.45 = [H_2] = [I_2]$$

Thus the two choices of unknown are equivalent. The point is that there are usually several correct ways to solve a problem. To use any of them, you must possess fundamental skills, such as being able to write down how many moles of each of the other ingredients will be used up or formed, given that x moles of some given component is used up or formed. A hint: the arithmetic is often simplified if you pick x to correspond to some ingredient that has a stoichiometric coefficient of 1; thus the text's choice of x in Example 8.15 works out a bit more easily than my choice of y. Likewise, Example 8.14 would have worked out a bit more easily had the unknown been defined as the number of moles O_2 reacting.

Example 8.16 involves the simple application of the same principles we have been discussing to a slightly different problem. Note that the text utilizes a "before-after" table. A good way for you to test your understanding would be to solve the same problem, using as the unknown not the increase in P_{H_2}, but the decrease in P_{HI}. You should come up with the same final answer as the text, though your table will look different.

Chapter 8 Introduction to Chemical Equilibrium

Questions

8.1 The ideal law of chemical equilibrium

1. Write the equilibrium constant expression for the following reactions; use P for gases, molar concentrations for substances in solution.

    (a) $HF(aq) = H^+(aq) + F^-(aq)$

    (b) $HF(aq) + H_2O(l) = H_3O^+(aq) + F^-(aq)$

    (c) $NH_3(aq) + CH_3COOH(aq) = NH_4^+(aq) + CH_3COO^-(aq)$

    (d) $2N_2O + O_2 = 4NO$

    (e) $N_2O + 1/2\ O_2 = 2NO$

    (f) $CaCO_3(s) = CO_2(g) + CaO(s)$

    (g) $4Fe^{2+}(aq) + O_2(g) + 4H^+(aq) = 4Fe^{3+} + 2H_2O$

    (h) $Hg_2Cl_2(s) = Hg_2^{2+}(aq) + 2Cl^-(aq)$

    (j) $Cu^{2+}(aq) + Zn(s) = Cu(s) + Zn^{2+}(aq)$

    (k) $H_2O(l) = H_2O(g)$

2. For the reactions in Question 1d, e, f, g and k, what would K_c have to be multiplied by to give K_p?

    (d)

    (e)

    (f)

    (g)

    (k)

8.2 The magnitude of the equilibrium constant and the direction of reaction

1. For the reaction

$$A = 2B$$

it is found that at equilibrium [A]=0.1M and [B]=0.01M

   (a) What is K_{eq}?

(b) Suppose initially a solution is made 1M in A and in B. What is Q?

(c) For the solution in (b), will the net reaction go to the left or the right?

(d) Suppose initially a solution is made 0.2M in A and 0.01M in B. What is Q?

(e) For the solution in (d) will the net reaction proceed to the left or the right?

(f) Suppose at equilibrium [A]=.001M. What is [B]?

2. For the dissolution of barium carbonate

$$BaCO_3(s) = Ba^{2+}(aq) + CO_2^{2-}(aq)$$

it is found that $K_{eq} = 1.6 \times 10^{-9}$. (This is the solubility product of $BaCO_3$).

(a) If pure water is added to solid $BaCO_3$, what will $[Ba^{2+}]$ and $[CO_3^{2-}]$ be in the solution after equilibrium has been achieved?

(b) If the same solution is now made 0.01M in CO_3^{2-}, what will Q be before the equilibrium shifts?

(c) Will the shift result in the dissolution or the precipitation of $BaCO_3$?

(d) What will be the new concentration of [Ba^{2+}] when equilibrium is again achieved?

8.3 Factors affecting the eqilibrium constant

1. Which of these will change the value of the equilibrium constant?

    _____(a) changing the solvent

    _____(b) changing the concentration of a reactant

    _____(c) changing the concentration of a product

    _____(d) changing the temperature

2. When reaction conditions change causing a change in Q, K_{eq} must also change. (True, False)

8.4 LeChatelier's principle

1. For the reaction

$$BaCO_3(s) = Ba^{2+}(aq) + CO_3^{2-}(aq)$$

    Tell whether adding each of the following to the solid in contact with its saturated solution will cause the equilibrium to shift to the left, the right, or not at all.

    _____(a) $BaCl_2$ (a soluble salt)

    _____(b) Na_2CO_3 (a soluble salt)

    _____(c) HCl

    _____(d) NaCl

2. Barium fluoride is slightly soluble in water. How would you exploit the common ion effect to cause more barium to precipitate in this form than from a saturated solution of this salt?

Chapter 8   Introduction to Chemical Equilibrium

3. Suppose the gas phase reaction

$$H_2(g) + Br_2(g) = 2HBr(g)$$

is at equilibrium in a cylinder fitted with a piston.

(a) If the pressure is doubled by moving the piston, in which direction will the equilibrium shift?

(b) If instead the pressure is doubled by adding an inert gas, in which direction will the equilibrium shift?

4. Now suppose the reaction of Question 3 is run at a lower temperature, so that HBr and Br_2 are both liquids.

(a) If the pressure is doubled by moving the piston, in which direction will the equilibrium shift?

(b) If instead the pressure is doubled by adding an inert gas, in which direction will the equilibrium shift?

5. The reaction

$$2H_2(g) + O_2(g) = 2H_2O(l)$$

is highly exothermic. Therefore, the higher the temperature, the *greater* the value of K_{eq}.

6. Lithium carbonate dissolves in water with the evolution of heat. Therefore it is *more* soluble in hot than in cold water.

Chapter 8 — Introduction to Chemical Equilibrium

8.5 Numerical values of equilibrium constants for reactions written in the reverse direction and for simultaneous equilibrium

1. Consider the following reactions in solution

$$A = B \quad \text{(I)}$$
$$2B = C \quad \text{(II)}$$

with corresponding equilibrium constants K_I and K_{II}. Write expressions for K of the following reactions, both in terms of concentrations and in terms of K_I and K_{II}; for example, K for B=A is $[A]/[B] = 1/K_I$.

(a) C = 2B

(b) 2A = 2B

(c) B = 1/2 C

(d) 2A = C

(e) A = 1/2 C

2. Given the following reactions

$$Zn(OH)_2(s) = Zn^{2+}(aq) + 2OH^-$$
$$K_{sp} = 5 \times 10^{-17}$$

$$Zn^{2+}(aq) + 4OH^-(aq) = Zn(OH)_4^{2-}(aq)$$
$$K_f = 3 \times 10^{15}$$

give an expression and the value of K for the overall reaction

$$Zn(OH)_2(s) + 2OH^-(aq) = Zn(OH)_4^{2-}(aq)$$

Chapter 8   Introduction to Chemical Equilibrium

8.6 Some typical problems in gas-phase equilibria

1. Consider the reaction

$$SO_2Cl_2(g) = SO_2(g) + Cl_2(g)$$

Suppose that 1atm SO_2 and 1atm Cl_2 are introduced into a sealed vessel. If we let x be equal to $P_{SO_2Cl_2}$ at equilibrium, then

(a) What is P_{SO_2} at equilibrium, in terms of x?

(b) What is P_{Cl_2} at equilibrium, in terms of x?

(c) If at equilibrium $P_{total}=1.76$ atm, calculate x

(d) What are the equilibrium values of

    (i) $P_{SO_2Cl_2}$

    (ii) P_{SO_2}

    (iii) P_{Cl_2}

(e) Calculate K_p

(f) Suppose that the system at equilibrium is compressed to half its volume. Tell whether each of the following will increase, decrease or remain the same.

    (i) moles SO_2Cl_2

    (ii) moles SO_2

    (iii) moles Cl_2

    (iv) K_p

    (v) K_c

Chapter 8

Introduction to Chemical Equilibrium

Answers

8.1 The ideal law of chemical equilibrium

1. (a) $K = \dfrac{[H^+][F^-]}{[HF]}$

    (b) $K = \dfrac{[H_3O^+][F^-]}{[HF]}$

    (c) $K = \dfrac{[NH_4^+][CH_3COO^-]}{[NH_3][CH_3COOH]}$

    (d) $K = \dfrac{P_{NO}^4}{P_{N_2O}^2 P_{O_2}}$

    (e) $K = \dfrac{P_{NO}^2}{P_{N_2O} P_{O_2}^{1/2}}$

    (f) $K = P_{CO_2}$

    (g) $K = \dfrac{[Fe^{3+}]^4}{[Fe^{2+}]^4[H^+]^4 P_{O_2}}$

    (h) $K = [Hg_2^{2+}][Cl^-]^2$

    (j) $K = \dfrac{[Zn^{2+}]}{[Cu^{2+}]}$

    (k) $K = P_{H_2O}$

2. (d) RT
    (e) $(RT)^{1/2}$
    (f) RT
    (g) 1/RT
    (k) RT

8.2 The magnitude of the equilibrium constant and the direction of reaction

1. (a) $K_{eq} = [B]^2/[A] = 0.001$
    (b) Q=1
    (c) left
    (d) Q = 0.0005
    (e) right
    (f) 0.001M

2. (a) 4.0×10^{-5}M for $[Ba^{2+}]$ and $[CO_3^{2-}]$ ($K = 1.6 \times 10^{-9} = 16 \times 10^{-10} = [Ba^{2+}][CO_3^{2-}] = x^2$, so $x = \sqrt{K} = 4 \times 10^{-5}$)
    (b) 4.0×10^{-7}
    (c) precipitation, since Q>K
    (d) 1.6×10^{-7}M ($[Ba^{2+}] = K/[OH^-] = 1.6 \times 10^{-9}/0.01$)

Chapter 8 Introduction to Chemical Equilibrium

8.3 Factors affecting the equilibrium constant

1. (a) and (d)
2. False

8.4 LeChatelier's principle

1. (a) left
   (b) left
   (c) right, due to $2H^+ + CO_3^{2-} = H_2CO_3 = H_2O + CO_2(g)$
   (d) no change
2. Add a soluble fluoride, such as NaF.
3. (a) Neither direction. (Same number of moles of gas on both sides.)
   (b) Neither direction. (Partial pressure of components don't change.)
4. (a) Right. (Only gaseous component is H_2, and it is on left.)
   (b) Neither direction. (As in 3(b).)
5. lower
6. less

8.5 Numerical values of equilibrium constants for reactions written in the reverse direction and for simultaneous equilibria

1. (a) $1/K_{II} = [B]^2/[C]$

   (b) $(K_I)^2 = [B]^2/[A]^2$

   (c) $(K_{II})^{1/2} = \sqrt{K_{II}} = [C]^{1/2}/[B]$

   (d) $(K_I)^2 K_{II} = [C]/[A]^2$

   (e) $K_I (K_{II})^{1/2} = [C]^{1/2}/[A]$

2. $K = [Zn(OH)_4^{2-}]/[OH^-]^2 = K_{sp}K_f = 0.15$

8.6 Some typical problems in gas-phase equilibria

1. (a) $(1-x)$
   (b) $(1-x)$
   (c) 0.24 ($P_T = 1.76 = x + (1-x) + (1-x) = 2-x$)
   (d)
       (i) 0.24
       (ii) 0.76
       (iii) 0.76
   (e) 2.4 ($K = P_{SO_2} P_{Cl_2}/P_{SOCl_2} = (7.6)^2/(2.4)$)
   (f)
       (i) increase
       (ii) decrease
       (iii) decrease
       (iv) no change
       (v) no change

CHAPTER 9: ACIDS, BASES AND SALTS

Scope

This and the next two chapters continue to develop methods of solving numerical problems involving chemical equilibrium. On the way, a number of new concepts are introduced, and some old ones elaborated. The prerequisites to assimilating this material are

1. the ability to write down the equilibrium constant expression starting with a balanced chemical equation;

2. the ability to write down new K's as reactions are reversed, multiplied and combined;

3. the ability to reason out changes in the concentration of various reaction components based on stoichiometry; for example, for the reaction

$$A + 2B \rightarrow 3C$$

   knowing that if x moles of C are formed, then $x/3$ moles of A and $2x/3$ moles of B must have reacted;

4. the ability to do algebra, such as solving the equation $4x^3 = 5 \times 10^{-17}$ for x, or using the quadratic formula to solve $x^2 - 4.4x + 1 = 0$.

These operations were introduced in the last chapter.

Another skill which will be useful, and which will be introduced in this chapter, is the ability to make simplifying algebraic approximations, so as to obtain easily solvable equations with answers good to two or three significant figures. Finally, amidst all these rules and techniques, there is no substitute for a good chemical understanding of what is going on in a particular reaction system. This gets more and more important as calculations get more difficult, because as this occurs it becomes easier to make numerical or algebraic mistakes. At the end of a lengthy calculation, you should always ask, "Does my answer make sense?" It takes chemical intuition to know whether it does or not.

Chapter 9 — Acids, Bases and Salts

Self-Ionization

Many solvents, but water in particular, *self-ionize* (or *auto-ionize*). To write the reaction in two equivalent forms

$$2H_2O(l) = H_3O^+(aq) + OH^-(aq)$$

$$H_2O(l) = H^+(aq) + OH^-(aq)$$

K for this reaction is termed K_w, the *ion-product of water*, and is equal to 10^{-14} at 25°C.

$$K_w = [H_3O^+][OH^-]$$

Of course, regardless of which way the equation is written, $[H_2O]$ does not appear in the K_w expression, for reasons you should know by now. In pure water, $[OH^-] = [H^+]$, from the reaction stoichiometry, so that each is at concentration $\sqrt{10^{-14}} = 10^{-7}$. Now, however, suppose we make up a solution 0.1M in strong acid, such as HCl. The solution will then be 0.1M in H^+ (or H_3O^+, if you prefer; recall, they are the same!). We can solve for $[OH^-]$ as follows

$$[OH^-] = \frac{K_w}{[H^+]} = \frac{10^{-14}}{0.1} = 10^{-13}$$

First, note that, as with other equilibrium constants, at constant temperature concentrations change, but K_w stays the same. This enables the calculation to be done in the first place. Second, we have made an approximation — two, in fact — in the assumption that $[H^+] = 0.1$M. The first approximation is that the HCl is 100% ionized. This is probably the more serious of the two. The second is more subtle. Recall that $[H^+]$ was 10^{-7}M even before adding the HCl; it stands to reason that after adding it there should still be some H^+ around due to the self-ionization of water; however, we are assuming our calculations are good to, say, three significant figures, and even if the H^+ from self-ionization amounted to 10^{-7}M, this would affect the concentration only in the seventh significant figure ($0.1 + 10^{-7} = 0.1000001$); therefore this contribution can be ignored. Furthermore, the H^+ from self-ionization would really amount to considerably less than 10^{-7}M, since adding extra H^+ reverses the self-ionization, by LeChatelier's principle. The concentration of H^+ present due to self-ionization must be equal to the concentration of OH^- present from the same process, due to the reaction stoichiometry, and we saw that this was about 10^{-13}M, not 10^{-7}M. In short, this second approximation is an excellent one.

Chapter 9 Acids, Bases and Salts

Before you are done with Chapter 11, you will learn to live with far worse.

pH

The term *pH* stands for the negative logarithm (to base 10) of $[H^+]$.

$$pH = -\log_{10}[H^+]$$

The term was invented and the notion popularized by the Danish chemist Sorensen in the early 20th century. Part of the utility of the pH notion is that it makes hydrogen ion concentration easy to talk about. In 0.1M strong acid $[H^+]$ is 0.1M, so that $pH = -\log_{10} 0.1 = 1$; in 0.1M strong base, $[OH^-]$ is 0.1M, so that $[H^+] = 10^{-14}/0.1 = 10^{-13}$, so that $pH = -\log_{10} 10^{-13} = 13$. It is easier, in common speech, to bandy about numbers like 13 than numbers like 10^{-13}. pH is also important because certain properties of solutions, such as the voltage across an electrochemical cell, respond to concentrations in a logarithmic fashion, and in solving certain problems it is easier to use pH directly than $[H^+]$. Note that at neutrality $pH = -\log_{10} 10^{-7} = 7$, and that pH's less than 7 are associated with acidic solutions ($[H^+] > 10^{-7}M$) and pH values greater than 7 are associated with basic solutions ($[H^+] < 10^{-7}M$). pH can be zero or negative, as well as positive; a 10M HCl solution has $pH = -\log_{10} 10 = -1$. Finally, note that K_w gives a relationship between pH and pOH ($pOH = -\log_{10}[OH^-]$)

$$K_w = 10^{-14} = [H^+][OH^-]$$

$$pK_w = -\log_{10} K_w = 14 = pH + pOH$$

In deriving the second equation we used the fact that $\log(xy)=\log(x)+\log(y)$ Thus, if pH is, say, 2, then pOH is 12. As should be clear from this example, the symbol "p" has been appropriated to mean "$-\log_{10}$", so that pAnything = $-\log_{10}$ (That Thing).

You should be aware that

$$10^{\log_{10}\text{Anything}} = \log_{10} 10^{\text{Anything}} = \text{That Thing}$$

This relationship allows one to convert pH values to $[H^+]$ values. For example, if pH = 2, then $[H^+] = 10^{-pH} = 10^{-2}$. This is easy enough, but if pH = 2.5 then clearly $[H^+]$ is between 10^{-2} and 10^{-3}, but where? The answer is $[H^+] = 3.16 \times 10^{-3}$, as can be verified on your pocket calculator. (If your

Weak Acids and Bases

For the ionization of an arbitrary acid, HA,

$$H + H_2O = H_3O^+ + A^-$$

The equilibrium constant is called the acidity constant, as discussed in the last chapter.

$$K_a = \frac{[H_3O^+][A^-]}{[HA]}$$

Likewise, for the ionization of a weak base

$$B + H_2O = BH^+ + OH^-$$

The equilibrium constant is called the basicity constant.

$$K_b = \frac{[BH^+][OH^-]}{[B]}$$

Examples of weak acids and bases are given in Tables 9.4 and 9.5, respectively. Note that K_a's and K_b's for common weak acids and bases are in the range 10^{-4} to 10^{-10}; of course, the weakest acids and bases are the ones with K's closer to 10^{-10}, the lower of these two values. Note as a matter of interest for now that the anion of a weak acid is a weak base (HCN has $K_a = 4.0 \times 10^{-10}$; CN^- has $K_b = 2.5 \times 10^{-5}$). This will be taken up again later in the chapter.

The *degree of dissociation*, α, is defined as the *fraction* of the original substance *ionized*. For example, if A 0.1M HA solution is made up, and it is found that $[H^+] = [A^-] = 0.01M$, then $\alpha = 0.01M/0.1M = 0.1$, so that 10% of the original molecules present have ionized. Note that if we have a pure HA solution, then $[H^+]$ must equal $[A^-]$, by the reaction stoichiometry

$$HA = H^+ + A^-$$

Again, this disregards any H^+ present due to the self-ionization of water. Note that α must be between 0 and 1; if $\alpha=0$ there is no dissociation (K=0) and the substance is not an electrolyte at all. If $\alpha=1$ all the molecules dissociate (K=∞) and the substance is a strong electrolyte.

Typical problems require you to calculate concentrations based on initial molarities and K's, or to calculate K's based on α's and concentrations. A relatively fancy kind of problem will give you a

Chapter 9 Acids, Bases and Salts

concentration and a freezing point (remember Chapter 5?), and ask you to calculate K and/or α. In all of these, skill in knowing when to make numerical approximations is often important.

Consider the reaction

$$H_2O + HA = H_3O^+ + A^-$$

$$K_a = \frac{[H_3O^+][A^-]}{[HA]}$$

Suppose that C, the *stoichiometric* or *nominal concentration*, is 0.01M. This is the concentration that one weighs out and dilutes in making up the solution. Suppose first that $K_a = 10^{-6}$, and that we wish to find the pH. Let $x = [H_3O^+]$; we know from the stoichiometry that then $x = [A^-]$ as well, and that $[HA] = C-x = 0.01-x$. This gives

$$K_a = \frac{[H_3O^+][A^-]}{[HA]}$$

$$10^{-6} = \frac{x^2}{0.01-x}$$

Now, we could multiply both sides of this equation by (0.01-x), and combine terms to get a quadratic in x, which we would then solve by the quadratic formula, but there is a simplifying approximation that can be made. Since K is so small, very little of the acid will ionize, and [HA] will be very close to C. This amounts to saying that x will be much smaller than 0.01, so that it can be ignored in the denominator of the last equation. This gives

$$10^{-6} \approx \frac{x^2}{.01}$$

$$x = [H^+] = \sqrt{10^{-8}} = 10^{-4}$$

$$pH = -\log_{10} 10^{-4} = 4.00$$

How good is this approximation? Let us solve the same equation exactly, using the quadratic formula, and see how much the result differs.

$$10^{-6} = \frac{x^2}{(0.01-x)}$$

$$10^{-8} - 10^{-6}x = x^2$$

$$x^2 + 10^{-6}x - 10^{-8} = 0$$

$$x = \frac{-b \pm \sqrt{b^2 - 4ac}}{2a}$$

$$a=1;\ b=10^{-6};\ c=10^{-8}$$

$$x = \frac{-10^{-6} \pm \sqrt{10^{-12} + 4 \times 10^{-8}}}{2}$$

$$x = [H^+] = 9.95 \times 10^{-5} \text{ (only positive solution)}$$

$$pH = 4.002$$

Our earlier result, using the approximation, gave $x = 10^{-4} = 10 \times 10^{-5}$ and therefore was about a half a percent higher than the true value. This much error is quite acceptable for this sort of calculation. The difference only occurs in the third significant figure, and a pH difference of 0.002 pH units is hard to measure reliably. Even the law of mass action, which declares K_{eq} to be independent of concentration, may not always be this good in this concentration range.

If we always had to apply the quadratic formula to make sure that our approximation was correct, there would be no point in applying the approximation — we would just use the quadratic formula from the start! Fortunately, we can check our approximation more easily. Note that our calculated value of x, using the approximation, was 10^{-4}. We can compare this with 0.01 in the denominator of the original expression, to see if our original guess that x could be neglected was justified. Note that 10^{-4} is one one-hundredth of 0.01. Whether that makes for insignificance depends on the context; for equilibrium problems, the text suggests that if x turns out to be less than 10% of C then the approximation was a good one; therefore, the approximation was certainly good. Generally, 10% criterion dictates that neglecting x in the denominator will end up being justified if $C > 100K$. This can be seen from the following. Assume that, in fact, $C = 100K$. Then our equation would look like

$$K = \frac{x^2}{C-x} = \frac{x^2}{100K-x}$$

Applying the approximation $x \ll C$ gives

$$K \approx \frac{x^2}{100K}$$

$$x^2 = 100K^2$$

$$x = 10K$$

Chapter 9 — Acids, Bases and Salts

Since 10K is just 0.1C, the approximation is just satisfactory when C=100K. When C is larger, the approximation is better. Incidentally, do not be troubled by the fact that to test the approximation you are comparing an approximate, rather than the true, value of x with C. The true value will always be less than the approximate value, in this sort of problem, making the approximation even better.

Now consider the same problem, in which $K=10^{-3}$, rather than 10^{-6}. We have

$$10^{-3} = \frac{x^2}{0.01-x}$$

Here, K is larger, and it may no longer be a good approximation to neglect x in the denominator; let us try doing so anyway. The result is $x=[H^+]=\sqrt{10^{-5}} = 0.0032$, pH = 250. In fact, the approximation is not good, since 0.0032 is not less than 0.001. We could apply the quadratic formula, but it is easier at this point to apply a *method of successive approximations*. We view our value of x as a "trial value," and solve the same equation again, but use (0.01-x) in the denominator instead of 0.01. This gives

$$10^{-3} = \frac{x^2}{.01-.0032}$$

$$x = \sqrt{(.0068)(10^{-3})} = 0.0026$$

This new x is again used in the denominator

$$10^{-3} = \frac{x^2}{0.01-0.0026}$$

$$x = \sqrt{(.0074)(10^{-3})} = .0027$$

And again

$$10^{-3} = \frac{x^2}{0.01-0.0027}$$

$$x = \sqrt{(.0073)(10^{-3})} = .0027$$

When successive values of x are the same to the desired number of significant figures, the process is terminated; thus $[H^+] = 0.0027M$, and pH = 2.57, rather than the 2.50 obtained from the original approximation.

A few words about the method. It may not be obvious to you why it should work. It does not, in fact, work on all algebraic equations, and to understand just when it will work requires more mathematics than this course presupposes. It will however, work in all the examples in this course where you might be tempted to use it, and in fact, if you try it in a non-applicable equation, the approximation will not *converge*; that is, successive values of x will not become closer and closer together. If the method converges, then it is applicable, and the final x can be believed to as many significant figures as it agrees with its predecessor. In other words, if it looks like it's working, then it's working.

Let us now calculate α for our two examples. When $K=10^{-6}$, we had $[H^+]=10^{-4}M$, so that $\alpha=[H^+]/C=10^{-4}M/0.01M = 0.01$, so that the material was 1% ionized. For $K=10^{-3}$, we had $[H^+]=0.0027M$, so that $\alpha = (0.0027M/0.01M) = 0.27$, or 27% ionized.

Recall that $\alpha = x/C$, so that $x=\alpha C$. This allows the expression for K to be restated in terms of α.

$$K = \frac{x^2}{C-x}$$

$$= \frac{\alpha^2 C^2}{C-\alpha C} = \frac{\alpha^2 C^2}{C(1-\alpha)}$$

$$= \frac{\alpha^2 C}{1-\alpha}$$

This is the *Ostwald Dilution Law*. From it it is easy to see that if C goes down, α goes up, so that a weak acid dissociates more as its concentration decreases. First consider a very weak acid, for which $K<<1$ and $\alpha<<1$. Here we have $K\approx\alpha^2 C$. Since K is a constant, when C goes down, α^2 must go up, and this requires that α go up. Even where we cannot make the approximation $(1-\alpha)\approx 1$, the equation says the same thing. Since K is constant, if C goes down, the quotient $(\alpha^2/(1-\alpha))$ must go up. If α increases, then the numerator gets bigger and the denominator gets smaller, both of which increase the quotient, so that the increase of α as the concentration goes down does not depend on very small values of K and α.

The increase of fraction dissociation with decreasing concentration is a direct consequence of LeChatelier's principle. For the reaction

$$HA = H^+ + A^-$$

There are more moles on the right than on the left; therefore, increasing the overall concentration shifts the reaction to the left, and decreasing C shifts it to the right. This is just like compressing or expanding a reacting gaseous system with more moles of gas on the right than on the left-hand side of the reaction. Even if the reaction is written

$$H_2O + HA = H_3O^+ + A^-$$

increasing C does not affect [H_2O], so that the analysis given above still holds. What is crucial is that there are more moles in the numerator than in the denominator of the expression for K_{eq}.

Even though our examples have referred to acids and K_a, the same methods are applicable to the analysis of ionization of bases, since this problem is algebraically identical. Example 9.8 in the text shows the application of the method of successive approximations to calculating α for a base, starting with the Ostwald dilution law. For methylamine, $K_b = 4.2 \times 10^{-4}$, and you are asked to find α for C = 0.02M. Denoting the amine by B,

$$B + H_2O = BH^+ + OH^-$$

$$K = 4.2 \times 10^{-4} = \frac{\alpha^2(0.02)}{(1-\alpha)}$$

We first try the approximation $(1-\alpha) \approx 1$, which gives $\alpha = \sqrt{(4.2 \times 10^{-4})/.02} = 0.145$, which is greater than 10% of 1, and therefore no good. Using this α in the denominator gives $\alpha = \sqrt{(1-0.145)(4.2 \times 10^{-4})/0.2} = 0.134$. The process converges on $\alpha = 0.135$. What makes the dissociation of a weak acid and a weak base algebraically identical is that for both the equilibrium constant expression has two moles in the numerator and one in the denominator; thus the same Ostwald law holds.

There is another sort of numerical problem which is related to this material, but which the text discusses later, in Example 9.10. Suppose I have a weak acid present in a solution also containing a strong acid. The strong acid will be completely dissociated, and by LeChatelier's principle the ionization of the weak acid will be somewhat suppressed. To use the text's example, suppose a solution is 0.120M in CH_3COOH and also 0.100M in HCl. What will the pH be?

$$CH_3COOH = CH_3COO^- + H^+$$

We can use a before-after table, letting x represent [CH_3COO^-].

	[CH_3COOH]	[CH_3COO^-]	[H^+]
before	0.120	0	0.100
after	0.120-x	x	0.100+x

Note that it is no longer true that $[H^+]=[CH_3COO^-]$, since most of the H^+ comes from the HCl.

$$K_a = 1.8 \times 10^{-5} = \frac{[CH_3COO^-][H^+]}{[CH_3COOH]} = \frac{(0.100 + x)x}{(0.120 - x)}$$

Now, reasoning as we have before, x should be small, and insignificant in comparison with either 0.100 or 0.120. This gives

$$1.8 \times 10^{-5} = \frac{0.100x}{0.120}$$

leading to $x = 2.16 \times 10^{-5}$, which is small enough to justify our approximation. In fact, $[H^+] = 0.100+x = 0.1000216 \approx 0.100$, so that x is entirely negligible, meaning that the ionization of acetic acid does not materially contribute to $[H^+]$; thus, pH=1.00. Incidentally, if in comparing x with 0.100 and 0.120 we had found that it was not negligible, we could have applied our method of successive approximation by inserting the value of x in (0.100+x) and (0.120-x), and solving for a new x interatively.

Conjugate Acids and Bases

We noted earlier that the anion of a weak acid is a weak base. According to the Lowry-Bronsted notion, if an acid dissociates according to

$$H_2 + HA = H_3O^+ + A^-$$

then HA is an acid because it is a proton-donor, and H_2O is a base because it is a proton acceptor. For the reverse reaction, however, note that H_3O^+ is an acid, and A^-, which accepts a proton to form HA, is a base. We say that A^- is the *conjugate base* of the acid HA, and that HA is the *conjugate acid* of the base A^-. Similarly, using ammonia as an example of a base,

$$NH_3 + H_2O = NH_4^+ + OH^-$$

NH_4^+ is the conjugate acid of NH_3.

Now consider putting the anion of weak acid in water; for example, making up a solution of $NaCH_3COO$, sodium acetate. The acetate anion is the conjugate base of acetic acid. Since acetic acid is more stable in its unionized form than in its ionized form ($K_a = 1.8 \times 10^{-5}$), it stands to reason that the following reaction should take place.

Chapter 9 — Acids, Bases and Salts

$$CH_3COO^- + H_2O = CH_3COOH + OH^-$$

This makes the solution basic. Thus the idea of a conjugate base is not just some abstract or formal notion; anions of weak acids really do act like bases! The reaction of a salt of a weak acid with water to form a a basic solution is sometimes called a *hydrolysis* reaction, and the equilibrium constant is sometimes denoted K_h. It is really just an ordinary base reaction, however, and to emphasize this we will use the symbol K_b.

Now, how can we calculate K_b for this reaction, based on what we already know? Consider the following two reactions, whose K's we know.

$$H_2O + CH_3COOH = H_3O^+ + CH_3COO^- \qquad K_a$$

$$2H_2O = H_3O^+ + OH^- \qquad K_w$$

If we substract the first reaction from the second, using the rules we described earlier for combining reactions, we get

$$CH_3COO^- + H_2O = CH_3COO^- + OH^-$$

which is the same reaction we had above for the action of CH_3COO^- as a base. Then, using the rules for calculating K_{eq} for combined reactions, K_b must be equal to K for the second reaction divided by K for the first; $K_b = K_w/K_a$. In this instance, $K_b = 10^{-14}/1.8 \times 10^{-5} = 5.6 \times 10^{-10}$. In general, a conjugate acid/conjugate base pair obeys the rule

$$K_a K_b = K_w$$

Note that the bigger K_a is, the smaller K_b is, so that the anions of stronger conjugate acids are weaker conjugate bases; for example, HCl is a very strong acid, and Cl⁻ is not basic at all. At the other extreme, OH⁻ is the strongest base that can exist in aqueous solution, and H_2O, its conjugate acid, is certainly a very weak acid. To pick some more conventional examples, we saw that K_a for acetic acid is 1.8×10^{-5}, so that K_b for the acetate ion is 5.6×10^{-10}. Now consider a weaker acid, HCN, $K_a = 4.0 \times 10^{-10}$. Its anion will be a stronger base than acetate: $K_b = K_w/K_a = 10^{-14}/4.0 \times 10^{-10} = 2.5 \times 10^{-5}$.

Note that some anions can themselves act like either acids or bases. For example, H_2S ionizes in two stages.

Chapter 9 — Acids, Bases and Salts

$$H_2S = H^+ + HS^- \qquad K_{a1} = 1.0 \times 10^{-7}$$

$$HS^- = H^+ + S^{2-} \qquad K_{a2} = 1.3 \times 10^{-13}$$

The HS^- ion is the conjugate base of H_2S, but is an acid in its own right, with $K_a = K_{a2} = 1.3 \times 10^{-13}$. For its action as a base

$$HS^- + H_2O = H_2S + OH^-$$

K_b is given by

$$K_w/K_{a1} = 10^{-14}/1.0 \times 10^{-7} = 1.0 \times 10^{-7}.$$

Thus HS^- is a stronger base than it is an acid, and solutions of NaHS in water will be basic. You should make sure you understand why K_{a2} gives the acidity of HS^-, whereas K_{a1} is used to calculate its K_b.

Weak Polyprotic Acids

H_2S is a *diprotic acid*, meaning that it has two acidic hydrogens. It is a *weak* diprotic acid because both K_a's are small. H_2SO_4 is a strong diprotic acid, because its first proton can be assumed to be completely dissociated. The text gives several other examples (besides H_2S) of weak diprotic acids, perhaps the most important of which is carbonic acid, H_2CO_3. A common *triprotic acid* is H_3PO_4, phosphoric acid.

For simple polyprotic acids it is found that $K_{a1} > K_{a2} > K_{a3} >$ etc. The reason is that when the first proton is removed, the energy of charge separation is compensated by ion-dipole interactions with the solvent. When subsequent protons are removed, the anion cannot achieve much additional charge stabilization, since it is already surrounded by water dipoles; it can hold its hydration shell more tightly, but there isn't room for more dipoles to crowd in. Since the di-anion cannot gain as much *additional* stabilization as the original anion had, the second ionization is suppressed; of course, the third (for a triprotic acid), is suppressed even more. For H_2S the sulfur atom acquires both the first and second charge on ionization, and $K_{a1}/K_{a2} = 1.0 \times 10^{-7}/1.3 \times 10^{-13} \approx 8 \times 10^5$. Tartaric acid, on the other hand, is a larger molecule, which has its ionizable hydrogens separated by six intervening atoms, and for it $K_{a1}/K_{a2} \approx 30$. Because its anionic sites are widely separated, they can each be hydrated separately, and

Chapter 9 — Acids, Bases and Salts

the relatively distant sites also do not experience charge-charge repulsion too strongly; therefore the second ionization is not suppressed too much.

Continuing to use H_2S as an example, we had, for the two ionizations,

$$K_{a1} = 1.0 \times 10^{-7} = \frac{[H^+][HS^-]}{[H_2S]}$$

$$K_{a2} = 1.3 \times 10^{-13} = \frac{[H^+][S^{2-}]}{[HS^-]}$$

For the total ionization process, we add the two reactions, giving

$$H_2S = S^{2-} + 2H^+$$

$$K_{overall} = K_{a1}K_{a2} = 1.3 \times 10^{-20}$$

The text states that it is not correct to look at either ionization or the total process alone to calculate $[H_3O^+]$; what is meant is that one usually does not have enough information about the problem to do so, even though all three equations are as valid at equilibrium as the law of mass action. For example, $K_{overall}$ does give the H^+ resulting from both the first and second ionizations, but to use it to calculate H^+ precisely one has to know S^{2-}, for which the other equations are necessary (see Chapter 10). Actually, for H_2S, K_{a2} is so much smaller than K_{a1} that the second ionization does not contribute significantly to $[H^+]$, so that one would be perfectly justified in assuming that H_2S was monoprotic, and calculating $[H^+]$ from C and K_{a1} only, as we did for monoprotic acids. This approach would not, however, suffice for tartaric acid, whose K_{a2} is almost as large as its K_{a1}.

The overall equation does have its uses, however. If $[H_3O^+]$ is maintained at some level by, say, addition of a strong acid (or by addition of a *buffer*; see next chapter), and if the solution is kept saturated with H_2S, so that its concentration is given by its solubility (0.10M), then there is only one unknown in the overall equation; namely, $[S^{2-}]$. In certain laboratory procedures the $[S^{2-}]$ concentration must be maintained at a certain level, and the overall equation tells what that level will be, for saturated H_2S solutions, as a function of $[H_2O]$; that is, of pH. This is straightforward, and Example 9.11 shows how it is done.

Chapter 9 — Acids, Bases and Salts

Salts

Earlier in the chapter we already discussed the fact that all salts are not neutral, because the conjugate base of a weak acid is a weak base. There is an easy way to figure out whether a salt will be acidic or basic. Recall the Arrhenius idea that a salt results from the neutralization reaction of an acid and a base.

$$\text{acid} + \text{base} = \text{salt} + \text{water}$$

Thus $NaCH_3COO$ can be thought of as the salt of sodium hydroxide and acetic acid. Sodium hydroxide is a strong base and acetic acid is a weak acid, and sodium acetate is basic. Similarly NH_4Cl is the salt of a weak base (ammonia) and a strong acid (hydrochloric), and is acidic.

This mnemonic can be justified by the Lowry-Bronsted theory. Recall from the discussion of conjugate acids and bases that the stronger the acid, the weaker the conjugate base, and vice versa. Using ammonium chloride as an example, since HCl is a very strong acid, Cl^- will not be appreciably basic; on the other hand, since NH_3 is a weak acid, NH_4^+ will be a somewhat strong acid; thus NH_4Cl is acidic. Of course, the salt of a strong acid and a strong base (like NaCl) will be neutral.

Now consider the salt of a weak acid and a weak base, such as NH_4F, the salt of HF and NH_3. K_b for NH_3 is 1.75×10^{-5}, so that K_a for NH_4^+ is $10^{-14}/1.8 \times 10^{-5} = 5.7 \times 10^{-8}$. K_a for HF is 7.2×10^{-4}, so K_b for F^- is $10^{-14}/7.2 \times 10^{-4} = 1.4 \times 10^{-11}$. Thus F^- is a weaker base than NH_4^+ is an acid, and the salt will be acidic.

It is easy to understand how to calculate the pH of a salt of a strong acid and a weak base or of a weak acid and a strong base, and we will go over these calculations. (To calculate the pH of a salt such as NH_4F is a bit harder, and we will show how to do it in chapter 11 of the Study Guide.) For a solution of NH_4Br with C=0.300M (Example 9.12), there is only one reaction of importance, since HBr is a strong acid, and Br^- does not hydrolize (does not act as a base).

$$NH_4^+ + H_2O = H_3O^+ + NH_3$$

$$K_a = K_w/K_b = 5.7 \times 10^{-10} = \frac{[NH_3][H_3O^+]}{[NH_4]}$$

Chapter 9 — Acids, Bases and Salts

We can use a before-after table, letting x = [NH$_3$]

	[NH$_4^+$]	[NH$_3$]	[H$_3$O$^+$]
before	0.300	0	0
after	0.300-x	x	x

Thus

$$K_b = 5.7 \times 10^{-10} = \frac{x^2}{0.300-x}$$

This can be solved by the methods discussed toward the beginning of the chapter, to give x = [H$^+$] = 1.3×10^{-5}, pH = 4.88.

Earlier we discussed the fact that the bisulfide ion, HS$^-$, can act either as an acid or as a base. Such a substance is called an *ampholyte*, and the behavior is called *amphiprotic*. We saw earlier that since K$_b$ for HS$^-$ (that is, for its hydrolysis to H$_2$S) is greater than its K$_a$ (that is, its ionization to S^{2-}), a salt like NaHS is basic. Some ampholytic salts are acidic, however, such as NH$_4$F; the appropriate K's have to be compared in each case.

Oxides of metals are usually, but not always, salts, and are usually, but not always basic. As an example, sodium oxide reacts with water to form the hydroxide.

$$Na_2O(s) + H_2O = Na^+ + 2OH^-$$

Oxides of the group IA and IIA metals are always basic. Oxides of non-metals, on the other hand, are usually, but not always, molecular, and are generally acidic. Thus SO$_3$ reacts with water to form sulfuric acid.

$$SO_3(g) + H_2O = 2H^+ + SO_4^{2-}$$

A substance which, upon addition of water, forms a strong acid is called an *acid anhydride*. SO$_3$ is an example. A non-molecular, non-metallic oxide is silica, SiO$_2$, common sand, which exists as a covalently bonded extended crystal lattice, rather than as discrete molecules. Common glass is mostly silica. Although silica does not dissolve in pure water to give acidic solutions, it does react with strong base, and thus is weakly acidic; in fact, strong base will corrode glass. Silica does not react with strong acids. A detailed discussion of the acidity of the oxides, based on the atomic structure of the elements, will be found in Chapter 15.

Leveling Effect of Water

Recall that HNO_3, HCl and H_2SO_4 are all strong acids in water; they dissociate completely, and the acidic species in solution — the species that actually participates in acid-base reactions — is the species we write as H_3O^+, the hydronium ion, but which is more accurately termed the hydrated proton. Thus, in aqueous solution, no species can exist that has a stronger tendency than H_3O^+ to donate protons. If pure nitric acid and sulfuric acid are mixed, however, the following reaction takes place to a certain extent.

$$HNO_3 + H_2SO_4 = H_2NO_3^+ + HSO_4^-$$

Because H_2SO_4 will protonate HNO_3, sulfuric is a stronger acid than nitric; in water, however, they have the same strength, because the acidic species is always H_3O^+. As another example, if pure $HClO_4$ and H_2SO_4 are mixed, the following reaction takes place:

$$HClO_4 + H_2SO_4 = H_3SO_4^+ + ClO_4^-$$

thus, perchloric is the stronger of these two acids. Note that $H_3SO_4^+$ is the most acidic species that can exist in pure H_2SO_4, and $H_2NO_3^+$ is the most acidic species that can exist in pure HNO_3.

Any solvent thus exerts a leveling effect on the strength of acids and bases dissolved in it. The more acidic the solvent, the more acidic the most acidic species, and the more basic the solvent, the more basic the most basic species. In more acidic solvents, one can distinguish among acidities of stronger acids. For example, if H_2SO_4 is the solvent, $HClO_4$ is an acid, but HNO_3 is a base (!) since it accepts a proton from H_2SO_4. In more basic solvents, one can distinguish among basicities of strong bases. A base as strong as $NaNH_2$ cannot exist in water, because of the reaction

$$NH_2^- + H_2O = NH_3 + OH^-$$

This reaction goes to completion, showing that NH_2^- is a stronger base than OH^-, and that H_2O is a stronger acid than NH_3. In liquid NH_3, however, NH_2^- can exist, and if methanol, CH_3OH, is added to a solution of $NaNH_2$ in liquid ammonia, the following reaction takes place

Chapter 9 — Acids, Bases and Salts

$$NH_2^- + CH_3OH = NH_3 + CH_3O^-$$

Methanol also acts like an acid in a strong solution of NaOH in water, but since OH^- is a weaker base than NH_2^-, here the methanol is not deprotonated to as great an extent. Thus, if one wishes to perform a reaction using methanol which requires its deprotonation, liquid NH_3 would be a better solvent than H_2O.

As the text mentions, knowing the relative strength of acids and bases in nonaqueous solvents is important in organic chemistry, where compounds first of all do not always dissolve in water, and, moreover, are often too hard to protonate (weak enough as bases) so that stronger acids than H_3O^+ might be needed, or too hard to abstract protons from (weak enough as acids) that stronger bases than OH^- might be needed. As a final note, you should realize that this entire systematization of acid and base strengths in non-aqueous media is made possible only by the Bronsted-Lowry concept of what an acid and a base are. Since the Arrhenius theory identifies an acid with H^+ and a base with OH^-, it is appliable only to aqueous solutions.

Chapter 9 — Acids, Bases and Salts

Questions

9.1 The ion-product of water and the pH scale

1. Write the equation for the self-ionization of methanol, CH_3OH.

2. Write the formula for a proton with four waters of hydration.

3. What are $[H_3O^+]$ and $[OH^-]$ in

   (a) a 0.01M HCl solution

   (b) A 0.01M NaOH solution

   (c) pure water

4. Give the pH of each of the solutions in Question 3.

   (a)

   (b)

   (c)

Chapter 9   Acids, Bases and Salts

5. What are pH and pOH in a 10M HNO_3 solution?

9.2 Weak acids and bases

1. Acetic acid has $K_a = 1.8 \times 10^{-5}$; formic acid has $K_a = 1.8 \times 10^{-4}$. Therefore a 0.1M acetic acid solution will have a *lower* pH than a 0.1M formic acid solution.

2. Ammonia has $K_b = 1.8 \times 10^{-5}$; methylamine has $K_b = 4.2 \times 10^{-4}$. Therefore a 0.1M ammonia solution will have a *lower* pH than a 0.1M methylamine solution.

3. The greater the value of K_a or K_b, the *greater* the value of α.

4. For an acid HA, $K_a = 10^{-5}$. For a 0.1M solution of HA

   (a) What is $[H^+]$?

   (b) What is the pH?

   (c) What is $[OH^-]$?

   (d) What is pOH?

   (e) What is $[A^-]$?

Chapter 9   Acids, Bases and Salts

(f) What is [HA]?

(g) What is α?

5. For a weak acid, the greater the value of C, the *greater* the value of α.

6. For a weak base, the greater the value of C, the *greater* the value of α.

9.3 Conjugate acids and bases and their relative strengths

1. Give the formula for the conjugate base of each of the following

   (a) acetic acid

   (b) H_2O

   (c) ammonium ion

   (d) H_2SO_4

   (e) HCN

2. Give the formula for the conjugate acid of each of the following.

   (a) ammonia

   (b) H_2O

   (c) CH_3NH_2

   (d) HPO_4^{2-}

   (e) S^{2-}

3. For an acid, HA, K_a is 10^{-6}. What is K_b for the conjugate base?

Chapter 9   Acids, Bases and Salts

4. For a base, B, K_b is 10^{-4}. What is K_a for the conjugate acid?

5. The *stronger* the base the weaker the conjugate acid.

9.4 Weak polyprotic acids

1. For sulfurous acid,

    (a) Write down the reactions corresponding to the first, second and overall ionization processes.

    (b) K_{a1} will be *greater* than K_{a2}.

    (c) Express $K_{overall}$ in terms of K_{a1} and K_{a2}.

9.5 Acidic, basic and neutral salts

1. Classify each of the following as acidic, basic or neutral.

    (a) sodium formate

    (b) calcium hydroxide

    (c) ammonium nitrate

    (d) potassium bromide

2. For H_2S, $K_{a1} = 1.0 \times 10^{-7}$ and K_{a2} is small enough to be ignored. What is the pH of a 0.1M solution of NaHS?

3. K_b for methylamine is 4.2×10^{-4}, and K_a for hydrocyanic acid is 4.0×10^{-10}. Will the salt methylammonium cyanide be acidic or basic?

4. For tartaric acid, $K_{a1} = 1.1 \times 10^{-3}$ and $K_{a2} = 4.3 \times 10^{-5}$. Is sodium hydrogen tartrate acidic or basic?

5. What acid is formed when each of the following anhydrides is added to water?

   (a) SO_2

   (b) P_2O_5

   (c) Cl_2O_7

6. What is the anhydride of each of the following acids?

   (a) H_2SO_4

   (b) H_3AsO_4 (arsenic acid)

   (c) $HMnO_4$ (permanganic acid)

9.6 The leveling effect of water on the strengths of strong acids

1. What is the most acidic and the most basic species that can exist in each of these solvents.

   (a) water

   (b) liquid ammonia

   (c) ethanol (CH_3CH_2OH)

Chapter 9 — Acids, Bases and Salts

Answers

9.1 The ion-product of water and the pH scale

1. $2CH_3OH = CH_3OH_2^+ + CH_3O^-$ or $CH_3OH = H^+ + CH_3O^-$
2. $H_9O_4^+$
3. (a) $[H^+] = 0.01M$; $[OH^-] = 10^{-12}M$
   (b) $[OH^-] = 0.01M$; $[H^+] = 10^{-12}M$
   (c) $[H^+] = [OH^-] = 10^{-7}M$
4. (a) 2.0
   (b) 12.0
   (c) 7.0
5. pH = −1; pOH = 15

9.2 Weak acids and bases

1. higher
2. True
3. True
4. (a) 0.001M
   (b) 3
   (c) $10^{-11}M$
   (d) 11
   (e) 0.001M
   (f) ≈ 0.1M
   (g) ≈ 0.01
5. lower
6. lower

9.3 Conjugate acids and bases and their relative strengths

1. (a) CH_3COO^-
   (b) OH^-
   (c) NH_3
   (d) HSO_4^-
   (e) CN^-
2. (a) NH_4^-
   (b) H_3O^+
   (c) $CH_3NH_3^+$
   (d) $H_2PO_4^-$
   (e) HS^-
3. 10^{-8} (= $10^{-14}/10^{-6}$)
4. 10^{-10}
5. True

9.4 Weak polyprotic acids

1. (a) $H_2SO_3 = H^+ + HSO_3^-$
   $HSO_3^- = H^+ + SO_3^{2-}$
   $H_2SO_3 = 2H^+ + SO_3^{2-}$
   (b) True
   (c) $K_{overall} = K_{a1}K_{a2}$

Chapter 9   Acids, Bases and Salts

9.5 Acidic, basic, and neutral salts

1. (a) acidic
   (b) basic
   (c) acidic
   (d) neutral

2. 10 ($K_b = 10^{-14}/10^{-7} = 10^{-7} = x^2/(C-x) \approx x^2/0.1$, so $x = [OH^-] = 10^{-4}$, so pOH=4, so pH=10)

3. basic (methylamine is a stronger base than HCN is an acid)

4. K_b acidic. ($K_b = 10^{-14}/1.1 \times 10^{-3} \approx 10^{-11} \ll K_{a2}$, so the acidity dominates)

5. (a) H_2SO_3 (sulfurous)
   (b) H_3PO_4
   (c) $HClO_4$ ($Cl_2O_7 + H_2O = "H_2Cl_2O_8" = 2HClO_4$

6. (a) SO_3
   (b) As_2O_5 ($As_2O_5 + 3H_2O = "H_6As_2O_8" = 2H_3AsO_4$)
   (c) Mn_2O_7 ($Mn_2O_7 + H_2O = "H_2Mn_2O_8" = 2HMnO_4$)

9.6 The leveling effect of water on the strengths of strong acids and bases

1. (a) most acidic: H_3O^+; most basic: OH^-
   (b) most acidic: NH_4^+; most basic: NH_2^-
   (c) most acidic: $CH_3CH_2OH_2^+$; most basic: $CH_3CH_2O^-$

Chapter 10: Buffer Solutions and Acid-Base Titrations

Scope

In the last chapter we concentrated on problems where a single acid or base or its salt is present in a system; for example, we discussed how to calculate pH and α for a weak base in water, given K_b and the initial concentration C. For a salt of the same weak base, we discussed how to do the same sort of calculation, using the relationship $K_a = K_w/K_b$. These generally led to equations like

$$K = \frac{x^2}{C-x}$$

which, depending on the relative values of C and K, could be solved either by assuming that $(C-x) \approx C$, or by the quadratic formula, or by a method of successive approximations. When a weak acid and its conjugate base are present in approximately equal concentrations, the calculation is somewhat different.

Such a solution, known as a *buffer*, has the interesting property that its pH changes little upon either dilution or addition of a small amount of strong acid or base. This makes buffers essential in physiological processes, since the body's chemicals are "tuned" to work best in certain narrow pH ranges, even though we ingest foods ranging from very acidic to mildly basic, and certain bodily processes, such as heavy exercise, release acidic substances. Without buffering, the pH of the blood, the gut, muscle and so on would undergo wild swings, and the proteins that carry oxygen, digest our food, and so on would exhibit coresponding swings in their behavior.

Buffers are also important in industrial processes, if only to protect against contamination. Suppose a process needs to run at pH=6. Without buffering, trace HCl present at 10^{-5}M would give the solution a pH of 5. This concentration amounts to only about 0.1g HCl in a 100-gallon tank! It is easy to imagine this degree of contamination occuring accidentally. Buffering the solution solves the problem.

Buffers

Let us see how the properties of buffers arise by examining a particular simple buffer system: a solution that is made up 0.1M in acetic acid (HOAc) and 0.2M in sodium acetate. The acid and its conjugate base react according to the equations

$$H_2O + HOAc = H_3O^+ + OAc^-$$

$$K_a = 1.8 \times 10^{-5} = \frac{[H_3O^+][OAc^-]}{[HOAc]}$$

$$H_2O + OAc^- = OH^- + HOAc$$

$$K_b = K_w/K_a = 10^{-14}/1.5 \times 10^{-5} = 5.6 \times 10^{-10} = \frac{[OH^-][HOAc]}{[OAc^-]}$$

We start by solving the equation for K_a for $[H_3O^+]$.

$$[H_3O^+] = K_a \frac{[HOAc]}{[OAc^-]}$$

Now, since K_a is small, the ionization of HOAc doesn't proceed very far, so that $[HOAc] \approx C_{HOAc} = 0.1M$. Also, since K_b is small, the hydrolysis of OAc^- does not proceed very far, so that $[OAc^-] \approx C_{NaOAc} = 0.2M$. Note that these are exactly the same approximations as we often made when dealing with an acid or its salt alone; namely, if K is small, then $(C-x) \approx C$, where $x = [H_3O^+]$ or $[OH^-]$. Recall that this approximation generally turns out to be justified when $C > 100K$. For reasons that will become clear below, most buffers are made up at reasonably high concentrations (like 0.1M), so that the approximation will usually work for the problems in this chapter.

Applying these approximations to our equation for $[H_3O^+]$

$$[H_3O^+] \approx K_a \frac{C_{HOAc}}{C_{NaOAc}} = 1.8 \times 10^{-5} \frac{0.1M}{0.2M} = 9.0 \times 10^{-6}$$

which gives pH = 5.05. Before examining the properties of this buffer solution, let us discuss an alternate way to calculate the pH. We take the negative logarithm of both sides of the equation for $[H_3O^+]$.

$$-\log_{10}[H_3O^+] = pH = pK_a - \log_{10} \frac{C_a}{C_b}$$

$$pH = pK_a + \log_{10}\left(\frac{C_b}{C_a}\right)$$

For generality, we have used C_a for C_{HOAc} and C_b for C_{NaOAc}. This equation is called the *Henderson-Hasselbalch equation*, but even though it has a name it does not really make solving buffer problems any easier (or any harder!). It is a matter of preference whether one calculates $[H_3O^+]$ directly, then takes the log to get pH, or whether one uses the Henderson-Hasselbalch equation to get pH directly. The equation does make it clear that the greater the ratio C_b/C_a, the greater the pH (since as this ratio goes up, so does its logarithm), but this is also clear from the original equation. Note that if $C_a = C_b$ then pH=pK_a; thus, if we made up a buffer that was 0.1M in both HOAc and NaOAc, we would have pH = pK_a = $-\log_{10}(1.8 \times 10^{-5}) = 4.74$.

Suppose now that we take 10mL of our buffer, pH=5.05, and dilute it up to 100mL. Note that the ratio C_b/C_a stays the same; therefore the pH stays the same. Let us now consider an unbuffered solution that has a pH of 5.05 because of the presence of a small amount of strong acid. Such a solution exhibits $[H_3O^+] = 10^{-5.05} = 8.9 \times 10^{-6}$. If this solution were diluted by a factor of 10, then the new value of $[H_3O^+]$ would of course be 8.9×10^{-7}, giving pH=6.05. Thus the pH of the unbuffered solution increases by a full pH unit on ten-fold dilution, whereas the buffered solution's pH remains constant.

It is instructive to compare these two results with what happens when a solution which contains just enough pure acetic acid to have a pH of 5.05 is diluted by a factor of 10. You can verify (in fact, you should!) that a solution 1.33×10^{-5}M in HOAc will exhibit this pH. If this solution is diluted by a factor of 10, we have

$$1.8 \times 10^{-5} = \frac{x^2}{1.33 \times 10^{-6} - x}$$

This must be solved by the quadratic formula (the method of successive approximation fails for C<K), giving $x = [H_3O^+] = 1.24 \times 10^{-6}$, pH = 5.91. Thus the pH of the HOAc solution increases by 0.86 pH units on ten-fold dilution. The reason it does not increase by a full pH unit is that α increases upon dilution, making the solution more acidic than it would otherwise be; nervertheless, the behavior of this solution on dilution resembles that of the pure strong acid much more closely than it does that of the buffer.

How far can the dilution of a buffer be carried out without an appreciable pH change? Recall that we assumed $[HOAc] \approx C_{HOAc}$ and $[OAc^-] \approx C_{NaOAc}$, and recall further that this should be fine, as long as C>100K; that is, as long as $C_{HOAc} > 100K_a$ and $C_{NaOAc} > 100K_b$. Our buffer started out with [HOAc] = 0.1M. 100K_a is 1.8×10^{-3}, and a 56-fold dilution of the system would give [HOAc] equal to this. At higher dilutions, the assumptions that went into the Henderson-Hasselbach relationship are no longer valid, and so the pH can no longer be assumed constant. (You can verify that it will take considerably more dilution for the assumption on C_{NaOAc} to fail; the first assumption to go spoils the relationship, however.) Note, incidentally, that if the buffer had started out 0.01M in HOAc, it would have been able to withstand only about a five-fold dilution before pH changes became significant; thus, to afford good protection against pH changes due to dilution, buffers have to be fairly concentrated.

Now let us consider adding 1mL of 0.1M HCl to 100mL of both our buffer and an unbuffered solution of pH 5.05. The solutions will both have 10^{-4}mol HCl in 101mL, giving $C_{HCl} \approx 0.001$. For the unbuffered solution this swamps out the low acid concentration already present, giving pH=3; thus the unbuffered solution's pH decreases by about two pH units. For the buffer, the HCl reacts with some of the sodium acetate, making acetic acid.

$$OAc^- + H_3O^+ = HOAc + H_2O$$

This reaction is the reverse of the ionization of HOAc, and has K = $1/K_a$ = 5.6×10^4, indicating that it is virtually complete. The net effect is that [HOAc] increases and [OAc$^-$] decreases. The following before-after table describes the effects of making the buffer 0.001M in HCl.

	[HOAc]	[OAc$^-$]	$\frac{[OAc^-]}{[HOAc]}$
before	0.100	0.200	2.00
after	0.101	0.199	1.97

The new C_b/C_a ratio gives pH=5.04; thus, the pH of the buffer has changed by only 0.01 pH unit on adding the acid, compared to a pH change of 2 for the unbuffered solution!

If, instead, 1mL of 0.1M NaOH is added, then $C_{NaOH} \approx 0.001$, and the reaction that takes place is

$$HOAc + OH^- = H_2O + OAc^-$$

This is the reverse of the hydrolysis of OAc$^-$, and has K = K_a/K_w = $1.8 \times 10^{-5}/10^{-14}$ = 1.8×10^9, so it, too, can be assumed to go to completion. This gives [OAc$^-$]=0.201, [HOAc]=0.099 and [OAc$^-$]/[HOAc] = 2.03, giving pH=5.05, unchanged, to two decimal places, from the starting value! For an unbuffered solution the pH would rise by 11 for a change of about 6 pH units!

Note that the reason the pH didn't change much on adding acid or base to our buffered solution is that acid could react with OAc⁻, and base could react with HOAc. Just as long as the concentration of added acid or base is low compared to the concentration of buffer ingredients, the ratio C_b/C_a will not change much, and the pH will not change much. Thus, to guard against pH changes due to addition of acid or base, as well as pH changes due to dilution, a buffer solution must be fairly concentrated.

Let us now consider how the effectiveness of a buffer varies with the ratio C_b/C_a. Recall that if this ratio is equal to 1, pH=pK_a, and that as this ratio varies from unity the pH of the buffer changes. We will see that the buffer is most efficient (offers the best protection) when this ratio is unity; therefore, when selecting a buffer for a particular application, one tries to pick an acid with a pK_a near the desired pH of the buffer.

Consider an acetate buffer with a *total* acetate concentration of 0.02M ($C_a + C_b$ = 0.02M). First consider the case C_b/C_a = 1 so that $C_a = C_b$ = 0.01M. If we make the solution 0.001M in HCl then the following before-after table will apply.

	[HOAc]	[OAc⁻]	$\frac{[OAc^-]}{[HOAc]}$	pH
before	0.010	0.010	1.00	4.74
after	0.011	0.009	0.82	4.65

Thus the pH decreases by 0.09 pH units upon addition of the acid. Now consider a solution with the same total acetate concentration, but with C_b/C_a = 1/9; we make C_a = 0.018 and C_b = 0.002. Again, let us examine the effectiveness of this buffer against 0.001M HCl.

	[HOAc]	[OAc⁻]	$\frac{[OAc^-]}{[HOAc]}$	pH
before	0.018	0.002	0.111	3.79
after	0.019	0.001	0.053	3.46

Here the pH changes by 0.33 pH units. Thus the buffer becomes less effective as C_b/C_a moves away from unity, at constant $(C_a + C_b)$. The reason should be clear from looking at the before-after tables. If the ratio is far from unity, then either C_a or C_b is small, and the relative change in this small value is large when a given amount of acid or base is added to the system. The result is that when the ratio starts far away from one, it changes a lot when acid or base is added, and this leads to a big pH change.

Just how serious this decrease in buffer efficiency is in a given case depends on the relative sizes of the smaller of C_a and C_b and the concentration of acid or base against which one is protecting. In general, one likes to keep the ratio C_b/C_a between 0.1 and 10. If it is necessary to buffer outside the pH range available within this range of ratios, it is best to select a different buffer system. Note that as C_b/C_a goes from 0.1 to 10, pH goes from (pK_a+1) to (pK_a-1), so that for an acid with a given pK_a, the useful buffer range is often quoted as $pH = pK_a \pm 1$.

There are several ways to prepare a buffer system of desired C_b/C_a and C_{total}. Suppose one wants a buffer 0.1M in HOAc and 0.2M in OAc$^-$. Here C_{total} is 0.3M. One can measure out HOAc and NaOAc at the requisite concentrations, or one can make up a solution 0.3M in HOAc, and then make the solution 0.2M in NaOH; likewise, one can make up a solution 0.3M in NaOAc and then make the solution 0.1M in HCl. Using this last method as an example, the 0.1M HCl reacts quantitatively with NaOAc to make 0.1M HOAc, so that C_b/C_a is just what it would have been if one had weighed out the HOAc and NaOAc separately. Note, however, that the solutions prepared by these methods will not be identical, since Cl$^-$ will be present if one starts out with NaOAc and adds HCl, but this does not affect the buffering action.

Acid-Base Titrations

We encountered the idea of a titration in the last chapter, when we discussed conductometric titrations. We also discussed the use of an indicator to determine the equivalence point of an acid-base titration. Now we will discuss in a detailed manner how the pH changes during a titration, and what implications this has for choosing an indicator. On the way, we will develop some new methods for calculating chemical equilibria.

Chapter 10 — Buffer Solutions and Acid-Base Titrations

For the titration of a strong acid with a strong base, the *titration curve* (that is, the plot of pH against volume of base added) is easy to understand. The calculation of this curve is well described in the text, and is illustrated in Figure 10.2. The basis of the calculation is that one starts with a certain number of moles of acid, and each mole of base added neutralizes an equivalent number of moles of acid. At the equivalence point the pH is 7, since here one has a solution of a neutral salt in water. Beyond the equivalence point the pH is greater than 7, since here one has a neutral salt plus excess base. Since the base is usually added in solution form, the volume continually rises as base is added, and this dilution effect must be taken into account; its net effect is to make the pH a bit closer to 7 (neutral solution) than it would be if base had been added with no volume change.

The most salient feature of Figure 10.2 is the rapid increase of pH in the vicinity of the equivalence point. Why this occurs is easy to understand. Suppose I have 100mL of 0.1M HCl. This solution has a pH of 1.00 and contains 0.01 moles of HCl. Suppose I add to the solution 0.001mol NaOH. The solution will then contain 0.009mol HCl. Assume the base was added as a solid, so that volume changes can be ignored. The new H^+ concentration will be 0.09M, and the pH will be $-\log_{10} 0.09 = 1.05$, so that the change in pH is only 0.05 pH units. Now consider adding the same amount of base to a neutral solution, whose pH is 7. After addition of 0.001mol NaOH, we have $[OH^-] = 0.01$, pOH=2 and pH=12. Thus making a 0.1M HCl solution simultaneously 0.01M in NaOH raises the pH by 0.05 pH units, but making a neutral solution 0.01M in NaOH raises the pH by 5 pH units! Except for minor corrections due to volume changes, this explains the overall shape of the curve in Figure 10.2.

Something you should realize is that you have been taught techniques for calculating the pH far from the equivalence point and right at it, but not very close to it. For example, when we calculated, above, the pH of a 0.1M HCl solution, we assumed that all the $[H_3O^+]$ came from the HCl, and none from the self-ionization of water, and, as discussed in the last chapter, this assumption is justified. At the equivalence point, all the $[H_3O^+]$ is from the self-ionization of water, and none from the HCl, since there is no HCl; therefore $[OH^+] = [H_3O^+] = 10^{-7}$, and pH=7. But very close to the equivalence point, at an HCl concentraation (C_{HCl}) of, say, 10^{-7}, clearly there must be some significant contribution to

Chapter 10 — Buffer Solutions and Acid-Base Titrations

$[H_3O^+]$ from both the acid and the self-ionization of water. Although the text discusses methods for solving this problem only in Appendix K, we will do so here. It is not difficult, and, moreover, it introduces the concepts of *charge balance* and *mass balance* which are employed later in the chapter.

The method of solving this problem is one example of the so-called "exact" calculation of equilibarium concentrations. It is exact in that it avoids the approximations that we have used before, but it is ultimately limited by how true the ideal law of mass action is. For highly dilute solutions, which is where exact calculations are often needed, the ideal law holds very well. Consider the following two processes taking place in solution

$$HCl + H_2O = H_3O^+ + Cl^-$$

$$2H_2O = H_3O^+ + OH^-$$

$$K_w = [H_3O^+][OH^-] = 10^{-14}$$

Note that there are three variables, $[H_3O^+]$, $[OH^-]$ and $[Cl^-]$, but only one algebraic equation relating them. In general, one needs three equations to solve for three unknowns. One of the missing equations, called the *charge balance equation*, comes from the fact that the overall solution must be electrically neutral; thus the sum of the positive charges must equal the sum of the negative charges.

$$[H_3O^+] = [Cl^-] + [OH^-]$$

The other missing equation is a *mass balance equation*. If we did not know that HCl was 100% dissociated, this would read

$$C_{HCl} = [HCl] + [Cl^-]$$

This says that all the HCl added initially winds up either in the ionized or the unionized form. Since HCl is a strong acid, however, [HCl]=0 and we can write:

$$C_{HCl} = [Cl^-]$$

This is called a mass balance because it relates the amount of a material added to a system (C_{HCl}) to the amounts of the various chemical species that can be derived from it ($[Cl^-]$ and $[HCl]$); actually, a more accurate term would be "mole balance." In any case, we now have three equations in three unknowns. To solve for $[H_3O^+]$ we can first substitute the value for $[Cl^-]$ from the mass balance equation into the charge

Chapter 10

Buffer Solutions and Acid-Base Titrations

balance equation

$$[H_3O^+] = [OH^-] + C_{HCl}$$

From the equation for the self-ionization of water, $[OH^-] = K_w/[H_3O^+]$, giving

$$[H_3O^+] = \frac{K_w}{[H_3O^+]} + C_{HCl}$$

This equation is rigorously true, and holds for *any* value of C_{HCl}. In our earlier work we always assumed a simpler relationship

$$[H_3O^+] = C_{HCl}$$

and stated that this ignores any H_3O^+ resulting from the self-ionization of water; the term $K_w/[H_3O^+]$ must then give the additional $[H_3O^+]$ present because of self-ionization.

Let us solve this equation for $C_{HCl} = 10^{-7}$, but first let us apply our chemical intuition to figure out about what the answer should be. If self-ionization were insignificant, then we would have $[H_3O^+] = 10^{-7}M$, but we know the answer must be bigger than this, because this value is characteristic of neutral solution, and we have, in fact, added some HCl. If self-ionization proceeded as fas as it does in neutral solution, then it would give an *additional* contribution to $[H_3O^+]$ (beyond that coming from the added HCl) of $10^{-7}M$, for a total $[H_3O^+]$ of $2\times10^{-7}M$, but we know that, by LeChatelier, the addition of HCl will suppress the self-ionization process, so that $[H_3O^+]$ must be less than $2\times10^{-7}M$. Thus our answer should lie between $1\times10^{-7}M$ and $2\times10^{-7}M$. This kind of "scoping out" of a problem before one begins is a useful way of checking one's answer when one is done. Now for the solution.

The equation can be solved in (at least!) two ways. One way is to use the method of successive approximations: pick an approximation to $[H_3O^+]$, say $10^{-7}M$, plug it into the expression on the right, giving a new value for $[H_3O^+]$, plug this into the right, and so on, until succesive values converge. You might try this, especially if you enjoy playing with your calculator. Alternatively, you can multiply both sides of the equation by $[H_3O^+]$, giving a quadratic, then use the quadratic formula. In either case, the result is $[H_3O^+] = 1.618\times10^{-7}$, giving pH=6.79. A final comment: methods of successive approximation are especially useful in solving exact equilibrium problems, because usually these problems give rise not to quadratic, but to cubic and even higher order equations, which are difficult or impossible to solve any other way. Recall again, though, that the exact solution is rarely needed; usually simplifying assumptions can be made. Thus, for the titration of a strong acid with a strong base, the exact method is only needed very close to the equivalence point.

Indicators

Because the titration curve is so steep going though the equivalence point, only a very small quantity of added base — a fraction of a drop — causes the pH to change dramatically in this region. A *pH indicator* (or *acid-base indicator*) is something that changes color over a particular pH range; thus (Table 10.2) phenolphthalein changes color between pH 8.0 and 9.8, bromthymol blue between pH 6.0 and 7.6 and methyl red betweeen 4.2 and 6.1. For the curve in Figure 10.2, all these ranges lie on the essentially vertical portion of the titration curve; thus all three indicators will all change color at essentially the same volume of base added. Thus any of them would suffice to indicate the equivalence point of the titration. On the other hand, an indicator whose range was near pH 2 would change color long before the equivalence point. The place the indicator changes color is called the *endpoint* of a titration; one uses an indicator whose endpoint occurs near the equivalence point. (It must be "near" in terms of amount of titrant added, not necessarily in terms of pH.) Note, incidentally, that more dilute solutions have titration curves with shorter vertical regions (Figure 10.4), so that the choice of indicator becomes more stringent.

An indicator is really an acid whose conjugate base has a different color from it. The color of at least one of the two forms must be very intense. If we represent the indicator as HIn, then we have

$$HIn + H_2O = H_3O^+ + In^-$$

$$K_{In} = \frac{[H_3O^+][In^-]}{[HIn]}$$

Taking the negative log of both sides and rearranging gives

$$pH = pK_{In} - \log_{10}\frac{[HIn]}{[In^-]}$$

$$\log_{10}\frac{[HIn]}{[In^-]} = pK_{In} - pH$$

This says that as the pH varies, the ratio $\frac{[HIn]}{[In^-]}$ varies. If the color of the indicator is very intense, then only a tiny amount need be present in order for the color change to be visible. Then the change in H_3O^+ due to the ionization of the indicator will also be tiny, and not contribute noticibly to the pH of the solution, so that we can view the ratio $[HIn]/[In^-]$ as being entirely driven by the pH changes of the substances being titrated.

As a general rule (though this will vary from indicator to indicator), if $[HIn]/[In^-] > 10$ then the solution has the color of the acid form of the indicator, HIn. By the above equation, this corresponds to pH $< pK_{In}-1$. Likewise, if $[HIn]/[In^-] < 0.1$, the solution has the color of $[In^-]$, and pH $> pK_{In}+1$. Over this range, $pK_{In}\pm 1$, we see a gradual color change. One cannot usually determine an endpoint to better than $pK_{In}\pm 1$, but as we have seen, this will not matter, if the titration curve is sufficiently steep, and the indicator is properly chosen.

Titration of a Weak Acid with a Strong Base

For the titration of a strong acid with a strong base (or a strong base with a strong acid) we saw that there are many good choices of indicator possible. We will see that for the titration of a weak acid with a strong base (or a weak base with a strong acid) the choices are far more limited. This is because the titration curve looks different; it does not have nearly as long a steep region.

The curve for the titration of HOAc with NaOH is calculated in Section 10.4, and shown in Figure 10.5. Note that unlike the calculation for HCl and NaOH, when NaOH is added to HOAc a buffer solution is formed. When enough NaOH has been added to neutralize half the HOAc originally present, then the ratio $(C_b/C_a) = 1$, and so pH $= pK_a$, based on our description of buffers. On either side of this point the buffer resists changes in pH as NaOH is added, so that the curve exhibits a fairly flat plateau in this region.

The equivalence point is the point where just enough NaOH has been added to fully neutralize the HOAc originally present.

$$OH^- + HOAc = H_2O + OAc^-$$

At this point the solution is identical to what would be obtained from adding pure NaOAc to water; there is some HOAc formed, however, due to the hydrolysis of OAc⁻, which is the reverse of the neutralization reaction. The net effect is that the solution is somewhat basic. In the last chapter we discussed how to calculate the pH of a pure NaOAc solution, using the fact that $K_b = K_w/K_a$.

Thus this titration differs from the titration of HCl with NaOH in two ways. The first way is that the pH at the equivalence point is greater than 7. (Of course if we had been titrating ammonia, a weak base, with HCl, a strong acid, we would have had NH$_4$Cl present at the equivalence point, and the solution would have been acidic, pH<7.) The second way is that the buffering action of the weak acid-weak base couple maintains a relatively high pH up to close to the equivalence point. For the example calculated in the book, 50mL of 0.1M HOAc titrated with 0.1M NaOH, the pH halfway to the equivalence point is 4.74, the pK_a of HOAc, and at the equivalence point it is 8.72, as shown in the text. If the acid had been HCl, then halfway to the equivalence point the pH would have been 1.48, and at equivalence the pH of course would have been 7.00. The total *rise* in pH from the half-equivalence point to the equivalence point is thus 3.98 pH units for the HOAc titration and 5.52 pH units for the HCl titration; thus, in the region approaching the equivalence point, the titration curve exhibits a shorter steep region for the HOAc titration than for the HCl titration. Because of this, it will *not* suffice to use a pH indicator, like methyl red, which changes color around pH 5, for the HOAc titration, even though this was entirely satisfactory for the HCl titration.

Of course, it is a good idea to use an indicator whose endpoint pH is near the equivalence point pH; phenolphthalein would be a good choice for the titration we have been discussing. On the other hand, if we had been titrating a weak base with a strong acid — NH$_3$ with HCl, for instance — the pH at the equivalence point would have been acidic, and an indicator such as methyl red would have been a good choice, whereas phenolphthalein would not.

Returning to the HOAc/NaOH system, what happens beyond the equivalence point? Further base has no more HOAc to react with, and except for the very small amount of OH⁻ present due to hydrolysis of OAc⁻, this is just like adding strong base to water. The result is that *after* the equivalence point the

Chapter 10 — Buffer Solutions and Acid-Base Titrations

titration of a weak acid with a strong base resembles that of a strong acid with a strong base, and the titration of a weak base with a strong acid resembles that of a strong base with a strong acid. Thus, for the HOAc/NaOH titration, the curve is steep after the equivalence point.

Diprotic Acids: Free Acid

The text describes how to calculate the pH of solutions of diprotic acids, and of their salts. For example, it describes how to calculate the concentrations of all species present in a solution of carbonic acid (H_2CO_3), or $CO_2 + H_2O$, of sodium bicarbonate ($NaHCO_3$) and of sodium carbonate (Na_2CO_3). This is of interest when discussing the titration of carbonic acid with base, since the pH will start out at the pH of a pure carbonic acid solution, rise, upon the addition of a single equivalent of base, to that of a $NaHCO_3$ solution (first equivalence point), and further rise, upon the addition of a second equivalent of base, to that of a Na_2CO_3 solution (second equivalence point). In addition, the method used for calculating the pH of a solution of $NaHCO_3$ can also be used to calculate the pH of a solution of the salt of a weak acid and a weak base, such as ammonium fluoride. These calculations use the concepts of mass balance and charge balance that we developed earlier.

Carbonic acid is really a solution of CO_2 in water. For its first ionization we can write

$$CO_2 + 2H_2O = H_3O^+ + HCO_3^-$$

the concentration of CO_2 in water obeys Henry's law; if a solution of CO_2 is in equilibrium with the gas above it, then

$$[CO_2] = 0.0337 P_{CO_2}$$

where P_{CO_2} is in atmospheres. In air, CO_2 is present at a level of 0.03 mole percent, so that $P_{CO_2} = 3 \times 10^{-4}$ atm, giving $[CO_2] = 1 \times 10^{-5}$ M. The two ionizations of carbonic acid are

$$CO_2 + 2H_2O = H_3O^+ + HCO_3^-$$

$$K_{a1} = \frac{[H_3O^+][HCO_3^-]}{[CO_2]} = 4.3 \times 10^{-7}$$

$$HCO_3^- + H_2O = H_3O^+ + CO_2^{2-}$$

Chapter 10

Buffer Solutions and Acid-Base Titrations

$$K_{a2} = \frac{[H_3O^+][CO_2^{2-}]}{[HCO_3^-]} = 4.7 \times 10^{-11}$$

The calculation of the pH of a carbonic acid solution in equilibrium with the air is given in Example 10.11. Note that $K_{a1} \gg K_{a2}$, and therefore let us start by assuming that all the H_3O^+ comes from the first ionization.

Let $x = [H_3O^+] = [HCO_3^-]$; let $C = [CO_2] = 1 \times 10^{-5}$. Then

$$K_{a1} = 4.3 \times 10^{-7} = \frac{x^2}{C}$$

Note that what is in the denominator of this expression is really C, not C-x. The reason is that the solution is assumed to be in equilibrium with the atmosphere, and thus more CO_2 dissolves to make up for any that ionizes. This equation gives $[H_3O^+] = [HCO_3^-] = 2.1 \times 10^{-6}$, pH=5.7.

To check the assumption that $[H_3O^+]$ from the second ionization is negligible, we can use the equation for the second ionization to calculate $[CO_3^{2-}]$. Note that our hypothesis that the second ionization can be ignored leads to the conclusion that $[H_3O^+] = [HCO_3^-]$. These cancel out the expression for K_{a2}, giving $[CO_3^{2-}] = K_{a2} = 4.7 \times 10^{-11}$. This is also an estimate of any additional H_3O^+ produced in the second ionization, since a mole of H_3O^+ ions is formed for each mole of CO_3^{2-} formed. The fact that 4.7×10^{-11}M $\ll 2.1 \times 10^{-6}$M justifies the assumption that only an insignificant amount of H_3O^+ comes from the second ionization. The result that, for a diprotic acid, H_2A, $[A^{2-}]$ is equal to K_{a2} is a general result that holds whenever $K_{a2} \ll K_{a1}$.

Diprotic Acids: Mono-hydrogen Salt

In the titration of a diprotic acid, H_2A, with NaOH, the solution at the first equivalence point is just like a solution of pure NaHA. In such a solution, the species Na^+, OH^-, H_3O^+, H_2A, HA^- and A^{2-} are to be found. Suppose we make up the solution at an initial concentration C. Then the following equations describe the system. The first is the charge balance equation, and the next two are mass balance equations. Note that in the charge balance, A^{2-} contributes $2[A^{2-}]$ negative charges.

$$[Na^+] + [H_3O^+] = [OH^-] + [HA^-] + 2[A^{2-}]$$

Chapter 10 — Buffer Solutions and Acid-Base Titrations

$$C = [Na^+]$$

$$C = [H_2A] + [HA^-] + [A^{2-}]$$

$$K_w = [H_3O^+][OH^-]$$

$$K_{a1} = \frac{[H_3O^+][HA^-]}{[H_2A]}$$

$$K_{a2} = \frac{[H_3O^+][A^{2-}]}{[HA^-]}$$

This is a system of six equations in six unknowns, which can solved numerically to any accuracy desired, but the methods of doing so are not straightforward. For this reason, we resort to approximations.

The main source of $[H_3O^+]$ in the solution is the ionization of $[HA^-]$, whose K is K_{a2}.

$$HA^- + H_2O = H_3O^+ + A^{2-}$$

The main source of OH^- is the hydrolysis of HA^-, whose K is K_w/K_{a1}

$$HA^- + H_2O = OH^- + H_2A$$

So far, all we have assumed is that any H_3O^+ or OH^- that results from the self-ionization of water is small compared to that which results from these processes. Now, however, note that the OH^- and the H_3O^+ resulting from these processes will react with each other to form water

$$H_3O^+ + OH^- = 2H_2O$$

K for this process is $(1/K_w) = 10^{14}$; the assumption that this reaction goes to completion is tantamount to the assumption that self-ionization supplies only negligible amounts of H_3O^+ and OH^-. If the equilibrium constants for the ionization process ($K = K_{a2}$) and for the hydrolysis process (K_w/K_{a1}) are "in the same ballpark," then the H_3O^+ and OH^- that they produce will neutralize each other, giving water, and perhaps leaving a small excess of one over the other.

In order to use this result, note that starting with the electroneutrality condition, substituting C for $[Na^+]$, then substituting the second mass balance expression for C, we get

$$[H_2A] + [H_3O^+] = [A^{2-}] + [OH^-]$$

The fact that the H_3O^+ and the OH^- formed in the ionization and hydrolysis processes neutralize each other means that their concentrations will be small compared to those of $[A^{2-}]$ and $[H_2A]$, which are

Chapter 10 — Buffer Solutions and Acid-Base Titrations

formed in the same processes. Thus, we make the approximation that

$$[A^{2-}] \approx [H_2A]$$

The text outlines the range of applicability of this assumption.

To use this approximation, consider the overall ionization of H_2A

$$2H_2O + H_2A = 2H_3O^+ + A^{2-}$$

$$K_{overall} = K_{a1}K_{a2} = \frac{[H_3O^+]^2[A_2^-]}{[H_2A]}$$

$$\approx [H_3O^+]^2$$

$$[H_3O^+] = \sqrt{K_{a1}K_2}$$

$K_{overall}$ was obtained by adding the chemical equations for the first and second ionizations, which necessitates taking the product of the two individual K's. Since $[A_2^-] \approx [H_2A]$, these cancel out of the expression for $K_{overall}$. Note that the expression for $[H_3O^+]$ does not contain C in it; therefore the pH of a salt like $NaHCO_3$ does not change upon dilution. For $NaHCO_3$ itself, the pH is 8.35, as shown in Example 10.12. Even though pH is invariant with concentration, however, the solution is not a buffer. Addition of small amounts of acid or base would change the material balance equations, and so the derivation we have given would no longer hold.

Salt of a Weak Acid and a Weak Base

Although not in the text, the calculation of the pH of the salt of a weak acid and a weak base, such as NH_4F, is conceptually similar to the calculation of the pH of an NaHA solution. One ion (NH_4^+) reacts with water as an acid to form H_3O^+, and the other ion (F^-) hydrolyses to form OH^-. This is what occurred for NaHA, but for NaHA it was the HA^- ion which underwent both processes. For the particular care of NH_4F, here are the equations describing the system. Again, the first equation is the electroneutrality condition, and the second and third are mass-balance conditions.

$$[NH_4^+] + [H_3O^+] = [F^-] + [OH^-]$$

$$C = [NH_4^+] + [NH_3]$$

$$C = [F^-] + [HF]$$

$$K_w = [H_3O^+][OH^-] = 10^{-14}$$

$$K_a(NH_4) = \frac{[NH_3][H_3O^+]}{[NH_4^+]} = \frac{10^{-14}}{1.75 \times 10^{-5}} = 5.7 \times 10^{-10}$$

$$K_b(F) = \frac{[HF][OH^-]}{[F^-]} = \frac{10^{-14}}{7.2 \times 10^{-4}} = 1.4 \times 10^{-11}$$

Before proceeding, note that $K_a > K_b$, so that the solution should turn out to be acidic. Again, we observe that if the acid-forming and base-forming processes take place to approximately the same extent, then the H_3O^+ and OH^- formed should react with each other, leaving only an insignificant residue of the one in excess — insignificant, that is, in comparison with the other species in solution. If we start with the electroneutrality condition, and use the mass-balance equations to eliminate $[NH_4^+]$ and $[F^-]$, we get, in analogy to the NaHA situation,

$$[HF] + [H_3O^+] = [NH_3] + [OH^-]$$

$$[HF] \approx [NH_3]$$

We can also do a similar substitution to eliminate [HF] and $[NH_3]$, giving

$$[NH_4^+] + [H_3O^+] = [F^-] + [OH^-]$$

$$[NH_4^+] \approx [F^-]$$

For NaHA, we used our approximation that $[H_2A] \approx [A^{2-}]$ in the equilibrium constant expression for the overall ionization of H_2A to give two hydronium ions. Here, the analogous reaction is the overall ionization of the acidic forms of both ions to give two hydronium ions.

$$HF + NH_4^+ + 2H_2O = 2H_3O^+ + F^- + NH_3$$

$$K_{overall} = \frac{[F^-][NH_3][H_3O^+]^2}{[HF][NH_4^+]}$$

$$\approx [H_3O^+]^2$$

$$[H_3O^+] \approx \sqrt{K_{overall}} = \sqrt{K_a(HF) K_a(NH_4^+)}$$

We have applied the approximations we derived to cancel terms from the expression for $K_{overall}$. The overall process is the sum of the ionization processes for NH_4^+ ($K = 10^{-14}/1.75 \times 10^{-5} = 5.7 \times 10^{-10}$) and for HF ($K = 7.2 \times 10^{-4}$), so that $K_{overall} = 4.1 \times 10^{-13}$, giving $[H_3O^+] = 6.4 \times 10^{-7}$, or pH=6.19.

Diprotic Acids: Fully Neutralized Salt

The fact that for the monohydrogen salt of a diprotic acid $[H_3O^+]$ is invariant with concentration means that the pH at the first equivalence point is the same, for a given diprotic acid, regardless of initial concentration. This pH is $-\log_{10}\sqrt{K_{a1}K_{a2}}$. We have also assumed that the acid is being titrated by a strong base. At the second equivalence point the solution is identical to a pure solution of a salt like Na_2A, for the generic diprotic acid H_2A; for carbonic acid, this is sodium carbonate, Na_2CO_3. The chief reaction going on in this solution is

$$CO_3^{2-} + H_2O = OH^- + HCO_3^-$$

$$K_b = \frac{K_w}{K_{a2}} = \frac{10^{-14}}{4.7 \times 10^{-11}} = 2.1 \times 10^{-4}$$

Thus the solution is basic. The HA^- can, in general, further hydrolyse to form H_2A (CO_2 and water for carbonic acid), and more OH^-, but for carbonic acid this second process has $K = K_w/K_{a1} = (10^{-14}/4.3 \times 10^{-7}) = 2.3 \times 10^{-8}$, which is 10,000 times smaller than that for the hydrolysis of CO_3^{2-}, and can be ignored, both because it has less of a tendency to take place and because its starting material, HCO_3^-, is present at a much lower concentrations than that of CO_3^{2-}. This is just an ordinary hydrolysis problem, then, and setting $x=[OH^-]\approx[HCO_3^-]$ gives

$$2.1 \times 10^{-4} = \frac{x^2}{C-x}$$

In Example 10.13 this is solved by successive approximation for C=0.05M to give $x = 3.1 \times 10^{-3} = [OH^-]$, so that pOH=2.51 and pH=11.49. The pH at this second equivalence point *will* depend on concentration. Note the overall shape of the titration curve of a diprotic acid (Figure 10.6). By a suitable choice of indicator, either endpoint can be located. Note, however, that if K_{a1} and K_{a2} are close together, the two sections of the curve will begin to coalesce, and the two separate endpoints will no longer be clearly separated.

In this chapter we have discussed how the simple ideas behind equilibrium calculations, introduced in Chapter 9, can be applied to rather complicated systems. Mass and charge balance were introduced as additional relationships that are often needed to complete the description of a set of simultaneous coupled equilibrium processes. Even when this is done, the system of equations that results may not always be easy to solve exactly, so that it is important to be able to make useful approximations. These approximations require quite a bit of chemical intuition to understand and use, but greatly enlarge the realm of problems that can be readily solved. In quantitative science in general, it is often easier to write down equations describing a system than it is to solve them; very often it is even harder to have enough insight into the workings of the system to know what simplifying approximations are justifiable. In this last process scientific intuition — understanding the system, physically or chemically — is extremely important. One's intuition must, however, be verified by checking one's assumptions numerically after the solution has been worked out.

Chapter 10   Buffer Solutions and Acid-Base Titrations

Questions

10.1 Buffer solutions

1. A buffer based on the weak acid HA will contain approximately equimolar amounts of _____ and _____.

2. For a buffer based on the weak acid HF,

    (a) write the main reaction which occurs when a small amount of strong acid is added.

    (b) write the main reaction which occurs when a small amount of strong base is added.

3. Suppose $K_a = 10^{-4}$ for the acid HA.

    (a) What is the pK_a?

    (b) In a 0.01M solution of HA

       (i) What is $[H_3O^+]$?

       (ii) What is the pH?

    (c) If a solution is made up 0.1M in [HA], and simultaneously 0.05M in strong base

       (i) What is [HA] equal to?

Chapter 10 — Buffer Solutions and Acid-Base Titrations

   (ii) What is [A⁻] equal to?

   (iii) What is the pH?

(d) If a solution is made up 0.11M in NaA and simultaneously 0.01M in strong acid

   (i) What is [HA] equal to?

   (ii) What is [A⁻] equal to?

   (iii) What is the pH?

4. The pH of a buffer is almost constant if (answer True or False)

   _____(a) the buffer is diluted

   _____(b) [HA] only is doubled

   _____(c) [HA] and [A⁻] are both doubled

   _____(d) small amounts of strong acid or base are added

10.2 Titration of strong acids vs. a strong base

1. In the titration of a strong base, B, with a strong acid,

   (a) the pH changes most rapidly *near the equivalence point.*

Chapter 10 — Buffer Solutions and Acid-Base Titrations

(b) the pH half-way to the equivalence point is given by the pH of *the buffer solution with equimolar [B] and [BH⁺]*.

(c) the indicator must be one which changes color *right at the pH of the equivalence point*

2. Suppose I have 1L of 0.10M strong base, B.

   (a) What is the pH?

   (b) Suppose I add to the solution 1L of 0.08M strong acid.

      (i) What is the new value of [B]?

      (ii) What is the new value of the pH?

      (iii) What was the change of pH as a result of this process?

3. Suppose I have 1L of pure water.

   (a) What is the pH?

Chapter 10   Buffer Solutions and Acid-Base Titrations

(b) Suppose I add to it 1L of 0.08M strong acid.

    (i) Estimate the pH (one or two significant figures is enough).

    (ii) What was the change of pH as a result of this process?

4. Describe what the pH changes you calculated in Question 2 and 3 have to do with some kind of acid-base titration.

10.3 Acid-base indicators

1. For an indicator HIn.

    (a) under what conditions will the solution generally take on the color of In^-?

    (b) under what conditions will the solution generally take on the color of HIn^-?

    (c) If the indicator has a pK_a of 4, over what pH range would you expect the color change to take place?

Chapter 10   Buffer Solutions and Acid-Base Titrations

2. Give the approximate pK_a value of an indicator which would be suitable for the titration of

   (a) a weak acid with a strong base.

   (b) a weak base with a strong acid.

10.4 Titration of a weak monoprotic acid vs. a strong base

1. In the titration of a weak acid ($K_a = 10^{-4}$) with a strong base,

   (a) what is the pH half-way to the equivalence point?

   (b) at what point in the titration is the pH that of a pure salt of the acid?

2. In the titration of a weak base with $pK_b = 10^{-4}$ with a strong acid,

   (a) what is the pH half-way to the equivalence point?

   (b) at what point in the titration is the pH that of a pure salt of the base?

10.5 Calculations involving weak diprotic acids and their salts

1. A solution of H_2S is made up at some concentration C.

   (a) List all species in solution

(b) Write the mass balance equation

(c) Write the charge balance equation

2. For a weak acid, H_2A, $K_{a1} = 10^{-4}$ and $K_{a2} = 10^{-10}$.

   (a) Estimate the pH of a 0.01M solution of H_2A.

   (b) Estimate the pH of a 0.01M solution of NaHA.

   (c) Estimate the pH of 0.01M Na_2A solution.

   (d) In the titration of H_2A with NaOH, what is the pH at the first equivalence point?

   (e) In the same titration, what is the pH at the second equivalence point?

Chapter 10 — Buffer Solutions and Acid-Base Titrations

Answers

10.1 Buffer Solutions

1. HA, A$^-$.

2. (a) F$^-$ + H$_3$O$^+$ = HF + H$_2$O
   (b) HF + OH$^-$ = F$^-$ + H$_2$O

3. (a) 4
   (b)
   - (i) 10^{-3}
   - (ii) 3

   (c)
   - (i) 0.05M
   - (ii) 0.05M
   - (iii) 4

   (d)
   - (i) 0.01M
   - (ii) 0.10M
   - (iii) 5

4. (a) True
   (b) False
   (c) True
   (d) True

10.2 Titration of a strong acid vs. a strong base

1. (a) True
   (b) a solution of pure B at somewhat less than half the initial concentration (due to dilution).
   (c) within about 2 pH units if the of the equivalence point.

2. (a) 13
   (b)
   - (i) 0.01M
   - (ii) 12
   - (iii) 1

3. (a) 7
   (b)
   - (i) ≈1.4
   - (ii) ≈6.6

4. pH changes more rapidly near the equivalence point than near the beginning of a titration.

10.3 Acid-base indicators

1. (a) [In$^-$] > 10[HIn]
   (b) [HIn] > 10[In$^-$]
   (c) 3—5

2. (a) pK_a ≈ 8.5
   (b) pK_a ≈ 5.5

10.4 Titration of a weak monoprotic acid vs. a strong base

1. (a) 4
   (b) equivalence point

2. (a) 10
   (b) equivalence point

Chapter 10 — Buffer Solutions and Acid-Base Titrations

10.5 Calculations involving weak diprotic acids and their salts

1. (a) H_3O^+, OH^-, H_2A, HA^-, A^{2-}
   (b) $C = [H_2A] + [HA^-] + [A^{2-}]$
   (c) $[H_3O^+] = [OH^-] + [HA^-] + 2[A^{2-}]$

2. (a) 3
   (b) 7 ($[H_3O^+] = \sqrt{K_{a1}K_{a2}}$)
   (c) 11 ($K_b = (K_w/K_{a2}) = 10^{-4}$)
   (d) 7 (from part (b))
   (e) 11 (from part (c))

CHAPTER 11: EQUILIBRIA INVOLVING SLIGHTLY SOLUBLE ELECTROLYTES

Scope

Solubility Product

The *solubility product*, K_{sp}, is nothing more than the equilibrium constant for the dissolution of a sparingly soluble electrolyte. You already encountered the concept in Chapter 8; in this chapter you will see how it is used in calculations. Much of the chemistry of *inorganic qualitative analysis* involves the precipitation of insoluble substances, and many "K_{sp} problems" have their origin in this technique.

Several K_{sp} expressions, and their corresponding reactions, are given in Table 11.1. A few are also given here, to illustrate how the solubility product is defined.

$$AgCl(s) = Ag^+(aq) + Cl^-(aq)$$

$$K_{sp} = [Ag^+][Cl^-] = 1.8 \times 10^{-10}$$

$$Bi_2S_3(s) = 2Bi^{3+}(aq) + 3S^{2-}(aq)$$

$$K_{sp} = [Bi^{3+}]^2[S^{2-}]^3 = 2 \times 10^{-72}$$

$$Hg_2Cl_2 = Hg_2^{2+}(aq) + 2Cl^-(aq)$$

$$K_{sp} = [Hg_2^{2+}][Cl^-]^2$$

Note that the stoichiometric coefficients in the chemical equations become the exponents in the K_{sp} expression, as they do in any equilibrium constant expression. Thus, Hg_2Cl_2, mercurous chloride, has the expression shown because the diatomic ion Hg_2^{2+} is the species in solution. Of course, the pure solid has a constant concentration and does not appear in the K_{sp} expression.

The text states as the *solubility product principle* that K_{sp} is a constant at a given T for a saturated solution of a given solid. This is just what we expect of an equilibrium constant. In order for an insoluble solid to be in equilibrium with its solution the solution must be saturated, as discussed early in Chapter 6, so that the stipulation that the solution be saturated is simply a statement of the fact the concentrations in the K_{sp} expression are equilibrium concentrations. Of course, if one puts some solid

AgCl into water and measures $[Ag^+]$ and $[Cl^-]$ before the solution becomes saturated, then one may still calculate the product $[Ag^+][Cl^-]$, but this product is not the value of K_{sp}; it is a value of Q, the reaction quotient (Chapter 8), and since the solution is not yet saturated $[Ag^+]$ and $[Cl^-]$ will be below their equilibrium concentrations, so that $Q < K_{sp}$. Recall that this implies a net forward reaction. All this may be obvious, but it is worth restating, if only to make it thoroughly clear that K_{sp} behaves just like any other equilibrium constant.

In Chapter 6 you learned about the different concentration units commonly employed for solutions: mole fraction, molality, molarity and so on. Next we will examine the relationship between K_{sp} and molarity, but before that we will note two things. First, any molarity we get starting with K_{sp} will be the molarity of the species at equilibrium; that is, in the saturated solution. For example, we will show how to calculate $[Ag^+]$ and $[CrO_4^{2-}]$ in a saturated solution of Ag_2CrO_4, given K_{sp} for this substance. Second, since you learned in Chapter 6 how to convert among the various concentration units, once you have calculated the molarity of a species you can re-express the result in any concentration unit you desire.

Ag_2CrO_4 dissolves according to

$$Ag_2CrO_4(s) = 2Ag^+(aq) + CrO_4^{2-}$$

$$K_{sp} = [Ag^+][CrO_4^{2-}] = 9.0 \times 10^{-12}$$

From the reaction stoichiometry it is clear that in a pure Ag_2CrO_4 solution if we let $x = [CrO_4^{2-}]$ then $[Ag^+] = 2x$. This gives

$$K_{sp} = 9.0 \times 10^{-12} = (2x)^2(x) = 4x^3$$

$$x = [CrO_4^{2-}] = 1.3 \times 10^{-4} M$$

$$[Ag^+] = 2x = 2.6 \times 10^{-4} M$$

Sometimes students are confused by the fact that we first multiply x by two for $[Ag^+]$ and then square this entire quantity; if you look carefully at the equations you can see why this is. The stoichiometry necessitates that $[Ag^+] = 2x$ if $[CrO_4^{2-}] = x$; you have encountered this before. Then the definition of K_{sp} requires that $[Ag^+]$ be squared; hence 2x must be squared.

Chapter 11 — Equilibria Involving Slightly Soluble Electrolytes

Common Ion Effect

Starting with K_{sp}, we have shown that in a saturated solution of Ag_2CrO_4 in pure water, $[Ag^+] = 2.6 \times 10^{-4}$M and $[CrO_4^{2-}] = 1.3 \times 10^{-4}$M. Note that this is equivalent to the statement that the solubility of Ag_2CrO_4 in water is 1.3×10^{-4}M. (Make sure you understand why before continuing!) Since a mole of Ag_2CrO_4 weighs 332g, the solubility of Ag_2CrO_4 in water is $(332\text{g/mol})(1.3 \times 10^{-4}\text{mol/L}) = 0.043$g/L. Suppose, however, we add to the liter of solution 0.1mol of the soluble strong electrolyte K_2CrO_4. By LeChatelier's principle, the common ion effect will suppress the solubility of Ag_2CrO_4. The value of $[CrO_4^{2-}]$ originally contributed by K_2CrO_4 is 0.1M, and since this is much larger than the 1.3×10^{-4}M originally contributed by the Ag_2CrO_4, the latter can be ignored. Then we have

$$K_{sp} = 9.0 \times 10^{-12} = [Ag^+]^2[CrO_4^{2-}]$$
$$= [Ag^+]^2(0.1)$$

so that $[Ag^+] = (9.0 \times 10^{-11})^{1/2} = 9.5 \times 10^{-6}$M. Note that $[Ag^+]$ has dropped by a factor of $(2.6 \times 10^{-4}/9.5 \times 10^{-6}) = 27$ due to the presence of 0.1M chromate.

Sometimes one is asked to do something like "calculate the solubility of Ag_2CrO_4 in a 0.1M potassium chromate solution." One considers the Ag^+ to come from dissolved Ag_2CrO_4, and, by the stoichiometry of dissolution, $[Ag_2CrO_4]$ is considered to be $(1/2)[Ag^+]$, or 4.7×10^{-7}M. This corresponds to 0.0016g/L. We still talk of the "solubility of Ag_2CrO_4" even though we know that what is really in solution are Ag^+ and CrO_4^{2-} ions, and even though a given CrO_4^{2-} ion in solution cannot be ascribed to either the potassium or the silver salt; you may view this word usage as an idiom of chemical discourse, and you must, of course, be able to recognize it and know what is meant by it. It does make sense, since the calculated amount of Ag_2CrO_4 is the number of grams that would dissolve in a liter of 0.1M K_2CrO_4. One more thing: you will note that when we went from pure water to 0.1M K_2CrO_4, K_{sp} stayed the same, but the solubility of Ag_2CrO_4 decreased. This is important: it shows that solubility may vary with composition (i.e., with concentration of components), but K_{sp} is constant at a given temperature.

Chapter 11   Equilibria Involving Slightly Soluble Electrolytes

Suppose now that we had asked for the solubility of Ag_2CrO_4 not in 0.1M K_2CrO_4 but rather in 0.0001M K_2CrO_4. This concentration is almost the same as that of CrO_4^{2-} in a pure aqueous solution of Ag_2CrO_4, so it no longer suffices to assume that all the chromate in solution comes from the potassium salt; therefore, the problem is harder. If we let x be equal to the molar solubility of Ag_2CrO_4 in the given solution, then the following before-after table applies. ("Before" means before adding the Ag_2CrO_4.)

	$[CrO_4^{2-}]$	$[Ag^+]$
before	1.0×10^{-4}	0
after	$1.0 \times 10^{-4}+x$	2x

$$K_{sp} = 9.0 \times 10^{-12} = (2x)^2(1.0 \times 10^{-4} + x)$$

This equation can be rearranged to read

$$x^2 = \frac{9.0 \times 10^{-12}}{4(1.0 \times 10^{-14} + x)}$$

and can be solved by our method of repeated substitution. We start by assuming $x = \sqrt{9.0 \times 10^{-12}} = 3.0 \times 10^{-6}$, substitute this value in the denominator on the right, giving $x = \sqrt{2.18 \times 10^{-8}} = 1.48 \times 10^{-4}$. Continuing the process gives the following successive values of x: 9.53×10^{-5}, 1.07×10^{-4}, 1.04×10^{-4}, 1.05×10^{-4}, 1.05×10^{-4}; thus, $x \approx 1.1 \times 10^{-4}$M, which corresponds to a solubility of about 0.035g/L, down from 0.043g/L in pure water.

Here is another sort of problem, not considered in the text, which I am presenting as a "brain-teaser." The method of solution and the answer can be found at the very end of this chapter. Consider the following two salts, with K_{sp}'s as shown.

$$CaCrO_4 \quad K_{sp} = 7.1 \times 10^{-4}$$
$$CaSO_4 \quad K_{sp} = 2.4 \times 10^{-5}$$

What are the solubilites of both salts in a solution saturated in both of them?

Example 11.8 in the text sets out still another sort of problem: will a precipitate form when two given solutions are mixed? In the example, the solutions are 70mL of 0.05M $Ba(NO_3)_2$ and 30mL of 0.02M NaF; K_{sp} of BaF_2 is 1.7×10^{-6}. The method is straightforward: calculate Q for the dissolution of

Chapter 11 Equilibria Involving Slightly Soluble Electrolytes

BaF_2, remembering to take dilution into account. If $Q<K_{sp}$, no precipitate will form (net process is dissolution). If $Q>K_{sp}$, then precipitate will form (net reaction is to the left, leading to precipitation). In this example, the total solution volume will be 0.1L, and the initial concentrations of Ba^{2+} and F^- are given by

$$[Ba^{2+}] = \frac{(0.05 mol/L)(0.07L)}{0.1L} = 0.035M$$

$$[F^-] = \frac{(0.02 mol/L)(0.03L)}{0.1L} = 0.006M$$

These are inserted into the expression for Q

$$Q = [Ba^{2+}][F^-]^2 = (0.035)(0.006)^2 = 1.3 \times 10^{-6}$$

Since $Q<K_{sp}$, no precipitate will form.

There are two possible sources of confusion in this problem. The reaction in question is

$$BaF_2 = Ba^{2+}(aq) + 2F^-(aq)$$

$$K_{sp} = 1.7 \times 10^{-6} = [Ba^{2+}][F^-]^2$$

One possibly confusing thing we did was to calculate the expression for K_{sp} and call it Q. Recall that the only difference between the expression for K_{sp} and that for Q is that for K_{sp} we know that the concentrations are equilibrium concentrations. Since here we are trying to find out whether the solution is at equilibrium with the solid (that is, it may not be), what we are calculating is Q. Second, we have chosen to write the equation in the forward direction, from the point of view of K_{sp}; that is, K_{sp} is K_{sp} for the *dissolution* of a solid, so we wrote the equation for dissolution. We said earlier that $Q<K_{sp}$ means that the net reaction proceeds to the right. In this problem, this must be understood to mean that there is a net *tendency* for the reaction to go to the right; in other words, if there were some solid already present, more would continue to dissolve. There is no precipitate, however, so that nothing happens; the solution is in an equilibrium state, but is not in equilibrium with the solid, there being no solid present.

Example 11.9 shows how to calculate the concentration of ions remaining in solution after a precipitation reaction occurs. This is quite straightforward, no more than a combination of the concept of limiting reagent (Chapter 1) with that of K_{sp}. In the example, a solution on mixing contains 0.072M

Zn^{2+} and 0.040M C$_2$O$_4^{2-}$ (oxalate); for the dissolution of ZnC$_2$O$_4$, Q is thus $(0.072)(0.040) = 2.9 \times 10^{-3}$. Since K_{sp} is 2.5×10^{-9}, precipitation will occur. Virtually all the oxalate will precipitate with an equimolar amount of Zn^{2+}, which is in excess, leaving [Zn^{2+}] = 0.032M; then [C$_2$O$_4^{2-}$] is given by

$$x = [C_2O_4^{2-}] = \frac{K_{sp}}{[Zn^{2+}]} = \frac{2.5 \times 10^{-9}}{.032 - x} = 7.8 \times 10^{-8} M$$

where we could neglect x in the denominator of the equation.

Dissolution of Precipitates

In Section 11.4 the text describes how precipitates contining the anion of a weak acid may sometimes be solubilized by addition of strong acid, as in Example 11.10.

$$MnS(s) + 2H^+(s) = Mn^{2+} + H_2S$$

If the dissolution of MnS in pure water is considered

$$MnS(s) = Mn^{2+} + S^{2-}$$

Then any process which removes S^{2-} (or Mn^{2+}) from solution will cause the reaction to go to the right — that is, will increase the solubility — by LeChatelier's principle. The addition of H$^+$ to sulfide ions in solution causes H$_2$S to form; because H$_2$S is a weak acid, both HS$^-$ and S^{2-} (especially) exist only at very low concentrations in an acidic medium. Note that when the equation for H$_2$S formation from H$^+$ and S^{2-} is added to the equation for the dissolution of MnS, what results is the equation for the overall process

$$MnS(s) = Mn^{2+} + S^{2-}$$

$$K = K_{sp} = 5 \times 10^{-15}$$

$$2H^+ + S^{2-} = H_2S$$

$$K = 1/K_{a1}K_{a2} = 1.3 \times 10^{20}$$

$$MnS(s) + 2H^+ = Mn^{2+} + H_2S$$

$$K_{overall} = \frac{[Mn^{2+}][H_2S]}{[H^+]^2} = \frac{K_{sp}}{K_{a1}K_{a2}} = 4 \times 10^5$$

Chapter 11 **Equilibria Involving Slightly Soluble Electrolytes**

The high value of $K_{overall}$ implies a strong tendency for the reaction to proceed to the right, *under standard conditions*; that is, when all dissolved substances are 1M in concentration. Later, in Example 11.12, the way to calculate the solubility of MnS in a solution of a given pH is described. Before going into that, however, you should understand two things. The first is that the MnS is dissolved not by the action of H^+ on the solid, but by the action of H^+ on S^{2-} in solution. Before H^+ is added, if the MnS is in equilibrium with its solution, precipitation and dissolution are occurring simultaneously at equal rates. When H^+ is added, $[S^{2-}]$ is diminished, due to H_2S formation, so that the rate of precipitation, which depends on $[S^{2-}]$, decreases. The rate of dissolution remains constant, however, and more MnS dissolves. This is in accord with the concept of dynamic equilibrium, as discussed in Chapter 5. Note in particular that it is the amount of S^{2-} per se in solution, not the total amount of dissolved sulfide, which includes H_2S, that determines the solubility of MnS.

The second thing to realize is that, although not discussed in the text, a precipitate can be dissolved by any reagent that results in a diminution in the concentration of one of the dissolved ions. For example, in Chapter 8 of the study guide, we gave an example involving zinc hydroxide.

$$Zn(OH)_2(s) = Zn^{2+} + 2OH^-$$

$$K_{sp} = [Zn^{2+}][OH^-]^2 = 5 \times 10^{-17}$$

Ordinarily, increasing $[OH^-]$ would be expected to decrease the solubility; however, at high enough $[OH^-]$ the complex $Zn(OH)_4^{2-}$ is formed, which increases the solubility by lowering $[Zn^{2+}]$. Similarly, AgCl can be solubilized by the addition of ammonia, which causes the stable complex ion $Ag(NH_3)_2^+$ to be formed. These solubilization mechanisms really are no different from the dissolution of MnS in strong acid, and we will work out the $Zn(OH)_2$ example below.

Now let us consider Example 11.12 in the text. We already calculated that for the reaction

$$MnS(s) + 2H^+ = H_2S + Mn^{2+}$$

$K_{overall} = [H_2S][Mn^{2+}]/[H^+]^2 = 4 \times 10^5$, and now we are told that $[H^+] = 0.20M$ (corresponding to pH=0.70) and $[H_2S] = 0.10M$. Clearly, $[Mn^{2+}]$ is the only unknown, and solving for it gives $1.6 \times 10^5 M$!! This would represent the equilibrium solubility of MnS in a solution with the given values of $[H_2S]$ and

Chapter 11 — Equilibria Involving Slightly Soluble Electrolytes

[H$^+$], and is preposterously high (the law of mass action would fail long before such a concentration was reached). The conclusion is that any reasonable amount of MnS will dissolve in a solution maintained at the given values of [H$^+$] and [H$_2$S].

When [H$^+$] is high, we can assume that virtually all soluble sulfide exists as H$_2$S. When S^{2-} is added to pure water, on the other hand, as you learned in Chapter 10, the predominant species is HS$^-$. At pH's somewhere between 7 and 1, both species will exist at comparable concentrations. This is not considered in the text, but Example 11.13 shows how to calculate the effect of S^{2-} hydrolysis on the MnS solubility in pure water. Without such hydrolysis it is easy to calculate that

$$\text{"[MnS]"} = [\text{Mn}^{2+}] = [\text{S}^{2-}] = \sqrt{K_{sp}} = \sqrt{3.8 \times 10^{-16}} = 7.1 \times 10^{-8}.$$

We use quotes as a reminder that it is really the ions that are in solution. Allowing a single step of hydrolysis gives

$$\text{S}^{2-} + \text{H}_2\text{O} = \text{HS}^- + \text{OH}^- \qquad K = K_w/K_{a2} = 7.7 \times 10^{-2}$$

Combining this with the K_{sp} reaction gives

$$\text{MnS(s)} + \text{H}_2\text{O} = \text{Mn}^{2+} + \text{HS}^- + \text{OH}^-$$

$$K_{overall} = \frac{K_{sp}K_w}{K_{a2}} = 7.7 \times 10^{-2}$$

Then [MnS]=[Mn^{2+}]=[OH$^-$]=[HS$^-$]=(K$_{overall}$)$^{1/3}$=7.3×10^{-6}, which is greater by a factor of 100 than the result of a naive calculation which doesn't include hydrolysis.

Once again we are shown that all relevant equilibria must be taken into account if a calculation is to be believed. Just as a peripheral note: in complicated naturally occurring equilibrium systems, such as ocean waters, there may be dozens of relevant equilibria, involving the various ionization states of species such as phosphate and carbonate, and solving them simultaneously can be a major chore, involving the use of high-speed computers. Verifying that the calculation is in accord with reality — that the equations really include all the pertinent information — is an even more formidable research task, involving much laborious experimental work. For many natural water systems there is still no system of equations that everyone believes; this contributes to the lack of policy consensus on environmental issues such as the abatement of acid rain and the use of phosphates in detergent. It takes considerable chemical insight to be able to know in advance what is relevant; very often new discoveries are made as a result of the inconsistency of an experimental result with the prediction of an existing model. This philosophical interlude will not help you realize that the hydrolysis of S^{2-} must be taken into account in calculating the solubility of MnS in water — only repeated exposure to the chemistry of the sulfides can do that — but at least it may make you realize that such complications are common, and help you be on your guard.

We skipped over Example 11.12 because it is self-explanatory, but it will be useful to summarize its significance. It is shown that $BaSO_4$ is not solubilized by acid. The reason is that the reaction

$$H^+ + SO_4^{2-} = HSO_4^-$$

has a small equilibrium constant; H^+ is not efficient in pulling sulfate out of solution; HSO_4^- is a fairly strong acid; the preceding three statements are equivalent. Since solubilization of MX by H^+ generally relies on formation of a species such as HX or H_2X, these must be *weak* acids if solubilization is to succeed. (For $BaSO_4$, since HSO_4^- tends not to be formed, naturally H_2SO_4, a very strong acid, tends even less to be formed.)

In the text it is shown that CuS will not be solubilized by H^+ even though S^{2-} is the anion of a very weak acid. Here the reason is that $K_{overall}$ for the dissolution is only 6.7×10^{-16} (instead of 4×10^5, as for MnS), due to very low value of K_{sp} for CuS (8.7×10^{-36}).

We now discuss the $Zn(OH)_2$ example, as promised (or perhaps we should say "threatened") earlier in this chapter and in Chapter 8. The two equilibria of interest are

$$Zn(OH)_2 = Zn^{2+} + 2OH^-$$

$$K_{sp} = [Zn^{2+}][OH^-]^2 = 4.5 \times 10^{-17}$$

$$Zn^{2+} + 4OH^- = Zn(OH)_4^{2-}$$

$$K_f = \frac{[Zn(OH)_4^{2-}]}{[Zn^{2+}][OH^-]^4} = 2.9 \times 10^{15}$$

(Recall from Chapter 8 that K_f is a *formation constant* for the complex ion $Zn(OH)_4^{2-}$.) Suppose, now, that we have a $Zn(OH)_2$ precipitate in water, and that we adjust $[OH^-]$ by changing the pH, using a pH

Chapter 11 — Equilibria Involving Slightly Soluble Electrolytes

meter. At any $[OH^-]$, we can calculate $[Zn^{2+}]$ from the K_{sp} expression

$$[Zn^{2+}] = \frac{K_{sp}}{[OH^-]}$$

This same value of $[Zn^{2+}]$ can then be used with the original value of $[OH^-]$ to calculate $[Zn(OH)_4^{2-}]$; since all the species are in solution together, all the equilibria must hold.

$$[Zn(OH)_4^{2-}] = K_f [Zn^{2+}][OH^-]^4$$

For example, if $[OH^-] = 10^{-7}$ (neutral solution), then, from the K_{sp} expression,

$$[Zn^{2+}] = \frac{4.5 \times 10^{-17}}{(10^{-7})^2} = 4.5 \times 10^{-3}$$

Then, for the K_f expression,

$$[Zn(OH)_4^{2-}] = (2.9 \times 10^{15})(4.5 \times 10^{-3})(10^{-7})^4 = 1.3 \times 10^{-15}$$

Now, the solubility of $Zn(OH)_2$ is considered to be equal to the solubility of zinc in all its forms, because this reflects how much weight the precipitate would actually lose if dissolved in a solution with the given $[OH^-]$. For pH=7 we have for the solubility of the precipitate

$$"[Zn(OH)_2]" = [Zn^{2+}] + [Zn(OH)_4^{2-}] \approx 4.5 \times 10^{-3}$$

Thus, almost all of the dissolved zinc is in the form of Zn^{+2}; the solubility does not differ from what would be found considering only the K_{sp} reaction. Now suppose instead that the pH is 13, so that $[OH^-]=0.1M$. Then we have

$$[Zn^{2+}] = \frac{4.5 \times 10^{-17}}{(0.1)^2} = 4.5 \times 10^{-15}$$

$$[Zn(OH)_4^{2-}] = (2.9 \times 10^{15})(4.5 \times 10^{-15})(0.1)^4 = 1.3 \times 10^{-3}$$

$$"[Zn(OH)_2]" \approx [Zn(OH)_4^{2-}] = 1.3 \times 10^{-3}$$

Note that raising $[OH^-]$ to a very high value *increases* rather than decreases the solubility of $Zn(OH)_2$, due to complex ion formation. The following table gives the results of the calculation for pH's between 7 and 13

Chapter 11 — Equilibria Involving Slightly Soluble Electrolytes

pH	$[Zn^{2+}]$	$[Zn(OH)_4^{2-}]$	"$[Zn(OH)_2]$"
7	4.5×10^{-3}	1.3×10^{-15}	4.5×10^{-3}
8	4.5×10^{-5}	1.3×10^{-13}	4.5×10^{-5}
9	4.5×10^{-7}	1.3×10^{-11}	4.5×10^{-7}
10	4.5×10^{-9}	1.3×10^{-9}	5.8×10^{-9}
11	4.5×10^{-11}	1.3×10^{-7}	1.3×10^{-7}
12	4.5×10^{-13}	1.3×10^{-5}	1.3×10^{-5}
13	4.5×10^{-15}	1.3×10^{-3}	1.3×10^{-3}

Note that as $[OH^-]$ increases over this range, the solubility of the precipitate decreases, goes through a minimum, then increases. Only at pH=10 do both dissolved forms of zinc significantly contribute to the solubility of the precipitate. This complicated behavior occurs only because the OH^- forms both the complex and the precipitate with Zn^{2+}. Most examples of complex ion formation, such as the dissolution of AgCl by NH_3, exhibit strictly increasing solubilities with increasing concentrations of complexing reagent. The method of solution, however, is exactly as we demonstrated it for $Zn(OH)_2$.

Separation of Precipitates

Another technique of qualitative analysis involves the separation of precipitates based on their K_{sp}'s. In Example 11.14 this is shown for magnesium and calcium carbonates, which exhibit K_{sp}'s of 4.5×10^{-5} and 4.8×10^{-9}, respectively. Rather than go through the text example, we will work a different but still typical problem involving these substances. Suppose a solution is 0.1M in both Mg^{2+} and Ca^{2+}. Suppose we add CO_3^{2-} until we just are about to precipitate $MgCO_3$ (the more soluble salt). By this time, how much Ca^{2+} remains in solution? In order to be just about to precipitate $MgCO_3$, we must have

$$[Mg^{2+}][CO_3^{2-}] = K_{sp}(MgCO_3)$$

where $[Mg^{2+}]$ is given by the initial concentration, 0.1M. This gives $[CO_3^{2-}]= (4.5\times10^{-5})/(0.1)= 4.5\times10^{-4}$M. At this value of $[CO_3^{2-}]$, $[Ca^{2+}]= K_{sp}(CaCO_3)/[CO_3^{2-}]= (4.8\times10^{-9})/(4.5\times10^{-4}\text{M})= 1.1\times10^{-5}$; thus, all of the Mg^{2+}, but only about 10^{-4}, or 0.01%, of the original calcium remains in solution. This is quite an efficient separation! Such a separation is the basis for a method of determining chloride ion concentration. Chromate, CrO_4^{2-}, a pale yellow ion, is added to the chloride solution. Silver chromate, Ag_2CrO_4, is bright red. $[Ag^+]$ is added to the solution as silver nitrate, precipitating AgCl, just until the red Ag_2CrO_4 precipitate begins to form. At this point virtually all the chloride has precipitated, and the amount originally present can be calculated from the concentration of the $AgNO_3$ solution and the volume added. Problem 11.19 involves a calculation on this system.

This chapter concludes the study of chemical equilibrium problems. In Chapter 18, which is concerned with thermodynamics, you will get a more profound appreciation of what an equilibrium constant really is; first, however, we return in earnest to an aspect of chemistry which we put aside after chapter 2; namely, the development of an understanding of why the different elements exhibit their characteristic chemical behaviors.

Chapter 11   Equilibria Involving Slightly Soluble Electrolytes

                                    Questions

11.1 The solubility product

1. For each of the following substances, write the K_{sp} expression and the chemical equation for which K_{sp} is the equilibrium constant.

    (a) $BaSO_4$

    (b) HgS

    (c) Hg_2S

    (d) CaF_2

    (e) Ag_2S

Chapter 11　　　　　　　　　　　　　　　Equilibria Involving Slightly Soluble Electrolytes

11.2 The relationship between the molar solubility and the solubility product

1. $BaSO_4$ has a K_{sp} of about 1×10^{-10}

    (a) What is its molar solubility in water?

    (b) The molecular weight is about 233. What is its solubility in grams per liter?

    (c) What is the molar solubility of $BaSO_4$ in a 0.001M solution of Na_2SO_4?

    (d) What is the solubility in grams per liter in 0.001M Na_2SO_4?

2. Tell whether the addition of acid would tend to increase, decrease or leave unaltered the solubility of the following insoluble salts.

    _____(a) $La(OH)_3$

    _____(b) Bi_2S_3

    _____(c) $AgCl$

3. Name a substance which, when added to each of the following, will decrease its solubility.

    _____(a) $AgCl$

    _____(b) $Al(OH)_3$

Chapter 11   Equilibria Involving Slightly Soluble Electrolytes

_____(c) BaSO$_4$

11.3 Rules for determining whether a solution is unsaturated, saturated or supersaturated

1. For each of the following salts, a table is given which contains the K_{sp}, the cation concentration and the anion concentration. Tell whether each solution is unsaturated, saturated or supersaturated.

	Salt	K_{sp}	[cation]	[anion]	Saturation
(a)	BaSO$_4$	1×10^{-10}	10^{-3}	10^{-4}	_____
(b)	AgCl	1.8×10^{-10}	10^{-6}	10^{-7}	_____
(c)	Hg$_2$Br$_2$	1.3×10^{-21}	10^{-8}	10^{-8}	_____
(d)	ZnC$_2$C$_4$	2.5×10^{-9}	5×10^{-4}	5×10^{-6}	_____

2. 50mL of 0.01M BaCl$_2$ is mixed with 50mL of 0.02M H$_2$SO$_4$. K_{sp} = (BaSO$_4$) = 1×10^{-10}. After precipitation, what are the concentrations of Ba^{2+} and SO$_4^{2-}$?

11.4 Equilibria involving both slightly soluble electrolytes and weak acids or bases

1. For a divalent cation M, the sulfide has the formula MS. Under standard conditions, MS will dissolve in acid if $K_{overall} > 1$ for the following reaction

$$MS + 2H^+ = M^{2+} + H_2S$$

(a) Write an equation for $K_{overall}$ in terms of K_{sp}(MS).

Chapter 11   Equilibria Involving Slightly Soluble Electrolytes

(b) If K_{sp} is large enough, MS will dissolve under standard conditions. How big does K_{sp} have to be for this to happen?

(c) How big does K_{sp} have to be for MS to dissolve to the extent of 1M in a saturated solution of H_2S ($[H_2S] = 0.1M$) maintained at a pH of 1?

2. In general, whether MX will dissolve in strong acid depends on $K_{sp}(MX)$ as well as the nature of the anion, X, and a calculation must in general be done to calculate whether MX will in fact, dissolve. For certain anions, however, you should be able to tell from your knowledge of chemistry that acid will not greatly enhance the solubility. For which of the following substances can you say, in advance, without doing a calculation, that strong acid will not materially increase the solubility.

(a) AgCl

(b) Al(OH)$_3$

(c) Bi$_2$S$_3$

(d) BaSO$_4$

(e) Ca$_3$(PO$_4$)$_2$

(f) AgBr

(g) Ca(H$_2$PO$_4$)$_2$

(h) Ba(CO$_3$)

3. It is found that the solubility of a sulfide, MS, in water is generally greater than the value predicted from its K_{sp}. Explain why this is.

Chapter 11   Equilibria Involving Slightly Soluble Electrolytes

11.5 Analytical separations based on a difference in solubility products

1. Suppose two insoluble salts share a common ion (for example, AgCl and AgBr, or $CaCO_3$ and $SrCO_3$). As the concentration of the common ion is increased, the member of the pair with the higher K_{sp} tends to precipitate (first, last).

2. $K_{sp}(AgCl) = 1.9 \times 10^{-10}$, and $K_{sp}(AgI) = 3.2 \times 10^{-17}$. Suppose a solution is 0.1M in both Cl^- and I^-.

    (a) What does $[Ag^+]$ have to be to just begin to precipitate out AgCl?

    (b) At this point, what is $[I^-]$ in solution?

Chapter 11 — Equilibria Involving Slightly Soluble Electrolytes

Answers

11.1 The solubility product

1. (a) $K_{sp} = [Ba^{2+}][SO_4^{2-}]$
   $BaSO_4(s) = Ba^{2+} + SO_4^{2-}$
   (b) $K_{sp} = [Hg^{2+}][S^{2-}]$
   $HgS(s) = Hg^{2+} + S^{2-}$
   (c) $K_{sp} = [Hg_2^{2+}][S^{2-}]$
   $Hg_2S(s) = Hg_2^{2+} + S^{2-}$
   (d) $K_{sp} = [Ca^{2+}][F^-]^2$
   $CaF_2(s) = Ca^{2+} + 2F^-$
   (e) $K_{sp} = [Ag^+]^2[S^{2-}]$
   $Ag_2S = 2Ag^+ + S^{2-}$

11.2 The relationship between the molar solubility and the solubility product

1. (a) $10^{-5} M$
   (b) $0.00233 g/L$
   (c) $10^{-7} M$
   (d) $2.33 \times 10^{-5} g/L$

2. (a) increase
   (b) increase
   (c) leave unaltered

3. Possibilities are
   (a) $AgNO_3$ or $NaCl$
   (b) $AlCl_3$ or $NaOH$
   (c) $BaCl_2$ or H_2SO_4

11.3 Rules for determining whether a solution is unsaturated, saturated or supersaturated

1. (a) supersaturated
   (b) unsaturated
   (c) unsaturated
   (d) saturated

2. $[SO_4] = 0.005$ (remember dilution!)
   $[Ba^{2+}] = (1/0.005) \times 10^{-10} = 2 \times 10^{-18}$

11.4 Equilibria involving both slightly soluble electrolytes and weak acids or bases

1. (a) $K_{overall} = K_{sp}/K_{a1}(H_2S)K_{a2}(H_2S) = K_{sp}/1.3 \times 10^{-20}$
   (b) $K_{sp} > 1.3 \times 10^{-20}$
   (c) $K_{overall} = K_{sp}/1.3 \times 10^{-20} = [M^{2+}][H_2S]/[H^+]^2$
   $K_{sp} = (1.3 \times 10^{-20})(1)(0.1)/(0.1)^2 = 1.3 \times 10^{-19}$ (ans.)

2. (a) no calculation needed
   (b) calculation needed
   (c) calculation needed
   (d) no calculation needed
   (e) calculation needed
   (f) no calculation needed
   (g) no calculation needed
   (h) calculation needed

3. S^{2-} acts like a base:

$$S^{2-} + H_2O = HS^- + OH^-$$

this lowers $[S^{2-}]$ in solution and increases solubility, by LeChatelier's principle.

Chapter 11 — Equilibria Involving Slightly Soluble Electrolytes

11.5 Analytical separations based on a difference in solubility products

1. last
2. (a) $[Ag^+] = 1.9 \times 10^{-10}/0.1 = 1.9 \times 10^{-9}$
   (b) $[I^-] = 3.2 \times 10^{-17}/1.9 \times 10^{-9} \approx 1.6 \times 10^{-8}$

Solution to "Brain-teaser" in Scope

Let $a = K_{sp}(CaCrO_4) = 7.1 \times 10^{-4}$; let $b = K_{sp}(CaSO_4) = 2.4 \times 10^{-5}$
Let $x =$ solubility of $CaCrO_4$; let $y =$ solubility of $CaSO_4$.

Then $[CrO_4^{2-}] = x$; $[SO_4^{2-}] = y$; $[Ca^{2+}] = x+y$.
Then the K_{sp} expressions give

$$(x + y)x = x^2 + xy = a = K_{sp}(CaCrO_4) \quad (x + y)y = xy + y^2 = b = K_{sp}(CaSO_4)$$

These are two equations in two unknowns. One way to solve them is to take their sum and difference, then factor, giving

$$x^2 + 2xy + y^2 = (x + y)^2 = a+b \quad x^2 - y^2 = (x + y)(x - y) = a-b$$

This gives

$$(x + y) = (a + b)^{1/2} = 2.71 \times 10^{-2} \quad (x - y) = (a - b)/(a + b)^{1/2} = 2.53 \times 10^{-2}$$

Then

$$y = \frac{1}{2}[(x + y) - (x - y)] = 9.00 \times 10^{-4} = \text{solubility of } CaSO_4$$

$$x = \frac{1}{2}[(x + y) + (x - y)] = 2.62 \times 10^{-2} = \text{solubility of } CaCrO_4$$

You can check the results by using x and y to calculate the two K_{sp}'s.

CHAPTER 12: ATOMIC STRUCTURE, ATOMIC SPECTRA, AND THE INTRODUCTION OF THE QUANTUM CONCEPT

Scope

Historical Overview

During the late nineteenth century, many physicists believed that all the laws of the universe had been discovered. There were two kinds of forces, gravity and electromagnetism. Newton's laws (1687) described what happened when forces were applied to matter ("Newtonian mechanics"), and Maxwell's equations (1856) described the mutual interactions of electrical and magnetic phenomena in detail; Newton had already described gravity. The young science of thermodynamics put forth the idea that heat, previously a mystery, was equivalent to work and energy, which were well understood mechanically. It seemed as if all that was needed to predict the course of any physical phenomenon was to write down the appropriate equations and solve them. Admittedly, this was not considered an easy task, but conceptually the universe seemed to be a deterministic — in fact, a determined — system, a big clock winding down, according to known laws.

There were a few flies in the unction — certain phenomena which classical physics could not explain. Their experimental and theoretical resolution, which are discussed in this chapter, led to quantum mechanics, which implies a very different view of the universe than that described in the last paragraph. Simultaneously with these developments, Einstein proposed and elaborated his theories of relativity, which in still different ways altered the way physicists — and perhaps even ordinary people — understand the universe. (Einstein also had a crucial role in the development of the quantum concept.) We understand the universe far better today than we did a hundred years ago, but it is possible that our belief that we understand it all will never again be as comfortably secure and final as it seemed to many at the end of nineteenth century.

Before discussing the specific topics treated in depth in the text, we will summarize the development of the new paradigms that led to our current understanding of what chemical bonding is. Hopefully this will convey an idea of where the theories that we now believe came from. It may be a good idea to come back and re-read this section after you have completed the chapter.

First we summarize several of the "paradoxes" that nineteenth-century physics could not explain, and how they were — tentatively, at the time — resolved.

Early Quantum Theory

Black-body Radiation (not in text)

Max Planck (1900) showed that the colors of glowing bodies at high temperatures ("black bodies") could be explained by assuming that they consisted of oscillators in equilibrium, with energies given by the formula $E=h\nu$, where ν is the frequency and h is a new constant, *Planck constant*.

Photoelectric Effect (Problem 12.8)

Albert Einstein (1905) showed that Planck's equation also explained the emission of electrons from a metal surface being illuminated with light, and introduced the idea that light itself — not, say, something in the glowing bodies described by Planck — is *quantized* according to $E=h\nu$.

Atomic emission spectra (Section 12-2, 12-3, Appendix H)

Niels Bohr (1913) derived an equation which quantitatively described the frequency of the spectral lines emitted by excited hydrogen atoms. To do this he introduced several *ad hoc* hypotheses, one of which was that the frequency of light emitted by an atom is given by Planck's formula, $\nu=\Delta E/h$, when an atom falls from an "excited state" to a "ground state" an energy ΔE beneath it. He also needed to postulate that the energy levels of the atom are quantized.

The basic model upon which Bohr built his theory was the idea of the *nuclear atom*; that is, the idea that the atom consists of a positively charged *nucleus* surrounded by negatively charged *electrons* in motion. Today, most of us learn in elementary school that that is what an atom is like, but when Bohr derived his theory, the idea was new. Here is how it came about.

The Nuclear Atom (Section 12-1)

Chapter 12 — Atomic Structure, Atomic Spectra, etc.

Electrical Nature of Matter (Chapter 16)

Michael Faraday (1830's) found that chemistry could be done by passing electricity through solutions (electrolysis), and that chemical reactions could produce electric current (as in batteries).

Cathode Rays (Section 12-1)

It was known that when a gas is subjected to a high voltage, it conducts electricity. J. J. Thomson (1894) explored the details of this phenomenon with experiments that showed that the *cathode* (the negatively charged electrode) emits small negatively charged particles (*electrons*), which then travel in the direction of the *anode* (the positively charged electrode). If there is no gas in the tube, this still occurs, but if there is a little bit of gas the electrons collide with the gas molecules and knock off some of their electrons, increasing the conductivity. Thomson found that all gases and cathode materials exhibit this behavior — so that all matter must contain electrons — and also that e/m, the *charge-to-mass ratio* is the same for all electrons, regardless of what the gas and cathode materials are.

Charge and Mass of an Electron (Section 12-1)

Robert A. Millikan (1910) determined the charge on the electron in his *oil-drop experiment*. He was able to ionize, then determine the charge on, tiny droplets of oil, and he found that the charge was always an integral multiple of 1.6×10^{-19} coulombs, which must therefore be the charge on an electron. Dividing Thomson's e/m ratio into this gives the mass of an electron.

The Nuclear Atom (Section 12-1)

Ernest Rutherford (1911) did experiments in which he fired *alpha particles* (helium nuclei, atomic mass 4 amu, charge +2) at very fine gold foil. He found that a surprising number passed completely through the foil undeflected, and that some bounced almost directly back. Assuming that the scattering was coulombic (repulsion of two positively charged bodies), Rutherford explained his results by saying that most of the mass and positive charge of an atom are in a very tiny nucleus. Many alpha particles travel straight through the foil because the nuclei are so tiny and far apart that there is only a small chance of hitting one. The alphas that bounce directly back do so because they happen to hit a nucleus almost head-on. Because the gold nucleus is so heavy (atomic mass 197 amu) it moves hardly at all in a head-on collision, and the light alpha particles recoil back. Because the gold foil itself is electrically neutral, electrons must in some way fill up the spaces between the nuclei.

The Bohr model was a great success in its day, since it described *hydrogen-like atoms* (atoms with a nucleus and a single electron, such as H and He^+) to a high degree of accuracy. Many attempts were made to describe multi-electron atoms and chemical bonds (as in H_2) using Bohr's ideas, but these did not succeed very well. In addition, as we shall see, Bohr had to make assumptions about atomic structure that violated the known physical laws of the day; the only justification for the assumptions lay in the fact that they worked — they gave the right answer for hydrogen. The inability of the theory to describe anything else meant that something was missing. Starting about ten years after Bohr developed his model, additional physical insights and experiments supplied the missing idea: *wave-particle duality*. This led, after only three years, to the formulation of the *Schroedinger equation*, which reproduces the Bohr results for hydrogen, but which also supplies an adequate framework for understanding multi-electron atoms and chemical bonding.

Wave Mechanics

Wave-particle Duality (Section 12-5)

Einstein, in his explanation of the photoelectric effect, said that light of frequency ν comes in little "bundles" of energy $E=h\nu$, so that light, previously considered a purely continuous wave, acts in certain experiments as if it were made up of particles. In 1924, Louis de Broglie, then a graduate student, put forth the astounding idea that anything we normally consider a particle can also act like a wave. De Broglie's relationship is that a particle of mass m moving at velocity v acts like a wave with wavelength $\lambda=h/mv=h/p$, where $p=mv$ is the momentum and h is Planck's constant. It turns out that λ for an electron is about the same length as the spacing between atoms in a crystal, so that if electrons can act as waves, crystals should diffract electrons, even as they do x-rays

(Chapter 21). This was found to occur.

The Uncertainty Principle (Section 12-6)

A direct consequence of wave-particle duality is that a particle's wave nature makes it "fuzzy". In optics, a wave of wavelenth λ cannot be used to resolve distances smaller than about λ; for example, if I illuminate an object under a microscope with green light (λ=500nm), I will not be able to distinguish details closer together than about 500nm. The "fuzziness" of a particle — the imprecision or *uncertainty* in its position — is also about as big as a de Broglie wavelength: $\Delta x \approx \lambda$. From de Broglie's relationship, $\lambda = h/p$, one can then write almost immediately $\Delta x \Delta p \approx h$, which is the *uncertainty principle* first stated by Werner Heisenberg (~1925). Δp is the uncertainty in the momentum. The more accurately one knows the position of a particle, the more uncertain one is of its momentum. (A more precise statement of the principle is $\Delta x \Delta P > h/4\pi$.)

The Schroedinger Equation (Section 12-6; see also **Perspective**)

The equations describing wave motion were well known in physics. Ernest Schroedinger (1927) took such a classical wave equation, and wherever the wavelength, λ, appeared, he substituted h/p from the de Broglie relationship. The resulting *Schroedinger Equation* describes the wave-like properties of particles. Since kinetic energy is given by $KE = mv^2/2 = p^2/2m$, the energy appears in the wave-equation. The Schroedinger equation can be solved — at least approximately — for a wide variety of systems of chemical interest, such as atoms and molecules. The solutions of the equation for the simple cases of the hydrogen atom, H, and the hydrogen molecule ion, H_2^+, give rise respectively to the ideas of *atomic orbitals* and *molecular orbitals*.

Atomic Spectra

When atoms of an element are heated, they emit light. As shown in Table 21.1 in the text, different elements exhibit different characteristic colors, and these form the basis of the "flame tests" for the elements. When this emitted light is passed through a spectroscope, it is found (Figure 12.9) that the spectra are *discrete*; only certain *lines*, corresponding to particular frequencies of light, are observed.

Each element gives its own characteristic line spectrum. The positions of the lines are described by either the frequency, ν, the wave-number, $\bar{\nu}$, or the wavelength, λ, which are related according to the equations

$$\nu = \frac{c}{\lambda} \quad \bar{\nu} = \frac{1}{\lambda}$$

where c is the speed of light.

As the text emphasizes, nineteenth-century physics was unable to explain why atomic spectra had the discrete nature that they do. The lines in the visible spectrum of the hydrogen atom, however, were found to form in a regular series, the *Balmer series*, given by Equation 12-9 in the text

$$\bar{\nu} = (109678 \text{cm}^{-1}) \left[\frac{1}{2^2} - \frac{1}{n^2}\right]$$

When n>2, the term in the brackets is positive; different values of $\bar{\nu}$ result from putting different values of n into the equation. It later turned out that other series of spectral lines also existed, given by

$$\bar{\nu} = \mathbf{R} \left[\frac{1}{n_L^2} - \frac{1}{n_H^2}\right]$$

where $\mathbf{R} = 109678 \text{cm}^{-1}$ is called the *Rydberg constant*, and where $n_H > n_L$. Thus, the Balmer series is given by $n_L = 2$, $n_H = 3, 4, 5...$, the Lyman series is given by $n_L = 1$, $n_H = 2, 3, 4...$, and there are other series for other values of n_L. The Balmer series was discovered first because it lies in the visible spectral region. For the Lyman series the spectrum lies in the ultraviolet, and for $n_L=3$ the spectrum lies in the infrared. Note that as n_L gets bigger, the first line in the series goes to lower and lower wave numbers, since the term in brackets gets smaller and smaller ([1/1-1/4]= .75 for Lyman, [1/4-1/9]= .14 for Balmer, etc.).

Niels Bohr came up with a mathematical model that explained all these series. The text shows how he did this in Appendix H, which we will now consider. As mentioned in the **Historical Overview**, certain of Bohr's postulates violated known laws of physics, but the fact that his model worked so well meant that there was some truth to it. The first basic idea that Bohr used was the idea of the nuclear atom; a hydrogen atom consists of a nucleus and an electron. He assumed that the electron could be in any of a number of fixed orbits, and experiences a centripetal force due to its Coulombic attraction to the nucleus, and a centrifugal force due to its motion in a circular orbit. These balance (Equation H-1).

Chapter 12 — Atomic Structure, Atomic Spectra, etc.

$$\frac{Ze^2}{r^2} = \frac{mv^2}{r}$$

Z is the nuclear charge and is equal to +1 for H, +2 for He$^+$, etc. Note that the idea of a fixed, stable circular orbit contradicts the expected behavior described at the end of Section 12.1 in the text; the electron "should" be drawn into the nucleus, emitting radiation, and the atom "should" cease to exist. Bohr simply postulated that, for unknown reasons, this does not occur.

The way Bohr arranged for discrete circular orbits was to postulate that the angular momentum be quantized (Equation H-2)

$$mvr = \frac{nh}{2\pi}$$

If n is an integer, then the angular momentum, mvr, can take on only certain values. Again, this postulate is purely *ad hoc*: integers are used because they work! But in any case, the second basic idea that Bohr used was this idea of quantization, discrete energy states. The n's are called quantum numbers. He also postulated that the energy difference between two states is given by Planck's equation, $\Delta E = h\nu$, and from these three equations Appendix H shows that

$$R = \frac{2\pi^2 me^4}{h^3 c} = 109737 \text{cm}^{-1}$$

in very close agreement with the experimentally observed value. The Bohr theory also allows the calculation of the radius of a Bohr orbit

$$r = n^2 \frac{h^2}{4\pi^2} me^2$$

By setting n=1, a_o, the radius of the first Bohr orbit can be found: a_o = 0.0529nm. This value turns out to be significant in the later Schroedinger treatment, but there it no longer refers to the radius of a fixed circular orbit.

Besides the spectrum, the Bohr model predicts the *ionization energy* of a hydrogen atom. This is the energy required to strip the electron completely off the atom, starting from the ground state (n=1). Stripping the electron off corresponds to an orbit of an infinite radius, which, according to the last equation corresponds to n=∞. Setting $n_L = 1$ and $n_H = \infty$, $\bar{\nu} = R$ for this stripping process; then

Chapter 12 — Atomic Structure, Atomic Spectra, etc.

$\Delta E = h\nu = hc\bar{\nu} = hc R = 1312$ kJ/mol corresponds to ionization, as the text shows (Equation 12-22). This matches the experimental value very well.

Wave Mechanics

Most of the material in Sections 12-5 and 12-6 has been discussed in the **Historical Overview**. The text gives examples (12.5, 12.6) of the calculation of the de Broglie wavelength for electrons and for macroscopic objects. It is both simple and instructive to see where de Broglie's relationship comes from. Consider a *photon* (quantum of light). It is described both by the Einstein equation and by the Planck equation.

$$E = mc^2 = h\nu = \frac{hc}{\lambda}$$

From this it follows inmediately that

$$\lambda = \frac{h}{mc} = \frac{h}{p}$$

Since c is a veleocity, mc=p is a momentum. Now a photon does not have a rest mass, but it does have a momentum; many of you have seen the delicately suspended pinwheel in an evacuated glass bulb that rotates when light hits the blackened blades. This rotation is due to the impact of the photons. De Broglie simply reasoned by analogy, and said that this relationship, $\lambda = h/p$, holds for anything, not just photons. Clearly, his contribution lay in his deep physical insight. The above derivation was the substance of de Broglie's doctoral thesis, which was only two pages long!

Although the formulations of quantum mechanics are theoretically correct even in the macroscopic world, quantum effects — such as diffraction — become insignificant for large objects. For large objects, quantum mechanics boils down to classical mechanics. (This is called the *correspondence principle*.)

As mentioned in the text and in the **Historical Overview**, and discussed in depth in the Perspective, the Bohr picture of the atom was eventually superseded by a model whose starting point was the de Broglie relationship. When the classical wave equation — which describes a standing wave, for example — is modified by substituting h/p for λ, the result is the *Schroedinger equation*. In the description of a classical wave, the square of the *amplitude* of the wave is the *intensity* of the wave. In quantum

Chapter 12 — Atomic Structure, Atomic Spectra, etc.

mechanics, the analog of intensity is *probability*, as the text describes. As you will see in the next chapter, the probability of finding an electron in certain regions of space is illustrated by drawing "electron clouds" which are densest where the electron spends the most time; again, this density is the direct analog of an intensity. It must be emphasized that this sort of behavior comes about when the Schroedinger equation is solved for a coulombically interacting system of nuclei and electrons; no new physical forces are involved. It is just that the wave nature of matter causes microscopic particles that interact coulombically to do strange things.

The chapter ends with a quote from P. A. M. Dirac, to the effect that if we could only solve the appropriate equations, we would be able to predict all of chemistry. This may sound similar to the earlier description of the state of physics in the late nineteenth century. Then, it seemed to be only the solution of certain equations — Newton's laws of motion — that lay between us and a complete understanding of the universe. There are several important differences, however, between the physicist's view of the universe then and now. The uncertainty principle says that we can never know both the position and momentum of a particle precisely. The positions and momenta of all the particles in a system, however, are needed for a deterministic prediction of how the system will evolve. Quantum mechanics predicts the diffraction pattern formed by a large number of particles passing through a slit, but cannot predict where in the pattern the next particle will fall. The most commonly accepted interpretation of the uncertainty principle (though there are physicists, including famous ones, who disagree) is that there is an inherently random, probabilistic element to the way the universe works, and that strict causality is no longer viable as a fundamental physical principle.

In addition, two entirely new forces, aside from electromagnetic interactions and gravity, have been discovered in this century. These are the "weak" and "strong" forces that operate within the nucleus, and between subatomic particles. Putting these, together with an electromagnetism and gravity, into a common framework is the major goal of theoretical physics today. And it is always possible, as physicists build bigger accelerators and probe more energetic interactions, that still more forces will be discovered.

Chapter 12   Atomic Structure, Atomic Spectra, etc.

Questions

12.1 The experimental basis for modern concepts of the atom

1. Cathode rays are composed of _____.

2. In a cathode-ray tube, the charge on the cathode is _____.

3. If the ends of the tube are coated with fluorescent material and an object placed in the beam, the shadow of the object appears at the (cathode, anode) end of the tube.

4. J. J. Thomson determined the e/m-ratio of the electron. e stands for _____, and m stands for _____.

5. Millikan, in his oil-drop experiment, determined the _____ of an electron.

6. If Millikan found droplets with charge 3.2×10^{-19}C, 4.8×10^{-19}C and 6.4×10^{-19}C, what is the largest value of e consistent with these results?

7. In Rutherford's scattering experiment, so many α-particles passed through the gold foil undeflected because

8. In Rutherford's scattering experiment, how was it determined that the nucleus must be massive?

9. From the point of view of classical physics, why was the picture of an atom as a planetary system unsatisfactory?

Chapter 12   Atomic Structure, Atomic Spectra, etc.

12.2 Atomic spectra

1. (a) Relate λ and ν for a light wave in a simple equation

   (b) Do the same for λ and $\bar{\nu}$

   (c) Do the same for $\bar{\nu}$ and ν.

2. What is meant by the statement that atomic spectra are "discreet?"

12.3 The emission spectrum of atomic hydrogen

1. Write an expression giving $\bar{\nu}$ for the lines in the Balmer series.

2. For this series, successive lines get (closer together, further apart).

3. What is the biggest possible value of $\bar{\nu}$ for a line in the Balmer series?

4. The *Rydberg Equation* gives the wave-number for all the lines in the hydrogen spectrum. Write this equation.

5. For $\bar{\nu}$ corresponding to ionization of atomic hydrogen from the ground state, n_L in the Rydberg equation is equal to _____, and n_H is equal to _____.

Chapter 12   Atomic Structure, Atomic Spectra, etc.

12.4 The Bohr theory of the hydrogen atom and the introduction of the quantum concept

1. In the Bohr model, the stable states of the hydrogen atom have energies proportional to what function of quantum number n?

2. The state of the hydrogen atom when the nucleus and electron are completely separated corresponds to

    (a) what value of the energy?

    (b) what value of n?

    (c) what radius of the orbit?

3. As n increases, successive Bohr states get (closer together, further apart, keep the same spacing)

4. Lines in the hydrogen spectrum correspond to *the energies of Bohr states.*

5. (a) The zero of energy in the Bohr atom corresponds to

    (b) With respect to this zero, the other states of H have (positive, negative) energies.

12.5 The dual nature of matter

1. A quantum of light is called a _____.

2. The equation relating the energy of a photon to its frequency is

3. In de Broglie's relationship,

    (a) the greater the momentum of a particle, the _____ its wavelength.

    (b) the greater the mass of a particle, the _____ its wavelength.

    (c) the greater the velocity of a particle, the _____ its wavelength.

(d) the greater the kinetic energy of a particle, the _____ its wavelength.

4. Name a wave property exhibited by electrons.

12.6 From Bohr's quantum theory to wave mechanics

1. In the Bohr model, if the charge on the nucleus increases, the spacing between the spectral lines gets (larger, smaller), and the lines are shifted to (greater, smaller) wave-numbers.

2. Give two phenomena that the Bohr model was unsuccessful in describing.

3. What do atoms and standing waves have in common?

4. A solution to the Schroedinger equation is called a _____.

5. Write the algebraic expression for the uncertainty principle.

6. Suppose a particle is confined to the x-axis, and I know that it is in between x=0 and x=2. (This is a "one dimensional box.")

    (a) What is Δx?

    (b) What is the minimum value of Δp?

(c) If I compress the box so that it now goes from x=0 to x=1, what is the new value of Δx?

(d) Express the new value of Δp in terms of the old one.

7. What function of Ψ gives the probability of finding an electron at a given point in space?

8. The phrase, "probability of finding an electron at a given point in space" refers to the particle nature of the electron. If instead we talk of the electron as a wave, what property of the wave at the corresponding point of space corresponds to this probability?

Chapter 12

Atomic Structure, Atomic Spectra, etc.

Perspective: The Schroedinger Equation

The Schroedinger equation (SE) is sometimes presented as having sprung full-grown from the brow of Schroedinger. Actually, its development is straightforward, given what came before, and how it arose is easier to understand than, say, how Bohr arrived at his postulates.

Classical Wave Equation:

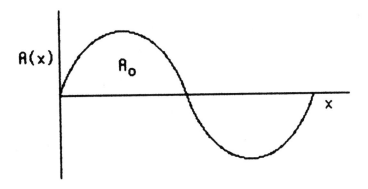

Consider a standing wave in some medium. A standing wave is one in which the nodes (positions of zero amplitude) do not move in time, and will prove to be a model for stable systems — such as atoms — which do not change in time; they are stable wave patterns. For such a wave, the amplitude is given by:

$$A(x) = A_o \sin \beta x$$

$$\beta = \frac{2\pi}{\lambda}$$

In this equation, A_o represents the maximum amplitude. Note that β is chosen so that when $x=\lambda$ the argument of the sine is 2π, or a full period. Using the fact that the derivative of $A_o \sin(\beta x)$ is $-A_o \beta \cos(\beta x)$ and that the derivative of that (the *second derivative* of the original function) is $-A_o \beta^2 \sin(\beta x)$, or $-\beta^2$ times the original function, we may write:

$$\frac{d^2 A(x)}{dx} = -\beta^2 A(x)$$

which is a classical wave equation. A *solution* to such a *differential equation* (so called because it contains derivatives) is any function which satisfies the equation; for example, the solution to this

12-15

Chapter 12 — Atomic Structure, Atomic Spectra, etc.

equation is a function whose second derivative is equal to minus a positive constant times the function.

We have already seen that the function $A_0 \sin(\beta x)$ is a solution of this equation; $B_0 \cos(\beta x)$ is another one, as is their sum. We will continue to use $A_0 \sin(\beta x)$ for illustrative purposes. It will be useful later to consider the geometric meaning of this equation. For any function, $y(x)$, recall that the derivative, dy/dx, stands for the *slope* of the function — that is, the slope of the tangent to the curve. Of course, dy/dx is a function of x, unless $y(x)$ is a straight line. In other words, the slope is, in general, changing with x. d^2y/dx^2 is the rate of change of the slope of the original function, just as dy/dx is the rate of change of the original function. d^2y/dx^2 tells how fast the slope is changing, and is called the *curvature* of the function. The figure below illustrates this concept for a few arbitrary curves. A solution to the classical wave equation, then, in words, is some function whose curvature is everywhere equal to minus a positive constant (β^2) times the value of the function itself. (You might like to sketch a sine function, and then its first and second derivatives, in order to convince yourself that it does indeed fulfill this condition.)

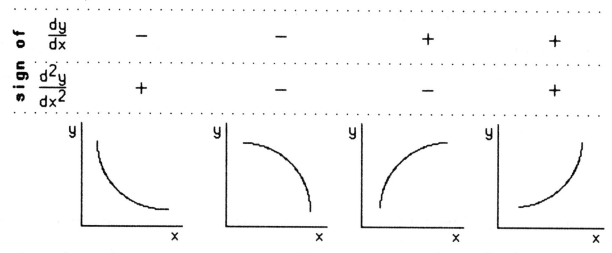

Schroedinger Equation:

Although the classical wave equation is mathematically satisfied by any function of the form $A_0 \sin(\beta x)$ (or $B_0 \cos(\beta x)$...), and is not dependent on the meaning of β, the discussion at the start of the previous section makes clear the fact that in the physical interpretation of wave motion, β is $2\pi/\lambda$. What Schroedinger did was to simply replace the wavelength, λ, with the de Broglie equivalent, h/p, giving $\beta = 2\pi p/h$. Since KE, the kinetic energy, is equal to $p^2/2m$, the Schroedinger Equation is:

Chapter 12 Atomic Structure, Atomic Spectra, etc.

$$\frac{d^2\Psi(x)}{dx^2} = -\beta^2\Psi(x) = -(2\pi\frac{p}{h})^2\Psi(x) = -(\frac{2\pi}{h})^2 2mKE\Psi(x) = -(\frac{2\pi}{h})^2 2m(E-V)\Psi(x)$$

$$\beta = \frac{2\pi}{h}\sqrt{2mKE}$$

where the last relationship in the first of the obove two equations comes from the fact that $E = KE + V$, where E is the total energy and V is the potential energy. We also used the relationship $KE = mv^2/2 = p^2/2m$. I have also replaced $A(x)$ in the classical equation by $\Psi(x)$, because that is the usual symbol for a quantum mechanical *wave function*. It is worthwhile to consider briefly what we have done. We have replaced a wavelength, λ, in the classical wave equation with its de Broglie particle equivalent, h/p. Since momentum is now in the equation, we can talk about kinetic energy. This, in turn, allows us to bring in potential energy. None of these concepts has any meaning for a classical wave. Similarly, the wave description that we started with has no meaning for a classical particle. Yet the Schroedinger Equation describes both aspects of a system. In this context, the wave function constitutes a complete description of the system.

Ironically, the value of $\Psi(x)$ itself has no straightforward physical interpretation, in the sense that $A(x)$ is the amplitude of a wave in the classical wave equation. $\Psi^2(x)$, however, may be interpreted as the probability of finding the particle at x. (More precisely, $\Psi^2(x)$ is the probability of finding the particle between x and x+dx.) This interpretation derives from the fact that for a classical wave $A^2(x)$ is the intensity of the radiation. In a quantum mechanical system the particle-like nature of the wave dicatates that its intensity must come in little packets; for a given quantum, either it is in the vicinity of x or it is not — there is no half-way. However, the result of observing a large number of quanta is a histogram or distribution of "hits" near different values of x, which is like an intensity, for all the measurements, taken together. This distribution is given by $\Psi^2(x)$. For a *single* measurement, however, $\Psi^2(x)$ gives only the probability that the particle will be found at x.

Another rewriting of the Schroedinger Equation gives:

$$\frac{-h^2}{8\pi^2 m}\frac{d^2\Psi}{dx^2} + V\Psi = E\Psi$$

Note that if I divide this form by Ψ, I get the statement that some funny term involving a second derivative, plus the potential energy, equals the total energy. This first term, involving the curvature of the wave function, must then be the kinetic energy, and the kinetic energy must get bigger as the curvature gets more negative. Note that the higher the frequency of a light wave, the greater its curvature, given the same amplitude. (If you do not believe this, draw and look at two sine waves with different frequencies.) This greater kinetic energy is in accord with the Planck equation, E=hν.

Schroedinger Equation for a Free Particle

A free particle is one which experiences no potential; that is, V = 0.

$$\frac{d^2\Psi}{dx^2} = -\beta^2\Psi = -\frac{8\pi^2 mE\Psi}{h^2}$$

Clearly, since any value of β will satisfy the equation, a solution can be found for any value of KE. In fact, since in this case KE = E, the particle can accomplish this by having the frequency given by E=hν. Thus, a free particle is not limited to discreet energy levels.

Schroedinger Equation for a Particle in a Box

Energy levels (*quantization*) arise from creating a *bound* (as opposed to a free) particle. This is done by setting up a potential which holds the particle in place. Consider an electron and proton interacting, for which the potential is given by $-e^2/r$, the Coulombic potential. Plugging this into the Schroedinger Equation (for V), and solving the equation, gives the wave functions for the hydrogen atom. These are known as *orbitals*. (Actually, our Schroedinger Equation is for only one dimension, x; for the H atom, it is necessary to use the three-dimensional Schroedinger Equation.)

Since the Schroedinger Equation for the H atom is hard to solve, we will use a simpler example,

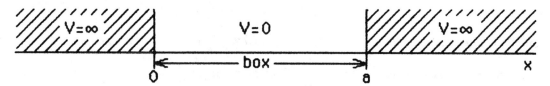

namely a particle confined to a one-dimensional "box;" that is, the portion of the x-axis between 0 and some point a. Inside the box, let us define V to be zero. In order to make sure that the particle can not get out of the box, V must be infinite for x<0 and for x>a. (To see this, note that V(x) is the potential energy experienced by a particle at position x. The greater the value of V, the less likely it is that the

Chapter 12   Atomic Structure, Atomic Spectra, etc.

particle will be at x, and if V is infinite, then the particle cannot be there at all.)

Wave Functions

Since the particle has zero probability of being outside the box, we know that $\Psi^2(x)=0$ for $x<0$ and $x>a$; this implies that $\Psi(x)$ is also zero there. We do not yet know what Ψ looks like inside the box, but we do know that it starts from zero at $x=0$. This is because it is zero at the "walls" of the box (the points 0 and a), and we assume that Ψ is continuous. We also know that Ψ must satisfy the Schroedinger Equation, with $V=0$, inside the box. We saw earlier that the functions $A_o\sin(\beta x)$ and B sub o $\cos(\beta x)$ will both satisfy the Schroedinger Equation with $V=0$, which is identical to the free-particle situation; however, of these, only A sub o $\sin(\beta x)$ is equal to zero at $x=0$, so it must be the correct solution.

We must also have, however, that $\Psi=0$ at $x=a$. The sine function evaluates to zero at $0, \pi, 2\pi$, etc., so βa must be equal to $n\pi$. In other words, $\beta=n\pi/a$. For any n, the function $A\sin(n\pi x/a)$ satisfies the Schroedinger Equation for the particle within the box. These n's are called *quantum numbers*, and note that they result from the mathematical process of solving the Schroedinger Equation for a bound particle; they do not result from *ad hoc* hypotheses, as in the Bohr atom.

Energies

Now that we know that β is equal to $n\pi/a$, we can use this relationship, squared, together with the fact that $\beta^2 = +(8\pi^2 mKE/h^2)$, to get:

$$E = \frac{n^2 h^2}{8ma^2}$$

This equation gives E as a function of quantum number, n. Note that for the particle in the box, only certain energies are allowed -- the ones given by this formula.

The Factor, A

We have not said anything about what A is equal to, in the expression $\Psi = A\sin(\beta x)$. Since $\Psi^2(x)$ is a probability, integrating $\Psi^2(x)$ from $-\infty$ to $+\infty$ must give unity; that is, the particle has 100%

Chapter 12 — Atomic Structure, Atomic Spectra, etc.

probability of being *somewhere*. This procedure turns out to yield $A=\sqrt{2/a}$ regardless of n. The wave functions and their squares have the general appearance shown here:

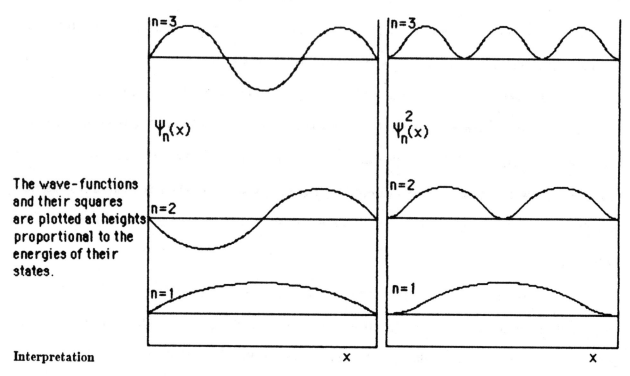

The wave-functions and their squares are plotted at heights proportional to the energies of their states.

Interpretation

Recall that we said that the curvature of $\Psi(x)$ is related to KE, and that the particle has only KE inside the box (V=0). The equation for E is in accord with this, since bigger n and smaller a, which both increase E, both give more curvature to the wave function. This should be clear from the way the shape of $\Psi(x)$ changes if n increases or if a decreases. Note that the number of nodes (points of zero amplitude) of the wave function (excluding the nodes at the boundaries) is given by n-1. This sort of behavior is typical to solutions to the Schroedinger Equation, and will be seen in the H-atom orbitals. For n=1, the particle is much more likely to be found in the middle of the box than toward the walls. This is contrary to classical behavior. In the classical system, the particle has an equal probability of being anywhere in the box. (Think of a marble in a shoebox.) For n=2, the particle has a high probability of being at a/4 and at 3a/4, and zero probability of being at a/2. Note, however, that the *average* position of the particle is at a/2; also, it is tempting to consider the question, "If the particle is found to be on the left, and on a subsequent measurement is found to be on the right, and it can never be in the middle, how did it get from the left to the right?" Of course, it is important to realize that the particle can never be located precisely, due to the uncertainty principle, and so the very idea that, in moving from left to right, the particle had to at one time be *exactly at* a/2 is not meaningful. These questions are mind-boggling, and fun.

Note that there is a finite energy even in the lowest state, n=1. (We can not have n=0, because then Ψ evaluates to zero everywhere, so that the particle does not exist!) This is the rule in quantum mechanics, and this energy is known as the *zero-point energy*. Quantum systems have zero-point energy even at absolute zero. This is required by the uncertainty principle. If there were no kinetic energy, there would be no momentum; that is, we would know that p=0, exactly. That would mean (because $\Delta x \Delta p > h/4\pi$) that the uncertainty in x would have to be infinite, but in our case it is not, since we know the particle is in the box, and not outside, so that x≈a. Therefore there must be a finite uncertainty in the momentum, and there must be at least enough kinetic energy to accommodate this amount of momentum. This is the zero-point energy.

Comments

Do not be concerned if some of this is very hard to grasp; people still argue about what it "really means." You are not responsible for knowing how to solve the Schroedinger Equation for any system, but you should know what it means for there to be a solution. You should understand the comments under **Interpretation**, above, and you should know about the meaning of Ψ(x) and Ψ²(x), and of the curvature of Ψ(x).

Just a few more things. In general, if you have j dimensions, a solution to the Schroedinger Equation will have j quantum numbers. A particle in a 3-dimensional box has three quantum numbers. Three quantum numbers also come out of solving the Schroedinger Equation for the H atom; they are n, l and m_l. The spin quantum number, m_s, does *not* come from solving the Schroedinger Equation, but comes from a relativistic treatment of the same problem (the Dirac equation).

One of the most important things to remember is that quantization (energy levels) appears naturally as a consequence of the mathematical solution to a Schroedinger Equation for a *bound* particle. This gives us some confidence that the Schroedinger Equation reflects physical reality, since quantization is experimentally observed (as in spectra). This way of arriving at quantization is far more convincing than by simply postulating that it must exist, as Bohr did. (This is not to deny Bohr's genius; in fact, it is all the more impressive that he saw the need for quantization before it could be justified from known physical laws.) The Schroedinger Equation reproduces all of Bohr's results for hydrogen, but also describes many-electron atoms, and provides us with a basis for understanding chemical bonding.

Chapter 12 — Atomic Structure, Atomic Spectra, etc.

Answers

12.1 The experimental basis for modern concepts of the atom

1. electrons
2. negative
3. anode
4. charge, mass
5. charge
6. 1.6×10^{-19} C (greatest common divisor)
7. most of the atom consists of empty space
8. A number of α-particles were scattered back toward their source
9. The electron circles the nucleus, and circular motion involves a continuous acceleration. A charged particle accelerating in an electric field should lose energy and emit radiation; thus the atom should not be stable.

12.2 Atomic emission spectra

1. (a) $\lambda = c/\nu$
   (b) $\lambda = 1/\bar{\nu}$
   (c) $\bar{\nu} = c\nu$
2. Only certain values of $\bar{\nu}$ appear.

12.3 The emission spectrum of atomic hydrogen

1. $\bar{\nu} = R[1/4 - 1/n^2]$ $n > 2$
2. Closer together. To satisfy yourself that this is true, calculate the expression in brackets in the answer to (1) for n=3, n=4 and n=5. You will see that the values for n=4 and n=5 are closer together than the values for n=3 and n=4.
3. $\bar{\nu} = R(1/4)$, corresponding to $n = \infty$
4. $\bar{\nu} = R[1/n_L^2 - 1/n_H^2]$
5. 1, ∞.

12.4 The Bohr theory of the hydrogen atom and the introduction of the quantum concept

1. $1/n^2$ ($E = -K/n^2$)
2. (a) 0
   (b) ∞
   (c) ∞
3. closer together
4. differences between the energies of Bohr states
5. (a) separated nucleus and electron
   (b) negative

12.5 The dual nature of matter

1. photon
2. $E = h\nu$
3. (a) shorter
   (b) shorter
   (c) shorter ($KE = p^2/2m$, so bigger KE means bigger p)

Chapter 12 — Atomic Structure, Atomic Spectra, etc.

12.6 From Bohr's quantum theory to wave mechanics

1. larger, greater
2. Possibilities are spectra of multi-electron atoms, chemical bonding, intensity of spectral lines for hydrogen.
3. They both are stationary (stable) states — they do not change in time.
4. wave function
5. $\Delta x \Delta p > h/4\pi$
6. (a) $\Delta x = 2$
   (b) $\Delta p = h/8\pi$
   (c) $\Delta x = 1$
   (d) $\Delta p_{new} = 2\Delta p_{old}$
7. $|\Psi|^2$
8. the intensity

CHAPTER 13: THE ELECTRONIC STRUCTURE OF ATOMS, THE PERIODIC TABLE, AND PERIODIC PROPERTIES

Scope

Wave Functions and Quantum Numbers

The Schroedinger equation can be solved exactly only for certain very simple systems, such as the hydrogen atom. Approximate solutions can be obtained for more complicated systems, however, such as many-electron atoms. The solutions for hydrogen are called the *hydrogen atomic orbitals*, and it turns out that the solutions for many-electron atoms can be understood as distorted versions of these. An understanding of the nature of the hydrogen atomic orbitals, plus some common sense and rules of thumb, can be used to infer many of the properties of the elements; for example, how chemical behavior is related to position in the periodic chart. This chapter develops this understanding, starting with a description of the hydrogen atomic orbitals. Chapter 14 then shows how these orbitals and other solutions for the Schroedinger equation lead to an understanding of chemical bonding.

Recall that Bohr had to introduce quantization of atomic energy levels as an *ad hoc* assumption; there was no underlying physical rationale. On the other hand, quantum numbers arise naturally in solutions of the Schroedinger equation. What makes them appear is what led to the Schroedinger equation in the first place; namely, the wave-particle duality introduced by de Broglie. De Broglie's relationship was not an *ad hoc* assumption; rather de Broglie had discovered a new law of nature, which was verified experimentally by the demonstration of electron diffraction. The remarkable success of the Schroedinger equation in rationalizing most of chemistry constitutes still further verification of de Broglie's principle.

Why should wave-particle duality lead naturally to quantization? Recall from Section 12.6 that a stable system can be compared to a standing wave, such as a plucked violin string. A plucked string gives not only the fundamental tone, but also *harmonics*, related to the fundamental by small whole-number ratios. For example, the fundamental of a plucked violin string (the predominant tone that is heard) has a wavelength twice the length of the string, the octave has a wavelength equal to the length of the string, and the perfect fifth has a wavelength 4/3 the length of the string. Small whole numbers arise naturally in descriptions of waves, and thus they also come about naturally in the wave description of matter, as embodied in the Schroedinger equation. These whole numbers are the quantum numbers.

Standing waves exhibit *nodes*, places where the amplitude goes to zero. A plucked string sounding the fundamental has a node at each end, where the string is clamped. The octave, with half the wavelength, has in addition a node in the middle. We will see shortly that the solutions to the Schroedinger equation also have nodes. As discussed in the last chapter, the intensity of a wave, defined as the square of its amplitude, has the quantum-mechanical interpretation of the probability of finding the system in a given configuration. Where a wave-function has a node, the amplitude, and therefore the intensity, is zero, and so in these places the particle is not found. If you read the **Perspective** in the last chapter you saw how the nodes come about for the quantum-mechanical particle in a box.

The Hydrogen Atomic Orbitals

Because the Schroedinger equation describes the particle-like as well as the wave-like properties of matter, it is possible to set up the equation for two particles which interact by means of some potential function, such as a proton and an electron interacting coulombically, which constitute a hydrogen atom. When this is done, a wave function, Ψ, can be obtained, whose absolute value squared, $|\Psi|^2$, gives the probability of finding the electron in any given position with respect to the proton. A whole series of Ψ's is in fact obtained, which correspond to the *ground state* of H (the state of lowest energy) and a series of *excited states*.

The Schroedinger equation for these interacting particles thus permits a number of related solutions, just as a vibrating string can exhibit a number of standing waves (the fundamental and the harmonics). Solving the Schroedinger equation in three dimensions gives three quantum numbers, sets of which characterize the respective solutions. These are n, *the principle quantum number*, l, *the azimuthal quantum number*, and m_l, *the magnetic quantum number*. A fourth quantum number, m_s, the *spin quantum number*, comes from a relativistic treatment of the same problem. Each wave function has a characteristic energy asssociated with it, and just as in the Bohr model. If the hydrogen atom makes a transition from a higher to a lower state, a photon of wavelength $\nu=\Delta E/h$ is emitted, where ΔE is the energy difference between the two states. The Schroedinger equation can also be used to calculate the energy associated with each state, and when this is done, it turns out that for the hydrogen atom only the quantum number n appears in the expression for the energy; in fact, the expression obtained for the energy as a function of n is identical to Bohr's: $E=-K/n^2$ (text, Equation 13-1). Bigger values of n imply higher (more positive) energies above the ground-state, and a greater average distance of the electron from the nucleus. n can take on values 1, 2, 3, etc. Thus when n=1 we have E=-K, the ground state, and when n=∞ we have E=0, the ionized atom.

The principle quantum number, n corresponds to orbital size. The bigger the value of n, the farther the electron is from the nucleus on average. l corresponds to orbital shape. The bigger the value of l, the more complex the shape. m_l corresponds to the orientation of the orbital in a magnetic field.

The azimuthal quantum number, l, is related to the angular momentum of the electron. l can range from 0 through (n-1), in steps of 1; thus if n=1, l=0 only; if n=2, l can be 0 or 1, and so on. For hydrogen, solutions with the same n but different l, such as (n=2, l=0) and (n=2, l=1), have the same energy, but for *many-electron atoms* they have different energies, with higher l always being associated with higher energy. The state (n=2, l=0) is called the 2s state, and (n=2, l=1) is called the 2p state. The 2 stands for n=2, and the s stands for l=0. l=1 is symbolized by the letter p, l=2 by d, and l=3 by f. Thus the state (n=1, l=0), the only set of (n,l) allowed for n=1, is symbolically written as 1s.

Each allowed pair of (n,l) is called a *subshell*. Each subshell includes a group of Ψ's, namely all the Ψ's with the given value of n and of l, and with the whole range of possible m_l's allowed for these (see below). In the same manner, all the Ψ's with a given value of n are considered to be in the same *shell*; thus the 1s subshell is in the shell characterized by n=1, called the *K-shell*, and the 2s and 2p subshells are both in the *L-shell*, characterized by n=2. Table 13.1 summarizes this nomenclature.

The *magnetic quantum number*, m_l, can range from -l to l, in steps of 1, so that for l=0 m_l must be 0, and for l=1 m_l can be -1, 0 or 1. For a given value of l, m_l has 2l+1 possible values, so that each subshell has 2l+1 "members". Each of these "members" is characterized by its own wave-function, Ψ, and corresponds to a possible state of the H atom. These states are termed *orbitals*. Thus, a subshell consists of one or more orbitals, and a shell consists of one or more subshells. Furthermore, it turns out that in the absence of an externally imposed magnetic field, all the orbitals in a given subshell have the same energy, even for many-electron atoms; they are *degenerate*. Since each subshell has 2l+1 orbitals, each subshell is 2l+1-fold degenerate. Thus the 2p subshell (recall that p means l=1) is 3-fold degenerate, as are the 3p, 4p, etc-subshells. For the hydrogen atom, in addition, the shells are degenerate, so that for it the L-shell, which contains the 2s orbital and the three 2p orbitals, is 4-fold degenerate.

Shapes of Orbitals

Recall that the Bohr theory allowed the calculation of the precise radius of an orbit. The Ψ's that come out of the Schroedinger equation, in contrast, give (through the function $|\Psi|^2$) not precise locations of electrons, but rather probability densities for them (intensities, in wave-language). Many of the figures in Chapter 13 are representations of these density patterns. Let us start by discussing the 1s orbital. Recall that 1s means n=1 and l=0; m_l can take on only one value, namely m_l=0. The text states that all s orbitals (l=0) are spherical. What this means is that, starting at the nucleus and going out in any direction, the orbital "looks" the same. The electron density changes as one gets further away from the nucleus (obviously, it must get very low very far away), but for any two points the same distance away, the density does not depend on direction; it is constant. Thus any sphere with the nucleus at its center will exhibit the same electron density (same value of $|\Psi|^2$) anywhere on it. (Again, this for an s orbital only.) Thus, $|\Psi|^2$ is a function only of r for an s orbital, and Ψ is also a function only of r. These functions are given by Equation 13-3

$$\Psi_{1s} = k e^{-Zr/a_o}$$

$$|\Psi_{1s}|^2 = k^2 e^{-2Zr/a_o}$$

Here, a_o is a constant, which turns out to be the same value (exactly!) as the radius of the first Bohr orbit. Z is the charge on the nucleus (Z=1 for H, Z=2 for He^+, etc), and k is another constant, whose value need not concern us at this moment.

Note that the equation for $|\Psi|^2$ says that the electron density decreases exponentially with distance from the nucleus. At $r=\infty$, $|\Psi_{1s}|^2=0$, so that the electron density goes to zero at great distances, and at $r=0$, $|\Psi_{1s}|^2 = k^2$, its maximum value, so that the electron has a high probability of being "at" the nucleus. Thus, in the stippled diagram 13-1a, which represents the electron density on a cross-section which cuts through the nucleus, the density of the stippling gets greater as one approaches the nucleus. Note that for hydrogen, at $r=a_o$, $|\Psi_{1s}|^2 = k^2 e^{-2} = 0.14k^2$, so that in the Schroedinger treatment a_o is not the radius of a fixed orbit, but rather the distance from the nucleus where the probability of finding a hydrogen 1s in a given volume of space has decayed to 14% of the probability of finding it in the same volume at the nucleus.

The text next proceeds to a discussion of the *radial distribution function*, $4\pi r^2|\Psi(r)|^2$. This is important, and students often have trouble understanding its meaning, even though it is, in fact, not really so difficult to grasp. Sometimes we want to know the probability of an electron's being a given distance from the nucleus. For a given distance, r, all the points that far away from the nucleus lie on a sphere of radius r. Since the area of this sphere is $4\pi r^2$, the total electron density on this sphere is given by $4\pi r^2|\Psi(r)|^2$. For the 1s orbital of hydrogen, this is equal to $4\pi r^2 k^2 e^{-2r/a_o}$. As r goes to zero, the exponential factor goes to 1, while r^2 goes to zero, so that this function tends to zero at the nucleus.

Another way of thinking of this function is to imagine an orange of radius r centered at the nucleus. $4\pi r^2$ is the area of the surface, and if the peel has a thickness Δr the volume of the peel is $4\pi r^2 \Delta r$. The total electron density that lies within this thickness (that is, within the *spherical shell* formed by the peel) is, to a good approximation, $4\pi r^2|\Psi(r)|^2 \Delta r$. If the orange shrinks, $|\Psi(r)|^2$ increases, but the total volume of the peel decreases faster, for small r. At large r, the volume of the peel is large, but $|\Psi(r)|^2$ has faded away to almost nothing. As Figure 13.3 illustrates, for hydrogen (Z=1) $4\pi r^2|\Psi(r)|^2$ goes through a maximum at a_o, the radius of the first Bohr orbit, 0.0529 nm. Thus, in the Schroedinger treatment, a_o is also the most probable distance of the electron from the nucleus in the 1s orbital of hydrogen.

The 2s orbital is also spherically symmetrical, but $|\Psi(r)|^2$ drops off in a more complicated fashion as r increases: there is a region of high electron density near the nucleus, then a *node* where $|\Psi(r)|^2$ goes to zero, and then another region of high density, before $|\Psi(r)|^2$ finally drops to zero again toward infinity. In general, an orbital with principle quantum numbre n has (n-1) nodes (excluding the node at r=∞ and the possible node at r=0); for the s orbitals these are all *radial nodes*, meaning that they are shaped like spheres: they are spherical surfaces where $|\Psi(r)|^2=0$ and $|\Psi(r)|=0$ (Figure 13.5).

From Equation 13-1, which, recall, applies to hydrogen only, for n=1, 2 and 3, $E_1 = -K_1$, $E_2 = -K/4$ and $E_3 = -K/9$. Referring now to Figure 13.5, which depicts the 1s, 2s and 3s orbitals, we can see that the reason E gets higher (more positive) as n increases is that the electron spends more of its time further and further from the nucleus. The figure shows that this is accompanied by the expansion of the electron cloud.

When l≠0 (p, d, f orbitals, and so on), Ψ is no longer spherically symmetrical. This means that going out from the nucleus the electron density as a function of distance looks different depending on direction. Consider, as an example, the p_z orbital (l=1; Figure 13.6). One passes through a region of high electron density as one goes out from the nucleus in the z-direction for this orbital, but not as one goes out in the x- and y-directions. In fact, the entire xy plane is a node of the $2p_z$ orbital, and thus no electron density is found there. Unlike the s orbitals, the p orbitals exhibit zero electron density at the nucleus. The directionality of the p orbitals is important in understanding bond angles in chemical compounds, as we shall see. The d-orbitals have more complicated nodal patterns, as illustrated in Figure 13.8.

The xy-plane is an *angular node* for the p_z orbital. Any node that is not radial (sphere-shaped) is termed an angular node. An orbital has the same number of angular nodes as the value of the azimuthal quantum number, l. Thus, for a 2p orbital, (n=2) there is (n-1)=1 node, of which (l=1) is angular; on the other hand, a 3p orbital (n=3, l=1) has (n-1)=2 nodes, of which (l=1) is angular, and therefore it has a spherical node as well. A 3d orbital (n=3, l=2) has (3-1)=2 nodes, of which (l=2) are angular; for all the d orbitals except the d_{z^2} (Figure 13.2) the nodes are planar. For d_{z^2}, the nodes are cone-shaped.

Chapter 13 — Electronic Structure of Atoms, Periodic Table, Periodic Properties

For any value of n, there is only one s orbital; p orbitals, however, come in sets of three, and d orbitals come in sets of five. There is one orbital for each allowed value of m_l; for example, for l=1 (a p orbital) m_l can be -1, 0 or 1, so that there are three p orbitals in a subshell. Similarly, for l=2 (d-orbital), m_l can range from -2 through +2, giving five d-orbitals in a subshell.

Electron Spin

As we have seen, much of the early confirmation of quantum theory came from the study of atomic spectra; the inability of the Bohr theory to fully explain the spectra of many-electron atoms led to the search for something better. The Schroedinger equation, though a great improvement, still left certain features of atomic spectra unexplained. These were resolved by assuming that the electron has an intrinsic angular momentum, in the same way that it has a charge and a mass. The electron posesses this property even in an s state, where we have noted that the *orbital angular momentum* is zero. This intrinsic angular momentum is termed the *spin*, in analogy to classical physics; a spinning classical charged body would exhibit a magnetic moment. The quantum number m_s, which can take on values +1/2 or -1/2, can be thought of as referring to the direction of the spin. m_s = +1/2 is considered to be *spin up* (spin state α) and m_s = -1/2 is considered to be *spin down* (spin state β). Thus, an electron in a given orbital characterized by values of n, l and m_l, can be in spin state α or β. The *Pauli exclusion principle* says that no two electrons can be in exactly the same state; however, two electrons in the same orbital with different spin states (different values of m_s) are in different states, so this situation is allowed. Thus each orbital can hold two electrons of opposite spin (Example 13.2 in the text).

Unpaired electrons result in *paramagnetism*, attraction to a magnetic field. Substances with no unpaired electrons are *diamagnetic*, and are repelled by a magnetic field. We shall see later (Chapter 20) that these properties are very important in understanding the chemistry of metal complexes. Although electron spin was first inferred from spectroscopic evidence, it was theoretically justified later by the application of the theory of relativity to quantum mechanisms (the Dirac equation).

The Periodic Table

So far we have been discussing the orbitals that arise from solving the Schroedinger equation for hydrogen. Consider briefly the 1s orbital. Imagine now increasing Z (the number of positive charges in the nucleus) to two. This corresponds to the ion He^+. This is still a one-nucleus, one-electron system (a hydrogen-like atom) and can be solved exactly using the Schroedinger equation; Equation 13-3 in the text gives the form of the 1s orbital for He^+, when Z=2. Due to the increased nuclear charge the orbital shrinks in toward the nucleus, compared to hydrogen. Now, however, imagine adding to the orbital a second electron (with spin opposite to that of the first.) This corresponds to a helium atom. One can imagine two main effects: first, the two electrons repel each other, so they tend to stay on opposite sides of the nucleus. This tends to expand the orbital compared with the shape it had in He^+; also, since the electrons cannot completely keep out of each others' way, there is some probability that one electron will be between the other and the nucleus. When this happens the second electron in effect "sees" a nuclear charge of about +1 instead of +2 (this is called *screening*), and will be less attracted to the nucleus. This also results in expansion of the orbital compared with the exact solution for He^+.

We have just taken the first step in what is called the *aufbau* (German for "building up") *process*: although the Schroedinger equation cannot be solved exactly for He, we can talk about the He orbitals as being expanded versions of the He^+ orbitals. This process can be continued to encompass the entire periodic chart. This can be done mathematically, but we will give only a qualitative, and hopefully intuitive, presentation of the results.

Suppose now we add another proton and electron to He, giving Li. The K shell is full with He, so the additional electron must go into the L shell (n=2), where it has the choice of a 2s or a 2p orbital. Recall that for hydrogen these two are equal in energy; however, recall that the 2s orbital (l=0) has high electron density at the nucleus, whereas the 2p has zero electron density at the nucleus. Recall also that there are two electrons in the 1s orbital close in to the nucleus. These will partly *screen* or *shield* the third electron, which is going to be further away from the nucleus than the 1s electrons, regardless of whether it goes into the 2s or the 2p orbital. (Recall that orbitals with the same value of n are, on average, approximately equally far from the nucleus.) If it goes into 2s, however, it *penetrates* the screen better than it would in the 2p, just because the 2s has higher electron density near the nucleus. Thus, in a many-electron atom, an electron in the 2s orbital has lower energy than one in the 2p; it can "feel" more of the nuclear charge. The *electronic configuration* of lithium is $1s^2 2s^1$. Adding additional protons (and possibly neutrons), and electrons gives first beryllium, $1s^2 2s^2$, then boron, $1s^2 2s^2 2p^1$, then carbon, $1s^2 2s^2 2p^2$.

Now, since there are three 2p orbitals, there is a choice as to how the two 2p electrons in carbon are to be arranged. They can be in the same orbital, with opposite spins, or in two separate orbitals, such as p_x and p_y. If they go into the same orbital the electrons tend to be closer, on average, than if they go into different ones, and since *electron-electron repulsion* is unfavorable, the latter course is chosen. This configuration results in two unpaired electrons (Figure 13.12). *Hund's rule* states that, other things being equal, electrons occupy different orbitals and have the same spin state. As we have seen, this is because the energy is usually lowest if they do; there are, however, exceptions. Hund's rule is really a rule of thumb — though a very good one — rather than a fundamental law of nature. It is sometimes violated. In contrast, the Pauli principle is a fundamental law, and is never violated.

Following carbon, we have nitrogen, $1s^2 2s^2 2p^3$ (three unpaired electrons), then oxygen, $1s^2 2s^2 2p^4$ (two unpaired electrons, plus two paired up in the remaining p-orbital), then fluorine, $1s^2 2s^2 2p^5$ (one unpaired electrons), and then neon, $1s^2 2s^2 2p^6$ (no unpaired electrons) completing the L-shell. The M shell then starts with sodium, $1s^2 2s^2 2p^6 3s$, which can be written (Ne)3s, where (Ne) indicates the neon electronic configuration. Note that lithium, (He)2s, and sodium, (Ne)3s, have similar configurations: an inert-gas *core* plus a single s electron. We will see that this accounts for their similar chemical properties. Likewise (Table 13.3) for fluorine, $(He)2s^2 2p^5$ and chlorine, $(Ne)3s^2 3p^5$. The electrons beyond the *core* are termed the *valence electrons*.

This filling-up or *aufbau* principle succesfully predicts the structure of the periodic chart, which had been deduced by chemists earlier from purely chemical evidence, mostly atomic weights and reactivity, such as the observation that Na, Li, and K all behave similarly. In particular, the *noble gases* (sometimes termed the *rare* or *inert gases*), He, Ne, Ar, and so on, are inert because all the electrons are in filled subshells. We will see a bit later that this is a specially stable condition. Much of the chemistry of the other elements can be explained by their tendency to achieve an inert-gas electronic configuration; for example, fluorine tends to form F⁻ ions, and sodium tends to form Na⁺ ions; both of these achieve the (Ne) configuration. As the text mentions, the noble gases are not completely inert; numerous compounds of xenon, Xe, are known, and can be understood using principles developed in this and the next chapter.

In applying the aufbau principle, electrons go into the orbitals in order of increasing energy; the energies of all the orbitals shift, however, with each stage of the process (that is, with each addition of a positive charge to the nucleus and a negative electron to the atom). For example, in hydrogen the 2s and 2p orbitals have equal energy; in helium the energy of the 2s is lower that of the 2p, but both of these are still lower in energy (more stable) than the 3s. Later on in the building-up of the periodic chart, energies of subshells within different shells begin to overlap, so that the 3d subshell is filled before the 4p. Within a subshell, Hund's rule provides a guide to the filling of orbitals: keep electrons in different orbitals with parallel spin wherever possible. Of course, all this must be done obeying the Pauli principle: no two electrons can have the same set of quantum numbers.

The *representative elements* are the elements whose s and p shells are in the process of being filled in. This occurs in the first two and last six columns of the periodic chart. Each column is labeled by the Roman numeral giving the number of electrons outside the previous inert gas configuration, followed by the letter A, so that Li and Na are in Group IA, F and Cl are in VIIA. Ne and Ar are in O, and are exceptionally stable.

As one applies the aufbau principle, successive *rows* (or *periods*) of the chart are filled. The third period terminates with Ar, which has electronic configuration $1s^2 2s^2 2p^6 3s^2 3p^6$. Recall that the K shell (n=1) can have only an s subshell, the L shell (n=2) can have s and p subshells, and the M shell (n=3) can have s, p and d subshells. After filling the 3p subshell (Ar), the next electron does not go into a 3d orbital, but rather into the 4s. Even though the most probable distance between the electron and the nucleus is greater in the 4s than in the 3d, the 4s has a greater electron density near the nucleus, as discussed earlier (Figure 13.4, 13.5). This lowers its energy below that of the 3d; thus Ca has the configuration $(Ar)4s^2$.

Beyond Ca the ten 3d electrons are added, forming, successively, the ten first-row *transition elements*, whose electronic configurations are given in Table 13.5. Through the third of them, V, vanadium, Hund's rule holds, but with the fourth, Cr, it breaks down, and chromium's configuration is $(Ar)3d^54s^1$. Here another rule-of-thumb is useful; namely, that half-full and full subshells have extra stability. Don't forget that with each application of the aufbau principle an extra charge is added to the nucleus in addition to the extra electron added to the next available orbital. Because the d orbitals are pointed in different directions in space, they do not screen the nuclear charge from each other very well. Thus by the time four d electrons are added, the fifth can experience more of the added nuclear charge by going into the 3d than into the 4s orbital. The same thing happens at Cu, the ninth first-row transition element, which has configuration $(Ar)3d^{10}4s^1$, and the same behavior is exhibited in the second transition period. When transition metal atoms are ionized, however, the 3d orbitals are always more stable than the 4s. This is because when electrons are removed to form the ions, there is less electron density to screen the outermost electrons, which then feel a greater nuclear charge. The ability of the 4s orbital to penetrate the inner electron screen then becomes less important.

After the 3d orbitals are filled, the 4p orbitals are filled, culminating in Kr. The following scheme can be used to remember the order of filling the orbitals. The arrows are followed from tail to head, starting at the top of the diagram.

Aufbau Chart

1s			
2s	2p		
3s	3p	3d	
4s	4p	4d	4f
5s	5p	5d	5f
6s	6p	6d	
7s			

Chapter 13 Electronic Structure of Atoms, Periodic Table, Periodic Properties

The filling of the f subshells corresponds to elements called the *second transition series*, or the *inner transition series*, or the *lanthanide* (row 6) and *actinide* (row 7) series. The chemistry of all these elements is very similar.

Ionization Energy (IE) and Electron Affinity (EA)

Ionization energies are also called *ionization potentials*. The ionization energy of an atom is ΔE for the reaction

$$X(g) \rightarrow X^+(g) + e^-(g)$$

and is always positive, since it always takes energy to separate charges. The ionization energy of an atom is a direct measure of the stabilization of the outermost electron; the higher the IE, the greater the stabilization. Ionization energy is a *periodic property*, one that is correlated with position in the periodic chart. The trends in IE can be explained using the concept of screening. As electrons and protons are added across a period (Table 13.9 and Figure 13.18), subsequent electrons are shielded only poorly by other electrons in the same shell, or especially by electrons in other orbitals of the same subshell. Thus Li (5.39 eV) has a lower IP than He (24.6 eV); Be's IP (9.32 eV) is higher than that of Li, because the new electron is also in 2s, but the first 2p electron (B, IE=8.30 eV) has a lower IP. IP climbs with C and N, because Hund's rule ensures that the first three 2p electrons go into separate orbitals, keeping screening to a minimum, but with the fourth 2p electron (C), IP drops again. This is another example of the "half-filled rule:" the fourth 2p electron fills an orbital already half occupied, and is relatively well screened by the first electron in it; therefore it does not experience as much of the added nuclear charge as the previous electron did. From here, IP rises again, reaching 21.6 eV at Ne. With Na, (Ne)3s^1, IP drops again, to 5.14 eV, and the same pattern repeats.

You may recall (Figure 2.2) that the atomic volume is a periodic property. Note, by comparison to Figure 13.18, that atomic volume is inversely correlated with ionization potential. Where the ionization energy is low, as in the alkali metals, the outermost electrons tend to be farther from the nucleus than in atoms whose ionization energy is high; for example, we noted earlier that the sodium 3s electron is well screened by the core electrons. Since it "sees" only low nuclear charge, it has a low IE; for the same reason, the electron tends to spend much of its time far from the nucleus. This enlarges the atom, leading to relatively high atomic (or molar) volume.

The *electron affinity* can be defined in words several different ways, which are all equivalent. The text defines it as the energy *released* in the following reaction

$$X(g) + e^-(g) \rightarrow X^-(g)$$

Since ΔE for a reaction is the energy *taken up*, EA is ΔE for the reaction

$$X^-(g) \rightarrow X(g) + e^-(g)$$

EA can thus be thought of as the ionization energy of a negative ion. EA is sometimes negative and sometimes positive, since there is no large positive energy of charge separation, as there is in IE. A positive electron affinity means that it takes energy to break the anion apart into the neutral atom plus a free electron, and thus large positive EA's are exhibited by atoms such as F and Cl which tend to form anions readily. As you can see (Table 13.10) this tendency is greatest as one approaches the rare gas configuration in each period. F has a high EA for the same reason that Ne has a high IP: it is hard to remove an electron from a filled shell (F^- or Ne). In fact, the periodic trends in IP and EA are directly related to the nature of the ions formed by the representative elements: Na has a low IP, so it easily loses an electron to form Na^+, whereas F has a high EA, so it easily gains an electron to form F^-. In both cases, an inert gas configuration (that of Ne) is reached. This sort of rationalization of the physical and chemical properties of the elements arises naturally from a consideration of atomic orbitals — the solutions to the Schroedinger equations for H — but was quite impossible with earlier pictures, such as the Bohr model of the atom.

Electronegativity, Ionic and Covalent Bonding

In a chemical bond between unlike elements, one element tends to at least partly donate electron density to the other. Since high IP and high EA both indicate the tendency of an atom to attract electrons, Mullikan originally defined electronegativity as their average: $X_M = (1/2)(IE+EA)$. An atom with a high value of X_M tends to attract electrons from an atom with a low value of X_M. IP's are easy to measure, but EA's are well-known only for a few elements; therefore other methods have been used to calculate electronegativities as well. Pauling's is the best known scale, and is based on measurements of bond strength. Fluorine has the highest electronegativity, 4.0; other halogens have electronegativities in the range 2 to 3; oxygen's is 3.5. Metals have low electronegativities, particularly those in Group IA and IIA. Electronegativity increases from the lower left to the upper right of the periodic table. An electronegativity of about 2.1 separates the metals and non-metals. Pairs of atoms with large electronegativity differences (say, greater than about 1.8) tend to form *ionic bonds*. NaF is an example: the F atom completely attracts the Na 3s electron, forming Na^+ and F^- ions with the (Ne) electronic configuration. On the other hand, when electronegativities are identical, as in F_2 or H_2, electrons are shared, forming *covalent bonds*. The empirical, pre-quantum-mechanical formulation of covalent bonding is still useful; it is embodied in the *octet rule*, which states that except for hydrogen, an atom "likes" to be surrounded by eight valence electrons; hydrogen "likes" to be surrounded by two. The valence electrons are represented by dots; thus, for the formation of H_2, HF and F_2 from atoms we write

$$2H\cdot \rightarrow H:H$$

$$H\cdot + :\ddot{F}\cdot \rightarrow H:\ddot{F}:$$

$$2:\ddot{F}\cdot \rightarrow :\ddot{F}:\ddot{F}:$$

Note that dots *between* the atoms are counted as belonging to both, so that the atoms in all these compounds obey the octet rule. The electron pairs that are not shared are called *lone pairs*; F has three lone pairs in both HF and F_2. In light of what we have learned of atomic structure, the octet rule simply reflects the tendency of atoms to achieve an inert gas configuration in chemical compounds.

In fact, ionic and pure covalent bonding form a continuum; the greater the electronegativity difference between the atoms the greater the ionic character of the bond. Covalent bonds with noticeable ionic character are called *polar covalent bonds*; in general, an increase in ionic character is accompanied by an increase in melting point, for solids, and in dipole moment, for isolated gas molecules.

These trends are exhibited by the data in Table 13.11, of melting points of the chlorides of the Group IIA metals. The greater the electronegativity difference, the higher the ionic character, and the

Chapter 13 Electronic Structure of Atoms, Periodic Table, Periodic Properties

greater the melting point. Table 13.12 shows that the same trend also holds across a period, as well as down a group. Note that electronegativity is not usually used to make exact predictions, but instead to correlate a great deal of chemical information in an economical though rough fashion, in a way that still has some connection to first principles. It is a key concept in the descriptive chemistry of the elements, as we shall see in subsequent chapters.

Group IA and IIA Metals

These are also called the *alkali metals* and *alkaline earth* metals respectively, and tend to form ionic compounds with +1 and +2 charge on their ions, respectively, with the exception of Be, which forms many covalent compounds. All react with water, often violently, to give the hydroxide and H_2. Electronegativity decreases as one goes down these groups, which accounts for Be's tendency to form many covalent compounds. Several important compounds and reactions involving these elements are given in the text.

Group VIIA and VIIA Elements

These are called the *chalcogens* and the *halogens*, respectively. The halogens tend to form ions with charge -1 (F^-, Cl^-, etc.), and the chalcogens tend to form ions with charge -2 (O^{2-}, S^{2-}, etc.), and also to enter into many covalent interactions. The compounds of hydrogen with these elements (HF, HCl, H_2S, etc.) tend to be acidic, with the exception of H_2O; this acidity occurs because the HX bond has much ionic character, and when surrounded with water tends to in fact break, giving salvated ions.

The chapter closes with a list of six *isoelectronic* ions — ions which have the same electronic configuration, in this case that of Ne. The ionic radius decreases as the nuclear charge increases; this is because higher nuclear charge binds the electron clouds more tightly, causing them to "shrink in", and it is the electron clouds which take up virtually all of the atomic volume (recall Rutherford's scattering experiment!). Since the radii of these ions varies so greatly (by more than a factor of 3, from 0.50 for Al^{3+} to 1.71 for N^{3-}), while the electronic configuration is constant (at $1s^2 2s^2 2p^6$), it is clear that the spacial arrangement of the electrons changes much from that given by the one-electron solutions to the

Schroedinger equations which are the hydrogen atomic orbitals. Nevertheless, we have seen in this chapter that the orbital concept allows us to rationalize many features of the physics and chemistry of the elements, and illuminates other chemical notions, such as electronegativity and ionic and covalent bonding, and is thus a powerful heuristic device in the understanding of chemical interactions.

Chapter 13 — Electronic Structure of Atoms, Periodic Table, Periodic Properties

Questions

13.1 Atomic orbitals and the quantum numbers n, l, and m_l

1. For hydrogen, as n increases, the energy of the system (increases, decreases, remains the same).

2. The ground state of hydrogen is given by n=_____.

3. As l increases, the energy of a hydrogen atom, at constant n, (increases, decreases, remains the same).

4. At constant n and l, states with different values of m_l have different energies only in the presence of _____.

5. What is the degeneracy of each of these subshells?

    _____(a) 2s

    _____(b) 3d

    _____(c) 4p

6. Fill in this table:

Subshell	Shell	n	l	Degeneracy	Possible Values of m_l
2s					
3p					
4f					
3d					

7. $|\Psi|^2$ is equal to _____ at a node.

8. Which of these orbitals has a finite electron density at the nucleus?

    _____ (a) 1s

    _____ (b) 2p

    _____ (c) 3s

    _____ (d) 4d

Chapter 13　　　　　　　　　Electronic Structure of Atoms, Periodic Table, Periodic Properties

9. As n increases, the electron spends more of its time (further from, closer to) the nucleus.

10. How many lobes does each of the following orbitals have?

　　_____ (a) 2s

　　_____ (b) 3p

　　_____ (c) $3d_{xy}$

11. How many nodes does each of the following orbitals have?

　　_____ (a) 2s

　　_____ (b) 2p

　　_____ (c) 3d

13.2 The spin quantum number and the Pauli exclusion principle

1. Two electrons in the same orbital must have different values of which quantum number?
   _____

2. This is required by which law of physics?

3. What determines whether a substance will be diamagnetic or paramagnetic?

13.3 The aufbau process and the periodic table

1. For a many-electron atom, as l increases, the energy of the subshell _____, at constant n.

2. Of the orbitals 3s, 3p and 3d, which

   (a) penetrates the screen of the inner orbitals the best? _____

   (b) the worst? _____

Chapter 13   Electronic Structure of Atoms, Periodic Table, Periodic Properties

3. Which of these species is "hydrogen like"?

   _____ (a) H^+

   _____ (b) H_2

   _____ (c) He

   _____ (d) He^+

   _____ (e) Li^+

   _____ (f) Li^{2+}

4. In applying the aufbau principle, which of the orbitals in each of the following pairs gets filled first?

   _____ (a) 1s, 2s

   _____ (b) 2s, 2p

   _____ (c) 2p, 3p

   _____ (d) 3s, 3p

5. Fill in the following table:

Atom	Electronic Configuration	Number of Unpaired Electrons
	$1s^2 2s^2 2p$	
C		
	(Ar)4s	
	(He)$2s^2$	
Ne		
P		

6. The elements whose s and p subshells are being filled are called the _____ elements.

Chapter 13   Electronic Structure of Atoms, Periodic Table, Periodic Properties

13.4 Group 0, the noble gases: the relationship between their chemistry and their electronic configurations

1. The inert gases are unreactive because

2. Which rare gases are known to form compounds?

3. What elements do these compound-forming rare gases most tend to combine with?

13.5 The relationship between the chemistry and the electronic configurations of the alkali and alkaline earth metals

1. Strictly speaking, an orbital is a

2. For many-electron atoms, a description in terms of orbitals is (exact, a useful approximation, a useless approximation).

3. Fill in this table:

Atom	Electronic Configuration	Number of Unpaired Electrons
V	$(Ar)3d^3 4s^2$	
Cr		
Ni		
Cu		
Mo		
Tc		

Chapter 13 Electronic Structure of Atoms, Periodic Table, Periodic Properties

4. Fill in this table:

Ion	Electronic Configuration	Number of Unpaired Electrons
V^{2+}		
Cr^{2+}		
Ni^{2+}		
Cu^{2+}		
Mo^{2+}		
Tc^{2+}		

5. Which subshell is being filled for each of the following:

   _____ (a) fourth row halogen

   _____ (b) lanthanide

   _____ (c) second period transition metal

13.6 Periodicity in bonding: ionic vs. covalent bonding across the short periods

1. For element X, write the chemical reaction whose ΔE is equal to

   (a) The first ionization energy

   (b) The second ionization energy

2. The more stable the outermost orbital, the (lower, higher) the IE.

3. For outermost electrons in the same shell, the higher the IE, the (lower, higher) the effective nuclear charge.

Chapter 13   Electronic Structure of Atoms, Periodic Table, Periodic Properties

4. When a new shell is started, the IE goes (up, down) compared to that of the previous element.

5. When a new subshell is started, the IE goes (up, down) compared to that of the previous element.

6. Tell whether each of these tends to increase or decrease the IE

   _____ (a) increased nuclear charge

   _____ (b) screening by inner electrons

   _____ (c) better penetration of outermost orbital

7. The higher the IE, the (greater, less) the tendency to form a positive ion.

8. Give an example of an element with a low IE.

13.7 The halogens: the relationship between their chemistry and their electronic configurations

1. For element X, write the chemical reaction whose ΔE is equal to the electron affinity.

2. The more positive EA, the (greater, less) the tendency to form a negative ion.

3. Give an example of an element with a high EA.

4. In general, more (IE's, EA's) are known accurately.

13.8 The chalcogens (oxygen family): the relationship between their chemistry and their electronic configurations

1. In a bond between dissimilar atoms, the atom with the (higher, lower) electronegativity tends to attract electron density.

2. An atom tends to be more electronegative the (higher, lower) its IE and the (higher, lower) its EA.

3. Where in the periodic chart do we find

   (a) The most electronegative elements?

Chapter 13　　　　　　　　　　Electronic Structure of Atoms, Periodic Table, Periodic Properties

(b) The least electronegative elements?

4. Which atom in each pair is the more electronegative?

　　_____ (a) Na, K

　　_____ (b) Na, Mg

　　_____ (c) F, Cl

　　_____ (d) S, Cl

13.9 The periods of the periodic table and the electronic configurations of the first series of transition metals

1. Write Lewis dot formulas for the following:

   (a) Na

   (b) P

   (c) S

   (d) K^+

   (e) Cl^-

   (f) O^{2-}

   (g) HCl

   (h) H_2S

2. Based on the data in Figure 13.19 in the text, predict whether each of the following compounds will be mostly ionic, polar covalent or pure covalent.

　　_____ (a) LiBr

　　_____ (b) LiI

　　_____ (c) HF

　　_____ (d) H_2O

　　_____ (e) Cl_2

　　_____ (f) PCl_5

Chapter 13　　　　　　　　　　Electronic Structure of Atoms, Periodic Table, Periodic Properties

13.10 A periodic property: the ionization energy

1. Most alkali metal compounds are (ionic, covalent).

2. Alkali metal ions tend to have a charge of _____.

3. Complete and balance:

    (a) $K + H_2O \rightarrow$

    (b) $NaHCO_3 + heat \rightarrow$

4. Most alkaline earth compounds are (ionic, covalent).

5. Alkaline earth ions tend to have a charge of _____.

6. Complete and balance

    (a) $Ca + H_2O \rightarrow$

    (b) $ + H_2O \rightarrow$ slaked lime

13.11 A periodic property: the electron affinity

1. Halide ions typically have a charge of _____.

2. For each of the following, on which atom does the positive end of the dipole reside?

    _____ (a) ClF

    _____ (b) IBr

    _____ (c) BrF

13.12 A periodic property: the electronegativity

1. Chalcogen ions tend to have a charge of _____.

2. Which of these exhibit mostly ionic, and which mostly covalent bonding?

    _____ (a) K_2O

    _____ (b) MgO

    _____ (c) Al_2O_3

    _____ (d) CO_2

_____ (e) H₂O

3. For a given electronic configuration, the more positive the ionic charge, the (greater, smaller) the ionic radius.

Chapter 13 — Electronic Structure of Atoms, Periodic Table, Periodic Properties

Answers

13.1 Atomic orbitals and the quantum numbers n, l and m_l

1. increases
2. 1
3. remains the same
4. an external magnetic field
5. (a) 1
   (b) 5
   (c) 3
6.

Subshell	Shell	n	l	Degeneracy	Possible Values of m_l
2s	L	2	0	1	0
3p	M	3	1	3	-1, 0, 1
4f	N	4	3	7	-3, -2, -1, 0, 1, 2, 3
3d	M	3	2	5	-2, -1, 0, 1, 2

7. zero
8. (a) and (c): s orbitals only
9. further from
10. (a) 1
    (b) 2
    (c) 4
11. (a) 1
    (b) 1
    (c) 2

13.2 The spin quantum number and the Pauli exclusion principle

1. m_s
2. Pauli exclusion principle
3. If it has unpaired electrons, it will be paramagnetic; otherwise, diamagnetic.

13.3 The aufbau process and the periodic table

1. increases
2. (a) 3s
   (b) 3d
3. (d, He^+) and (f, Li^{2+}). Each has only a single electron.
4. (a) 1s
   (b) 2s
   (c) 2p
   (d) 3s

Chapter 13 — Electronic Structure of Atoms, Periodic Table, Periodic Properties

5.

Atom	Electronic Configuration	Number of Unpaired Electrons
B	$1s^2 2s^2 2p$	1
C	$1s^2 2s^2 2p^2$	2
K	$(Ar)4s$	1
Be	$(He)2s^2$	0
Ne	$1s^2 2s^2 2p^6$	0
P	$(Ne)3s^2 2p^3$	3

6. representative

13.4 Group 0, the noble gases: the relationship between their chemistry and their electronic configurations

1. They exhibit complete s and p subshells.
2. Xe and Kr.
3. F and O

13.5 The relationship between the chemistry and the electronic configurations of the alkali and the alkaline earth metals

1. one-electron wave function
2. a useful approximation
3.

Atom	Electronic Configuration	Number of Unpaired Electron
V	$(Ar)3d^3 4s^2$	3
Cr	$(Ar)3d^5 4s^1$	6
Ni	$(Ar)3d^8 4s^2$	3
Cu	$(Ar)3d^{10} 4s^1$	1
Mo	$(Kr)4d^5 5s^1$	6
Tc	$(Kr)4d^5 5s^2$	5

4.

Ion	Electronic Configuration	Number of Unpaired Electrons
V^{2+}	$(Ar)3d^3$	3
Cr^{2+}	$(Ar)3d^4$	4
Ni^{2+}	$(Ar)3d^8$	3
Cu^{2+}	$(Ar)3d^9$	4
Mo^{2+}	$(Kr)4d^4$	4
Tc^{2+}	$(Kr)4d^5$	5

5. (a) 4s
   (b) 4f

(c) 4d

13.6 Periodicity in bonding: ionic vs. covalent bonding across the short periods

1. (a) $X \rightarrow X^+ + e^-$
   (b) $X^+ \rightarrow X^{2+} + e^-$
2. higher
3. higher
4. down
5. down
6. (a) increase
   (b) decrease
   (c) increase
7. less
8. Na, Li, Mg, Ca, etc.

13.7 The halogens: the relationship between their chemistry and their electronic configurations

1. $X^- \rightarrow X + e^-$
2. greater
3. O, F, Cl, etc.
4. IE's

13.8 The chalcogens (oxygen family): the relationship between their chemistry and their electronic configurations

1. higher
2. higher, higher
3. (a) top right
   (b) bottom left
4. (a) Na
   (b) Mg
   (c) F
   (d) Cl

13.9 The periods of the periodic table and the electronic configurations of the first series of transition metals

1. (a) Na

   (b) :P̈·

   (c) :S̈:

   (d) K^+ or :K̈:$^+$

   (e) :C̈l:$^-$

   (f) :Ö:$^{2-}$

Chapter 13 Electronic Structure of Atoms, Periodic Table, Periodic Properties

(g) H:Cl̈:

(h) H:S̈:H

2. (a) ionic
   (b) polar covalent
   (c) polar covalent
   (d) polar covalent
   (e) covalent
   (f) polar covalent

13.10 A periodic property: the ionization energy

1. ionic
2. +1
3. (a) $2K + 2H_2O \rightarrow 2KOH + H_2$
   (b) $2NaHCO_3 \rightarrow Na_2CO_3 + CO_2 + H_2O$
4. ionic
5. +2
6. (a) $Ca + 2H_2O \rightarrow Ca(OH)_2 + H_2$
   (b) $CaO(lime) + H_2O \rightarrow Ca(OH)_2$ (slaked lime)

13.11 A periodic property: the electron affinity

1. -1
2. (a) Cl
   (b) I
   (c) Br

13.12 A periodic property: the electronegativity

1. -2
2. (a) ionic
   (b) ionic
   (c) ionic
   (d) covalent
   (e) covalent
3. smaller

CHAPTER 14: THE NATURE OF THE CHEMICAL BOND

Scope

The concept of the chemical bond predates quantum mechanics, and some of the pre-quantum-mechanical ways of describing bonding are still useful. These include *ionic bonding, Lewis structures* and *VSEPR (Valence Shell Electron Pair Repulsion)*[*], which involve quantum mechanics only indirectly. Then *molecular orbitals*, an inherently quantum mechanical description of bonding, are taken up.

Ionic Bonding

The text starts with a discussion of ΔH for the overall reaction

$$Na(g) + Cl(g) \rightarrow Na^+Cl^-(g)$$

where the product is the NaCl ion pair in the gas phase. It is important to remember that for

$$Na(g) \rightarrow Na^+(g) + e^-$$

ΔH is equal to the ionization energy of Na, but for

$$Cl(g) + e^- \rightarrow Cl^-(g)$$

ΔH is the *negative* of the electron affinity of Cl, based on the definitions of these terms in Chapter 13. Therefore, the energy of the reaction

$$Na(g) + Cl(g) \rightarrow Na^+(g) + Cl^-(g)$$

is given by IE(Na)-EA(Cl). Then, if the Na-Cl distance in the ion pair is known, the energy can be calculated for

$$Na^+(g) + Cl^-(g) \rightarrow Na^+Cl^-(g)$$

It is simply the Coulombic energy, kq_1q_2/Dr, where k and D are constants, and q_1 and q_2 are opposite in sign, but equal in magnitude to the charge on an electron, and r is the known Na-Cl distance. What makes the ionic bonding model a good one for NaCl is that the energy for the overall reaction can actually be measured, and the result agrees with this calculation. Note also that r, IE(Na) and EA(Cl) are all taken as experimental quantities. Of course, to calculate these from first principles would require the use of quantum mechanics.

---

[*] VSEPR was devised long after quantum mechanics (it dates from about 1957), but it takes no quantum mechanical expertise to use it.

Chapter 14 The Nature of the Chemical Bond

Lewis structures

Ionic bonding provides an appropriate description of a chemical bond when the electronegativities of the two elements involved are sufficiently dissimilar (say, an electronegativity difference of 1.8 or greater). For many chemical bonds between atoms of similar electronegativity, the bonding scheme can be described by the *octet rule*, which was discussed in the last chapter; it should be borne in mind that the octet rule does not always work, and that sometimes Lewis structure do not give a good picture of the bonding; however, several additional ideas (beyond the simplest applications of the octet rule) can be used in conjunction with Lewis diagrams; these make them applicable to the vast majority of chemical compounds.

The first additional concept is *multiple bonding*; for example, $:\ddot{O}=C=\ddot{O}:$ for CO_2. Note that all atoms still exhibit complete octets. The notion of multiple bonds is firmly rooted in reality, since when the octet rule requires a double bond between two elements, the bond is in fact found to be both shorter and harder to break (greater bond energy) than a single bond between the same two elements.

Formal charge is the second additional concept. Sometimes the octect rule can be satisfied only by putting a charge on the atoms. It is important to note that electrons are counted differently for the calculation of formal charge than for the fulfillment of the octet rule. In both cases lone pairs of electrons are assigned only to the atom on which they reside; however, for the octet rule, each shared electron is assigned to both atoms, whereas for formal charge the shared electrons are divided up between the two atoms. For example, in $:C\equiv O:$, there are six shared electrons. All of them contribute to filling the octet around C, and all of them also contribute to filling the octet around O. On the other hand, to calculate the formal charges, three electrons are assigned to C and three to O. Then C and O each have five valence electrons each (these three, plus two in the lone pair). Since a neutral C atom has only four valence electrons, whereas neutral O has six, the formal charges on C and O in CO are -1 and +1, respectively. Formal charge extends the applicability of the octet rule beyond simple compounds such as Cl_2.

Formal charge is a largely fictitious notion. Since O is more electronegative than C, it attracts much of the electron density in the bond; however, the relatively positive and negative ends of the molecule are where the formal charge idea predicts them to be, though the magnitude of the charge separation is much less than a full electronic charge..

The third notion that can be added to the octet rule to make Lewis structures work is that of *resonance*. With basic resonance we admit that G. N. Lewis was not completely correct when he said that a covalent bond is a pair of electrons shared between two atoms. A bond therefore can not always be represented by one or more lines connecting two atoms. For SO_2 we write the following resonance structures:

$$:\ddot{O}::\ddot{S}:\ddot{O}:$$

$$:\ddot{O} \quad :\ddot{S}::\ddot{O}:$$

The double bond and one lone pair are *delocalized* across all three atoms. The real structure is in some sense the average of the resonance structures. Delocalized bonds are actually fairly common. They exist in many organic compounds (like benzene), in the boron hydrides (see Section 14.9), and are responsible for the properties of metals (Chapter 21). Note that in SO_2, each S-O bond is neither a single nor a double bond, but rather a "one-and-a-half" bond. Also note that these resonance structures allow the octet rule to be fulfilled for all the atoms, but also allow the two oxygen atoms to be equivalent.

The fourth idea which we can bring in to make Lewis structures work is to allow more than eight electrons in the valence shell, thus violating the octet rule. This is often done for resonance structures describing species such as CrO_4^- (Examples 14.2 and 14.4 in the text); each double bond introduced into the structure expands the Cr valence shell by two electrons, but diminishes by one the formal charge on Cr and removes the formal charge on an oxygen.

Lewis structures are hardest to write for species with unpaired electrons. For example, a perfectly reasonable structure for O_2 would appear to be $:\ddot{O}=\ddot{O}:$, which satisfies the octet rule, but which campletely fails to predict that oxygen is paramagnetic. This result is explained using molecular orbital theory (see below). In summary, Lewis structures are easy to use, and can usually be made to apply, but not always.

VSEPR

This method starts with counting the number of lone and bonded electron pairs around a central atom, and then assumes that all these pairs give an overall geometry which allows the pairs to get as far away from each other as they can; thus two pairs gives a linear overall geometry, three pairs gives trigonal (three bonds in a plane at 120° angles), four gives a tetrahedron with the central atom in the middle, five gives a trigonal bipyramid (see Figure 14.5 in the text) and six gives an octahedron.

Lone pairs take up more space than bonded pairs, and so they tend to take up the less hindered positions where there is a choice — as in the trigonal bipyramid. Here, the lone pair takes an equatorial position, where it has to interact with only two other pairs at 90° (plus two at 120°), rather than an axial position, where it would have three 90° interactions.

The VSEPR method is best understood by example. First consider some simple molecules. CH_4 has eight electrons in the valence shell (four from C, $2s^2 2p^2$, and four from the four H's, each $1s^1$), thus CH_4 has four electron pairs in the valence shell; therefore the overall geometry is tetrahedral, with C in the middle. The H-C-H bond angles are all $\approx 109°$, the tetrahedral angle. For NH_3, there are also eight valence electrons (five from N and three from the three H's), so the overall geometry is still tetrahedral; however, one corner of the tetrahedron consists of a lone pair, so that the atoms form a trigonal pyramid, with N at the apex. Furthermore, since the lone pair takes up more space than a bonded pair, the H-N-H angles are compressed to about 107°. It is the same story for H_2O, which is tetrahedral overall, but whose atoms exhibit a bent geometry; two lone pairs squish the H-O-H angle to 105°. It is sometimes hard to understand why a lone pair takes up more space than a bonded pair, with its associated atom; it should help to remember that the nucleus takes little space itself, and tends to attract and compress the bonded pair. In VSEPR language, CH_4 is an AX_4 species, NH_3 is AX_3E and H_2O is AX_2E_2; X represents a bonded pair and E a non-bonded pair.

When dealing with charged species, it is easiest to start by ascribing the charge to the central atom; thus NH_4^+ has four electrons from N, and one each from the four H's, and is AX_4, tetrahedral both in overall and in atomic geometry. Now consider some more complex examples, first I_3^-, which you need to know has one central and two "outer" I's (that is, it is a chain, not a three-membered ring). The central I is viewed as I^-, and has eight valence electrons; the outer two each contribute one, for a total of ten. This gives five pairs total, so I_3^- is AX_2E_3, which is a trigonal bipyramid overall; however, the lone pairs take up the equatorial positions, as discussed above, so that the atomic geometry is linear. Now consider TeF_5^-. Te is below S in the periodic chart, so Te^- contributes seven electrons to its valence shell, and each F contributes one; the compound thus has six electron pairs total, and is AX_5E, octahedral overall. The atomic geometry is roughly a square pyramid (like the top half of AX_6 in Figure 14.5), but the lone pair takes up more space than an F, so that the Te is actually *below* the planes of the four F's.

Several more examples are instructive. Consider XeF_6. Xe contributes eight electrons to its valence shell, and each F contributes one, for a total of 14, so the compound is AX_6E. Its true geometry is not known, but it is known experimentally not to be octahedral, and although VSEPR does not have a rough-and-ready geometry for seven-coordinate species, it does correctly predict that XeF_6 is not octahedral, since only species with six pairs are octahedral. Finally, consider the three species NO_2^+, NO_2^- and NO_2. NO_2^+ is AX_2, and is in fact linear, as predicted. NO_2^- is AX_2E, and should be trigonal overall, with bent atomic geometry. One also predicts the O-N-O angle to be less than 120°, due to the extra space taken up by the lone pair. The angle is in fact 115°. NO_2 has an unpaired electron, so I will represent the compound as $AX_2E_{1/2}$. This electron must go into some orbital, so overall geometry should be trigonal, but a lone unpaired electron should take up less space than a lone pair, so we predict trigonal geometry overall, with an O-N-O angle certainly greater than 115°. In fact, the angle is 135°, so that a lone unpaired electron takes up less space than a *bonded* pair as well. Clearly, the VSEPR picture correlates and predicts much data starting from very simple assumptions.

Molecular orbitals

The Schroedinger equation can be solved exactly for two nuclei a fixed distance apart interacting coulombically with a movable electron. This corresponds to H_2^+, the hydrogen molecule ion. The resulting solutions are called *molecular orbitals*. (Note that an orbital is still a one-electron wave function!) These molecular orbitals can also be constructed mathematically by combining atomic orbitals on adjacent atoms. Once this is done, a kind of molecular aufbau principle is applied to define the electronic configurations of diatomic molecules with more than a single electron. The molecular orbital approach can also be applied to polyatomics, but then becomes much more complicated.

When two H's combine, their two 1s orbitals coalesce to form two σ molecular orbitals (Figure 14.17). A sigma orbital is one with axial symmetry along the bond, which means that if the molecule is rotated around the bond the orbital looks the same. Sigma orbitals do not have to come from s atomic orbitals, as we shall see later. One of the two σ orbitals formed, written 1s, is a *bonding orbital*: its electron density is concentrated between the two nuclei, and thus serves to pull them together. The other, written σ^*, is an *antibonding orbital*. Its electron density is concentrated outside the two nuclei, and electron density in this orbital pulls the molecule apart. (Again, see Figure 14.17.) In H_2^+, the ground state is σ^*1s^1. If the electron is excited to σ^*1s by a photon of appropriate energy, the molecule will become unstable and begin to fly apart into a hydrogen atom and a naked proton, H^+. In order for a molecule to be stable, there have to be more electrons in bonding than in antibonding orbitals. Starting with H_2^+, we can add electrons to form first H_2, which has electronic configuration $\sigma 1s^2$, then H_2^-, $\sigma 1s^2 \sigma^* 1s$, then H_2^{2-}, $\sigma 1s^2 \sigma^* 1s^2$. By the above criterion, all these should be stable except H_2^{2-}, even though H_2^- has not been found experimentally. Using similar criteria, He^+ ($\sigma 1s^2 \sigma^* 1s$) is stable, whereas He_2 ($\sigma 1s^2 \sigma^* 1s^2$) is not. *Bond order* is defined as the number of electrons in bonding molecular orbitals (MO's) minus the number in antibonding MO's, all divided by two. Thus H_2^+ has a bond order of 1/2, H_2 has a bond order of 1 and H_2^- has a bond order of 1/2. If the bond order is greater than zero, the species should be stable.

It is a good approximation to treat only the valence atomic orbitals as combining to form MO's. Thus, in the second row of the periodic chart, two Li's can form Li_2 (which exists in the gas phase) by each contributing an electron to a $\sigma 2s$ MO (formed from the 2s atomic orbitals), while each Li is viewed as retaining its own atomic 1s orbital. This makes sense because the 1s orbitals are very tightly bound by the +3 charge on the Li nucleus, and therefore tend not to be shared. This scheme works through Be_2 ($\sigma 2s^2 \sigma^* 2s^2$, unstable), but with boron the p atomic orbitals also came into play. Two p orbitals directed along the bond axis combine to form a $\sigma 2p$ bonding orbital and a $\sigma^* 2p$ antibonding orbital. These orbitals share the axial symmetry of the $\sigma 2s$ and $\sigma^* 2s$ orbitals, but have nodes at the nuclei (Figure 14.19, 14.20). Each atom also has two p orbitals perpendicular to the bond axis, and these overlap "sideways" to form new kinds of orbitals, π molecular orbitals. If we call the bond axis the z-direction, then the p_z orbitals of each atom combine to form the $\sigma 2p$ and $\sigma^* 2p$ orbitals, the p_x orbitals of each atom combine to form the $\pi 2p$ and $\pi^* 2p_x$ orbitals, and the p_y orbitals of each atom combine to form the $\pi 2p_y$ and $\pi^* 2p_y$ orbitals, which are equal in energy to their $\pi 2p_x$ counterparts.

Note from Figure 14.20 that π orbitals do *not* have axial symmetry. A π orbital has a planar node which includes the bond axis, and electron density "above" and "below" this node. This is true for both the bonding and the antibonding π orbitals. The $\pi 2p_x$ and $\pi 2p_y$ orbitals differ only in that, referring to Figure 14.20, for $\pi 2p_y$ the node is "horizontal," sticking out of the page, so that the lobes of electron density are above and below it, whereas for $\pi 2p_x$ the node is "vertical", in the plane of the page, so that the electron density is in front of and behind the page. The $\pi^* 2p$ antibonding orbitals have these same nodes as well, in addition to the node between the two bonded atoms which is characteristic of all antibonding orbitals.

We can now continue to apply the molecular aufbau principle to describe the bonding in the first row diatomics. This is complicated, however, by the fact that some of the molecular orbitals switch relative energies as electrons are added. (This complication should not be too surprising; recall that when we applied the atomic aufbau principle such shifts also occured; for example, as we went from vanadium to chromium to manganese, the 4s orbital went from being more to less to more stable than the 3d.) $\sigma 2s$

and σ^*2s are always the lowest in energy, and $\pi^*2p_{x,y}$ and σ^*2p are always the highest (Figure 14.21); however, the $\pi 2p_{x,y}$ are lower than $\sigma 2p$ for B_2, C_2 and N_2, whereas this ordering is reversed for O_2 and F_2. Remember that $\sigma 2p$ has its highest electron density on the bond axis, and so does $\sigma 2s$. For B_2, C_2 and N_2 $\sigma 2p$ is destabilized by electron-electron repulsion with $\sigma 2s$, and so $\pi 2p$ is more stable. With increasing nuclear charge, however, the $\sigma 2s$ electrons tend to spend more time near the nuclei (they become more like electrons in inner orbitals), and so there is more "room" for the $\sigma 2p$ orbital, which then becomes more stable.

With this in mind, B_2 has valence electronic configuration $\sigma 2s^2 \sigma^*2s^2 \pi 2p^2$, where the $\pi 2p$ electrons are in two separate orbitals (x and y) with parallel spin, as predicted by Hund's rule. B_2 is thus paramagnetic. With C_2 the two additional electrons fill the $\pi 2p_x$ and $\pi 2p_y$ orbitals, and C_2 is diamagnetic. The bond order of B_2 is 1 and of C_2 is 2. With N_2 the $\sigma 2p$ orbital is filled, and the bond order is 3. Thus the valence electronic configuration of N_2 is $\sigma 2s^2 \sigma^*2s^2 \pi 2p^4 \sigma 2p^2$. For O_2, the order of $\pi 2p$ and $\sigma 2p$ changes, and the two additional electrons go into π^*2p_x and π^*2p_y, for a configuration of $\sigma 2s^2 \sigma^*2s^2 \sigma 2p^2 \pi 2p^4 \pi^*2p^2$. The bond order is 2, and Hund's rule is again obeyed, successfully predicting that O_2 is paramagnetic, with two unpaired electrons. Recall that Lewis structures and the octet rule could not predict this.

For heteronuclear second-row diatomics, the story is much the same. The orbitals generally occur in the same energetic sequence as in B_2, C_2 and N_2: $\pi 2p$ below $\sigma 2p$. Thus CO is isoelectronic with N_2: they both have ten valence electrons, and the same electronic configuration. In general (Figure 14.26), when two atomic orbitals combine to form a pair of MO's, the bonding MO will be closer in both energy and shape to the lower-lying atomic orbital, and the antibonding MO will be closer to the atomic orbital with higher energy. Oxygen's atomic orbitals are lower in energy than carbon's, due to higher nuclear charge, and so the electrons in CO's bonding orbitals tend to spend most of their time near the oxygen, whereas those in CO's antibonding orbitals tend to spend most of their time near the carbon. This is also in accord with what we predict from the electronegativities of C and O.

The molecular orbital method is capable of making better predictions about some of the properties of molecules — paramagnetism, for example — than any of the other methods described so far. It is also capable of being carried out mathematically to predict more complicated properties, such as the electronic absorption spectra of molecules, and is the method that gives the truest picture of what a chemical bond really is. On the other hand, it is quite cumbersome to use, and a detailed molecular orbital calculation for even a small molecule takes many hours of computing time. For this reason, chemists usually use the simplest description of a molecule that still is likely to predict the desired property; for example, if molecular geometry is desired, then VSEPR is likely to be used. Furthermore, the fundamental ideas of atomic and molecular orbital theory can be used in a qualitative (non-mathematical) manner to describe molecules too large to perform detailed MO calculations on.

Directed Valence*

This is an approach to bonding which starts out with atomic orbitals on each atom, and uses them to discuss the way each atom bonds with its neighbors in a molecule. It is a simplification of the MO method in two ways; first of all, it does not include the idea of antibonding orbitals; second of all, most bonds are viewed simply as pairwise interactions of atomic orbitals. Molecular orbitals which span several atoms still came about for delocalized bonds, thus giving a good description of what we described earlier as resonance. The directed valence method is widely used in organic chemistry, and successfully describes many of the properties of even rather large molecules. These properties include geometry and, to a certain extent, chemical reactivity. The method is really adequate only for the ground-states of stable molecules, however, so that it cannot be used to predict spectra, or the properties of highly reactive species such as C_2; it is an excellent example of how the ideas of molecular orbital theory can be used in a qualitative way to describe important features of chemistry.

The key concept in the directed valence method is the idea of *hybrid orbitals*. Recall that VSEPR successfully predicts that when carbon is bonded to four nearest neighbors, it has four pairs of bonding electrons, which are in tetrahedral orientation (bond angle $\approx 109°$). Recall as well that atomic carbon has

---

* This is another term for the method discussed in Section 14.7 ("Hybrid atomic orbitals and the molecular geometry of

valence electronic configuration $2s^2 2p^2$, with the only two unpaired electrons, and that the p orbitals are 90° apart. How can this be reconciled with the observed tetrahedral geometry, and valency of four? The answer is that atomic orbitals can *hybridize* or mix, giving a new set of atomic orbitals. For example, the 2s and the three 2p orbitals of C can hybridize to form four identical so-called sp^3 orbitals, tetrahedrally oriented. Note that the number of orbitals which results is the same as the number which one starts with (four in this case), and the relative proportions of the kind of orbitals in the hybrid are the same as in the starting set.

$$1s + 3p \rightarrow 4sp^3$$

Mathematically, this corresponds to taking different combinations of the s, p_x, p_y and p_z orbitals to form each of the sp^3 orbitals.

Thus carbon in its valence state (the state in which it enters into bonds with four neighbors) does not have the electronic configuration $2s^2 2p^2$, but rather one electron in each of four equivalent sp^3 orbitals. These orbitals can each *overlap* with a half-filled 1s orbital of H, forming four σ bonds, giving rise to the tetrahedral compound methane, CH_4. Even though the sp^3-hybridized valence state is higher in energy than carbon's ground state, the energy decrease on forming four bonds more than compensates for this.

In some compounds, such as formaldehyde, $CH_2=O$, carbon is bonded to three atoms, and VSEPR successfully predicts trigonal geometry. Here the s and two of the p orbitals on carbon hybridize to form three sp^2 orbitals, and the other p orbital is left alone. The three sp^2 orbitals are in a plane, 120° apart, and the p orbital is perpendicular to the plane; each contains one electron. Two of the sp^2's can overlap with hydrogen 1s orbitals to form σ bonds. Oxygen's ground state valence electronic configuration is $2s^2 2p^4$, with two unpaired electrons in perpendicular p orbitals. One of these can overlap end-wise with the remaining sp^2 from carbon, forming a σ bond, and the other can overlap sideways with carbon's p orbital, forming a π bond. The situation is similar for ethylene, Figure 14.38 in the text.

---

polyatomic molecules") and 14.8 ("Multiple Bonding").

Carbon also forms linear compounds, such as acetylene, HC≡CH. Here the hybridization on the carbon is sp, with two p orbitals left alone. On each carbon, the two sp orbitals are 180° from each other; one forms a σ bond with the other carbon's sp, and the other sp forms a σ bond with an H's 1s. Both p orbitals on each carbon overlap with p orbitals on the other carbon, forming two perpendicular π bonds.

For benzene, C_6H_6, each carbon is bonded to two other carbons and a hydrogen, and hybridization is trigonal (sp^2). The resulting 120° degree angle is what leads to benzene's hexagonal geometry. The p orbitals are perpendicular to the plane of the molecule (Figure 14.41), and the six electrons which they contribute are free to move anywhere within the delocalized twin-donut shaped delocalized π orbital formed by the overlap of the p's. This picture of resonance is certainly superior to that given by a pair of resonance structures, as it demonstrates that all these six electrons are shared by all the atoms. It also allows bond order to be calculated. Each C-C single bond contributes one to the bond order; looking at the delocalized π orbital as six π bonds, each contains only a single electron, half its capacity, so this contributes 1/2 to the bond order of each bond. Thus a C-C bond in benzene has a total bond order of 1.5.

Hydrogen Bridge Bonds

These can be seen as delocalized σ bonds, as illustrated in Figure 14.43 for B_2H_6. B is sp^3 hybridized; two sp^3's of each B overlap with the 1s's of the H's they share, and the other two overlap with the 1s's of the H's they have to themselves. The bond order of the bond from B to a shared H is 1/2, since two electrons make two such bonds. B_2H_6 can also be depicted by the following resonance structures.

Chapter 14 — The Nature of the Chemical Bond

Questions

14.1 Ionic vs. covalent bonding

1. Consider elements M and X, which react in the gas phase to form an ion pair:

$$M(g) + X(g) \rightarrow M^+X^-(g)$$

Tell whether each of the following will tend to increase, decrease or have no effect on ΔH for this reaction. (Note: "Increase" means "make more positive".)

_____(a) IE of M increases.

_____(b) EA of M increases (becomes more positive).

_____(c) IE of X increases.

_____(d) EA of M increases (becomes more positive).

_____(e) Bond length of ion pair increases.

14.2 Lewis structures

1. Draw Lewis structures satisfying the octet rule for the following. Indicate all non-zero formal charges.

   (a) NH_3

   (b) NH_4^+

   (c) H_2O

   (d) H_3O^+

   (e) OH^-

Chapter 14   The Nature of the Chemical Bond

    (f) N_2

    (g) NO_2^-

    (h) HCN

    (j) HCl

    (k) PO_4^{3-}

    (l) H_2SO_4

    (m) ClO_3^-

    (n) ClO_4^-

2. Draw additional resonance structures (not satisfying the octet rule) for the compounds in Question 1 which require them. Also indicate non-zero formal charges.

Chapter 14　　　　　　　　　　　　　　　　　　　　　　　The Nature of the Chemical Bond

14.3 Valence-shell electron-pair repulsion (VSEPR) theory

1. Characterize each of the following as AX_nE_m (for example, SO_2 is AX_2E), and give for each both the overall geometry of all the electron pairs (trigonal for SO_2) and the geometry of the atoms themselves (bent for SO_2).

    (a) PO_4^{3-}

    (b) SCl_2

    (c) SCl_4

    (d) SCl_6

    (e) ClO_4^-

    (f) ClO_3^-

    (g) $HgCl_2$

    (h) BF_3

    (j) IF_5

    (k) XeF_2

    (l) XeF_4

    (m) XeO_3

14.4 Molecular orbitals of H_2

1. Refer to Figure 14.16 in the text.

    (a) Two hydrogen atoms closer together than R_e experience an (attractive, repulsive, zero) force.

    (b) Two hydrogen atoms farther apart than R_e experience an (attractive, repulsive, zero) force.

    (c) Two hydrogen atoms a distance R_e apart experience an (attractive, repulsive, zero) force.

    (d) If the potential energy of two hydrogen atoms is positive, then they must be (closer together than R_e, farther apart than R_e, cannot tell).

2. Give the electronic configuration and the bond order for each of the following, and predict whether it will be a stable or unstable molecule.

Chapter 14   The Nature of the Chemical Bond

   (a) H_2^+

   (b) H_2^-

   (c) He_2^+

   (d) He_2

14.5 Homonuclear diatomic molecules of the second period

1. Give the valence electronic configuration, the bond order and the number of unpaired electrons for each of the following.

   (a) Li_2

   (b) C_2

   (c) N_2

   (d) F_2

   (e) Ne_2

14.6 Heteronuclear diatomic molecules of the second period

1. For second-period atoms A and B, suppose the 2s and 2p of B are lower in energy than the corresponding atomic orbitals of A.

   (a) Which atom has the higher IE?

   (b) In the molecule AB, tell which atom the electrons are closer to, on average, in each of the following orbitals.

   i. $\sigma 2s$

   ii. $\sigma^* 2s$

   iii. $\pi 2p$

   iv. $\pi^* 2p$

2. Give the valence electronic configuration, the bond order and the number of unpaired electrons for each of the following

   (a) CO

14-15

(b) CO⁺

(c) CN

(d) CN⁻

(e) NO

(f) NO⁺

(g) NO⁻

14.7 Hybrid atomic orbitals and the molecular geometry of polyatomic molecules

1. A hybrid orbital is a (molecular, atomic) orbital.

2. Given an s orbital and three p orbitals, give all the possible hybridization schemes, and tell what sort of bonding geometry is associated with each.

14.8 Multiple bonds

1. For each of the following, tell how many σ and how many π bonds there are between the indicated atoms.

    (a) C and C in ethylene

    (b) C and O in carbon dioxide

    (c) C and O in carbon monoxide

    (d) C and O in formaldehyde

2. What is the bond order in each of the following

    (a) benzene

    (b) SO_2

    (c) graphite

Chapter 14 　　　　　　　　　　　　　　　　　　　　　　　　The Nature of the Chemical Bond

14.9 Hydrogen-bridge bonds; the boranes

1. What does the bonding in B_2H_6 have in common with that in benzene?

Chapter 14 — The Nature of the Chemical Bond

Answers

14.1 Ionic vs. covalent bonding

1. (a) increase (M^+ formed less easily)
   (b) no effect (M^- not involved)
   (c) no effect (X^+ not involved)
   (d) decrease (X^- formed more easily)
   (e) increase (less Coulombic attraction)

14.2 Lewis structures

1. (a)

   (b)

   (c)

   (d)

   (e)

   $\overset{\ominus}{:}\!\ddot{O}\!-\!H$

   (f)

   $:N\!\equiv\!N:$

(g) [Lewis structure: O=N-O⁻ ↔ ⁻O-N=O resonance structures for NO₂⁻]

(h)

H–C≡N:

(j)

H–Cl:

(k) [Lewis structures of PO₄³⁻ resonance forms] etc.

(l) H–O–S(2+)(O⁻)₂–O–H ↔ H–O–S(+)(=O)(O⁻)–O–H ↔ H–O–S(=O)₂–O–H etc.

(m) [ClO₃⁻ resonance structures] etc.

(n) [ClO₄⁻ resonance structures] etc.

14.3 Valence-shell electron-pair repulsion (VSEPR) theory

1.

Species	AX_nE_n	Overall Geometry	Atomic Geometry
(a) PO_4^{3-}	AX_4	tetrahedral	tetrahedral
(b) SCl_2	AX_2E_2	tetrahedral	bent
(c) SCl_4	AX_4E	trig. bipyr.	lone pair-equatorial
(d) SCl_6	AX_6	octahedral	octahedral
(e) ClO_4^-	AX_4	tetrahedral	tetrahedral
(f) ClO_3^-	AX_3E	tetrahedral	tri. pyr.
(g) $HgCl_2$	AX_2	linear	linear
(h) BF_3	AX_3	trigonal	trigonal
(j) IF_5	AX_5E	octahedral	~square-pyramid
(k) XeF_2	AX_2E_3	trig. bipyr.	linear
(l) XeF_4	AX_4E_2	octahedral	sq. planar
(m) XeO_3	AX_3E	tetrahedral	tri. pyr.

14.4 Molecular orbitals of H_2

1. (a) repulsive
   (b) attractive
   (c) zero
   (d) closer together than R_e

2.

Species	Electronic Configuration	Bond Order	Stable?
(a) H_2^+	$\sigma 1s^1$	1/2	yes
(b) H_2^-	$\sigma 1s^2 \sigma^* 1s^1$	1/2	yes, in theory
(c) He_2^+	$\sigma 1s^2 \sigma^* 1s^1$	1/2	yes
(d) He_2	$\sigma 1s^2 \sigma^* 1s^2$	0	no

14.5 Homonuclear diatomic molecules of the second period

1.

Species	Electronic Configuration	Bond Order	Number of Unpaired Electrons
(a) Li_2	$\sigma 2s^2$	1	0
(b) C_2	$\sigma 2s^2 \sigma^* 2s^2 \pi 2p^4$	2	0
(c) N_2	$\sigma 2s^2 \sigma^* 2s^2 \sigma 2p^2 \pi 2p^4$	3	0
(d) F_2	$\sigma 2s^2 \sigma^* 2s^2 \sigma 2p^2 \pi 2p^4 \pi^* 2p^4$	1	0
(e) Ne_2	$\sigma 2s^2 \sigma^* 2s^2 \sigma 2p^2 \pi 2p^4 \pi^* 2p^4 \sigma^* 2p^2$	0	0

Chapter 14 — The Nature of the Chemical Bond

14.6 Heteronuclear diatomic molecules of the second period

1. (a) B
   (b)
       i. B
       ii. A
       iii. B
       iv. A

2.

Species	Electronic Configuration	Bond Order	Number of Unpaired Electrons
(a) CO	$\sigma 2s^2 \sigma^* 2s^2 \pi 2p^4 \sigma 2p^2$	3	0
(b) CO$^+$	$\sigma 2s^2 \sigma^* 2s^2 \pi 2p^4 \sigma 2p$	2.5	1
(c) CN	$\sigma 2s^2 \sigma^* 2s^2 \pi 2p^4 \sigma 2p$	2.5	1
(d) CN$^-$	$\sigma 2s^2 \sigma^* 2s^2 \pi 2p^4 \sigma 2p^2$	3	0
(e) NO	$\sigma 2s^2 \sigma^* 2s^2 \pi 2p^4 \sigma 2p^2 \pi^* 2p$	2.5	1
(f) NO$^+$	$\sigma 2s^2 \sigma^* 2s^2 \pi 2p^4 \sigma 2p^2$	3	0
(g) NO$^-$	$\sigma 2s^2 \sigma^* 2s^2 \pi 2p^4 \sigma 2p^2 \pi^* 2p^2$	2	2

14.7 Hybrid atomic orbitals and the molecular geometry of polyatomic molecules

1. atomic

2.

Scheme	Geometry
four sp^3	tetrahedral
three sp^2+p	trigonal
two sp + two p	linear

14.8 Multiple bonds

1.

	Sigma	Pi
(a)	1	1
(b)	1	1
(c)	1	2
(d)	1	1

2. (a) 3/2
   (b) 3/2
   (c) 4/3

14.9 Hydrogen-bridge bonds; the boranes

1. Both compounds exhibit delocalized bonds. In benzene the π-bonds are delocalized, and in diborane the σ-bonds are delocalized. In both, some electrons are shared by more than two atoms.

Chapter 15 — Oxidation States and Oxidation-Reduction Reactions

CHAPTER 15. OXIDATION STATES AND OXIDATION-REDUCTION REACTIONS

Scope

Oxidation and Reduction

This chapter introduces oxidation-reduction reactions, and also includes descriptive material on the redox chemistry of several groups of elements.

A reaction such as

$$2MgO + O_2 \rightarrow 2MgO$$

is historically called an oxidation because it involves the reaction of oxygen with something. In modern terms, we know that oxygen is electronegative and tends to attract electrons, giving it some degree of negative charge in most compounds; similarly, magnesium is electropositive. In this particular reaction, the electronegativity difference between oxygen and magnesium is so great that the compound is ionic; the magnesium oxide crystal consists of Mg^{2+} and O^{2-} ions.

Thus, the oxidation of magnesium results in the transformation of Mg to Mg^{2+}, and of O_2 to O^{2-}. If we temporarily define *oxidation state* or *oxidation number* as the charge on an atom or monatomic ion, we can see that the oxidation state of magnesium has gone from 0 to +2, so that oxidation corresponds to an increase in oxidation number, which must correspond to a loss of electrons. Oxygen has decreased in oxidation number, so it must have gained electrons; this we term *reduction*. Historically, the reduction of an ore meant reacting the ore with a substance which freed the metal, as in:

$$2FeO(s) + C(s) \rightarrow CO_2(g) + 2Fe(l)$$

The term "reduction" was probably used because the ore is reduced in both size and mass when this process takes place. But regardless of the historical origin of the terms, you must know that *oxidation is the loss of electrons* and *reduction is the gain of electrons*. Some more definitions: an *oxidizing agent* (or *oxidant*), such as O_2 above, is the species that is reduced in a redox reaction; a *reducing agent* (or *reductant*), for example Mg, is the one that is oxidized.

Chapter 15   Oxidation States and Oxidation-Reduction Reactions

Half-Reactions

A redox reaction can be conceptually broken up into separate oxidation and reduction parts. Consider the two reactions:

$$2Mg(s) + O_2(g) \rightarrow 2MgO(s)$$

$$4Na(s) + O_2(g) \rightarrow 2Na_2O(s)$$

They have in common the reduction of $O_2(g)$ to O^{2-}. This can be denoted by the *half-reaction* (or *half-cell reaction* or *redox couple*):

$$O_2 + 4e^- \rightarrow 2O^{2-}$$

The $4e^-$ are put there to denote that four negative electrons must be gained by O_2 to reduce it. What we do when we balance redox reactions is to "dissect" the overall process into a reduction half-reaction, such as the above, and an oxidation half-reaction. For the first reaction above, the oxidation half-reaction is

$$Mg \rightarrow Mg^{++} + 2e^-$$

Note that for these, and all other couples, the total charge is the same on both sides of the equation, once electrons have been added.

To reconstitute the overall reaction from the two couples, the general rule is that the number of electrons lost in oxidation must equal the number of electrons gained in reduction; therefore the oxidation half-reaction (Mg to Mg^{++}) has to occur twice each time the reduction half-reaction occurs:

$$O_2 + 4e^- \rightarrow 2O^{2-}$$

$$2(Mg \rightarrow Mg^{2+} + 2e^-)$$

This means that all stoichiometric coefficients in the oxidation couple get multiplied by two. If the couples are then added, the balanced overall equation results.

$$O_2 + 2Mg \rightarrow 2Mg^{2+} + 2O^{2-}$$

Note in particular that the electrons disappear; of course, that's because we chose our multiplier so that they would.

Chapter 15 — Oxidation States and Oxidation-Reduction Reactions

Oxidation State

We temporarily defined the oxidation state as the charge on an atom or simple ion. When we have covalent bonds, electrons have been shared, rather than completely transferred, so we do not have ions. As a simple example, consider the following reaction:

$$2C + O_2 \rightarrow 2CO$$

Stoichiometrically, this is identical with the oxidation of Mg. And it is certainly an oxidation, even in the primitive sense of a "reaction with oxygen". Clearly, it is carbon that is being oxidized. Now, we know that CO is a covalent molecule, rather than an ionic crystal like MgO; nevertheless, we still ascribe oxidation states of +2 and -2 to the carbon and oxygen, respectively, in CO. This illustrates both the great strength and the great weakness of the oxidation-state concept:

Weakness: Oxidation numbers rarely give a true picture of "where the electrons are" in a compound.

Strength: You don't have to know about molecular structure in great detail to use oxidation states. This is, in fact, a great strength, since, as you saw in the last few chapters, chemical bonding can be subtle, and, in fact, researchers are continually discovering new surprises in that realm.

In Section 15.2 the text gives a detailed set of rules for determining oxidation number. We will summarize the most useful of them here. A pure element has an oxidation number of zero. In a compound or ion, the sum of the oxidation numbers on the atoms must be equal to the charge on the species (zero if a compound); thus MgO is neutral overall, so that if O has oxidation number -2, Mg must have oxidation number +2. Similarly, of course, a monatomic ion, like Cl^-, has an oxidation state equal to the charge on the ion (-1 for Cl^-).

Oxygen has an oxidation state of -2, except in the peroxide ion (O_2^{2-}), where it has oxidation number -1, and in the rarely encountered superoxide ion (O_2^-), where its oxidation number is -1/2; thus, in CO, carbon must have oxidation number +2, and in CO_2 carbon must have oxidation number +4. In the carbonate ion, CO_3^{2-}, carbon must also have oxidation number +4, so that the sum of the oxidation numbers comes out to -2 (that is, $3(-2)+4=-2$).

The alkali metals (Li group) and alkaline earth metals (Ca group) have oxidation numbers +1 and +2 respectively in compounds. This corresponds to their ionic nature. H has oxidation number +1, unless it is bonded to a metal, in which case it has oxidation number -1. Thus in CH_4, C has oxidation number +4; in LiH, lithium hydride, H has oxidation number -1.

Where the above rules are in conflict, the more electronegative element keeps its usual oxidation number. Here are a few examples of the applications of these rules.

Compound	Oxidation Numbers
H_2S	H: +1, gives S: -2
SO_2	O: -2, gives S: +4
NH_3	H: +1, gives N: -3
NI_3	I: -1, gives N: +3
SO_4^{2-}	O: -2, gives S: +6
$Cr_2O_7^{2-}$	O: -2, gives Cr: +6
ClO_4^-	O: -2, gives Cl: +7
ICl	Cl: -1, gives I: +1
$FeCl_2$	Cl: -1, gives Fe: +2
$FeCl_3$	Cl: -1, gives Fe: +3

Note that certain atoms, such as S and Fe, are capable of a range of oxidation numbers. Also note that oxidation numbers can be fractional: for propane, C_3H_8, H: +1 gives C: -8/3. (For organic compounds, there is a way to ensure that carbon will never take on a fractional oxidation state. This is described in Chapter 23).

For simple (monatomic) ions recall that the oxidation number is given by the ionic charge. Since the formation of many simple ions (Na^+, F^-, etc.) is easy to rationalize in terms of atomic structure, it stands to reason that for these species, at least, oxidation number is a periodic property. For other simple ions, one has to dig a bit deeper for an explanation of the observed oxidation states. For example, why does Fe commonly exhibit the +2 and +3 oxidation states? The ground state of the neutral atom is $(Ar)4s^2 3d^6$, and the +2 oxidation state corresponds to stripping off the 4s electrons. Then stripping off a single additional electron gives the stable $3d^5$ configuration (half-filled subshell), corresponding to the +3 oxidation state.

Not all oxidation states are this easy to rationalize, however, even among the metals. For the non-metals, except for oxygen and fluorine, a bewildering variety of oxidation states is exhibited. For example, nitrogen exhibits all oxidation states between -3 and +5. It is possible to make some sense of this, and we try to do so in the **Perspective** at the end of this Study Guide chapter. For now, realize that elements of intermediate electronegativity, such as N, tend to form covalent, rather than ionic bonds. The oxidation number of N in a compound is dictated by the other atoms to which it is bound; for example, in NH_3, H: +1 gives N: -3, and in NI_3, I: -1 gives N: +3. This is in a sense an artifact of the rules for determining oxidation number, since the bonding is similar for these two compounds, as it is for CH_4 (C: -4) and CI_4 (C: +4).

Example 15.2 in the text discusses the relationship between oxidation number and formal charge. In CrO_4^{2-}, chromium exhibits the +6 oxidation state, but the formal charge is +2, +1 or 0, depending on the resonance structures one draws. The formal charge gives a better picture than oxidation number of "where the electrons are" in a covalently bonded system, but recall that one of the weaknesses of the oxidation number concept is that it is impervious to the subtleties of chemical bonding. In order to use formal charge you have to know about bonding; to use oxidation numbers you do not.

Balancing Redox Equations

How are oxidation numbers used? One important application is in the balancing of redox reactions. Recall that earlier we showed how the reaction of Mg with O_2 could be divided into two half-reactions which could be balanced separately, then combined to give an overall reaction. Balancing redox equations in aqueous solution is somewhat complicated by the fact that H^+ (in acid solution), OH^- (in

Chapter 15 — Oxidation States and Oxidation-Reduction Reactions

basic solution) and H_2O (in either case) can be among the reactants. For example, the permanganate ion, MnO_4^-, is a strong oxidizing agent in basic solution, where it is reduced to manganese dioxide, $MnO_2(s)$. In the corresponding half-reaction, somehow two oxygens must be disposed of. A balanced half-reaction that achieves this is:

$$2H_2O + MnO_4^- + 3e^- \rightarrow MnO_2(s) + 4OH^-$$

Note that this reduction half-reaction is balanced with regard to all elements and overall charge; it is unbalanced with regard to electrons, but when combined with an oxidation half-reaction the electrons will disappear in the overall reaction, as they did in our MgO example. The different methods of balancing redox equations differ only in the procedure used to make charge, H and O all balance at once.

The text gives several methods of doing this, and describes them in such detail that it makes little sense to repeat the description here. Instead, we will give a few extra examples. First, we will show how to balance the reaction

$$Zn(s) + NO_3^- \rightarrow Zn^{2+} + NH_4^+$$

in acid solution, using the *ion-electron method*. As we do this, you can follow the numbered steps of the method as given in the text. It should be obvious that Zn is being oxidized from the 0 to the +2 oxidation state, so that the oxidation couple is straightforward:

$$Zn \rightarrow Zn^{2+} + 2e^-$$

The $2e^-$ are added to achieve charge balance. The reduction couple must then be

$$NO_3^- \rightarrow NH_4^+$$

The N must be changing oxidation state, but it is not necessary at this point to calculate oxidation number; it is, however, necessary to make sure that N is balanced, which it is (one N on each side). Now we balance O using H_2O. This gives

$$NO_3^- \rightarrow NH_4^+ + 3H_2O$$

Next, we balance H using H^+. There are 10 H's on the right, and none on the left, so we must add 10 H^+ to the left, giving

$$NO_3^- + 10H^+ \rightarrow NH_4^+ + 3H_2O$$

Chapter 15 — Oxidation States and Oxidation-Reduction Reactions

Now we balance charge with e^-. The charge is +9 on the left and +1 on the right, so we must add $8e^-$ to the left, giving

$$NO_3^- + 10H^+ + 8e^- \rightarrow NH_4^+ + 3H_2O$$

The result is checked by making sure that the number of electrons corresponds to the change in oxidation number. For NO_3^- we have O: -2 gives N: +5, and for NH_4^+ we have H: +1 gives N: -3, so that the oxidation state of N is reduced by 8; this corresponds to a gain of $8e^-$, which is what we obtained, so it appears that we've done things correctly.

So far we have

$$(Zn \rightarrow Zn^{2+} + 2e^-) \times 4 \quad \text{(oxidation)}$$

$$NO_3^- + 10H^+ + 8e^- \rightarrow NH_4^+ + 3H_2O \quad \text{(reduction)}$$

each half-reaction is balanced in terms of all elements and overall charge, but we must combine the half-reactions in such a way that the electrons disappear from the final result. This is done by multiplying the first couple by 4, so that $8e^-$ are produced, which can be used up in the second reaction. Doing this, then adding the two couples, gives

$$4Zn(s) + NO_3^- + 10H^+ \rightarrow 4Zn^{2+} + NH_4^+ + 3H_2O$$

As a final check, we make sure that the overall reaction is balanced both in terms of charge and in terms of each element. The charge is +9 on both sides, and all elements balance, so that the overall equation is balanced.

Hint: check charge balance first. If you did something wrong, it will usually show up here. *Another hint*: there are so many little calculations to make in balancing redox equations that you should expect to come up with an incorrectly balanced final result pretty often. It is *very important* to check the final result. If it doesn't balance, go back and check whether the half-reactions balance. If they do, then obviously the mistake was in combining them. If a half-reaction doesn't balance, your best bet is to pull out a fresh sheet of paper and start over on it, meticulously, step by step. It's usually easier to do this than to check your first attempt.

Chapter 15 Oxidation States and Oxidation-Reduction Reactions

As an additional example we balance the *disproportionation* of chlorine dioxide, ClO_2, in basic solution.

$$ClO_2 \rightarrow ClO_2^- + ClO_3^-$$

A disproportionation is where two or more molecules of one species react to form two or more separate species. Here, for ClO_2, we have O: -2 gives Cl: +4. For ClO_2^-, O: -2 gives Cl: +3, and for ClO_3^-, O: -2 gives Cl: +5. Thus a species with Cl in the +4 state disproportionates to species with Cl in the +5 and +3 states. ClO_4 must therefore be on the left-hand side of both the oxidation and reduction couple.

We will go through the balancing for the oxidation couple, for which the skeletal equation is

$$ClO_2 \rightarrow ClO_3^-$$

One first makes certain that the atom undergoing redox (Cl) is balanced, which it is. Then one calculates the oxidation numbers, which we already did (+4 → +5), and adds electrons to balance this, giving

$$ClO_2 \rightarrow ClO_3^- + e^-$$

Charge is then balanced with OH^-, giving

$$ClO_2 + 2OH^- \rightarrow ClO_3^- + e^-$$

Next oxygen is balanced with H_2O, resulting in

$$ClO_2 + 2OH^- \rightarrow ClO_3^- + H_2O + e^- \qquad \text{(oxidation)}$$

If everything was done correctly, H should balance, which it does.

The reduction couple is easy.

$$ClO_2 + e^- \rightarrow ClO_2^- \qquad \text{(reduction)}$$

Since both the oxidation and the reduction half-reactions are one-electron processes, they are added just as they are, to give

$$2ClO_2 + 2OH^- \rightarrow ClO_2^- + ClO_3^- + H_2O$$

Charge and all atoms balance, so the result is correct.

Note that there is nothing special about balancing a disproportionation reaction. Once you realize that the same species is both reduced and oxidized, and therefore appears on the left in both half-reactions, the balancing process proceeds just as for any other redox reaction.

Chapter 15   Oxidation States and Oxidation-Reduction Reactions

Descriptive Redox Chemistry

Oxidation numbers can also be used to correlate and understand a fair amount of descriptive chemistry This subject is covered in some detail in the text. Not all the phenomena described can be easily explained in terms of the fundamentals of atomic structure and bonding; much of the material in these sections must be memorized if it is to be learned. In the **Perspective** we will concentrate on making sense of those parts of the material which can be readily understood, or at least readily classified. Here we summarize only the most important material in Sections 15.4, 15.5 and 15.6. Of course "importance" is in the eye of the beholder, and you should make sure you know what your instructor feels is important!

Nitrogen exhibits all the oxidation states from -3 to +5, and in virtually all its compounds the bonding is covalent. The most commonly occurring oxidation states are -3, as in ammonia, NH_3, 0, as in N_2, +5, as in nitric acid, HNO_3, and +3 as in nitrous acid, HNO_2. You should, however, also be familiar with the oxides of nitrogen: nitrogen dioxide, NO_2 (+4), nitric oxide NO (+2) and nitrous oxide (laughing gas), N_2O (+1). In addition, you should know that N_2O_5 (+5) is the *acid anhydride* of nitric acid; adding one molecule of water gives "$H_2N_2O_6$", which corresponds to $2HNO_3$.

Phosphorus, beneath N on the periodic chart, exhibits mainly the +3 and +5 oxidation states, as in PCl_3 and PCl_5. P_2O_5 is the anhydride of phosphoric acid, which you should be able to show by adding $3H_2O$ to the formula. As the text mentions, usually higher oxidation numbers give stronger oxyacids (HNO_3 stronger than HNO_2, H_2SO_4 stronger than H_2SO_3), but phosphorus provides an exception: H_3PO_3 (phosphorous acid) is stronger than H_3PO_4 (phosphoric). The reason for the more common trend is discussed later, in Section 15.5, for the oxyacids of chlorine, and also in the **Perspective**; the reason phosphorous acid is an exception is that one of the hydrogens is bonded directly to the phosphorous, whereas the general rule presumes that all the H's are bonded to the O's, as you shall see, below.

As one goes down Group VA, +3 and +5 remain the only common oxidation states, but +3 becomes more and more predominent as the elements become more metallic and the bonding becomes more ionic.

Chapter 15 — Oxidation States and Oxidation-Reduction Reactions

Among the halogens, fluorine exhibits only the -1 oxidation state in compounds, but the other halogens exhibit in addition oxidation states ranging up to +7. The odd-numbered states are by far the most common, and correspond, for chlorine, to the following well-known ions.

Formula	Name	Oxidation State
Cl^-	chloride	-1
ClO^-	hypochlorite	+1
ClO_2^-	chlorite	+3
ClO_3^-	chlorate	+5
ClO_4^-	perchlorate	+7

Among the oxyacids, hypochlorous is the weakest and perchloric is the strongest. Formal charge explains this. If comparable resonance structures are drawn (such as the ones obeying the octet rule, with no double bonds), then the greater the oxidation state, the greater the formal charge (see Section 15.5). Thus in perchlorate the formal charge on Cl is +3. This attracts electron density in the bonds connecting it to oxygen, helping the highly electronegative chlorine shed its uncomfortable positive charge. If a hydrogen is bound to an oxygen it competes with the Cl for electron density, so that loss of hydrogen helps satisfies the chlorine's thirst for electrons, contributing to the acidity. Where the formal charge on the chlorine is less, the Cl is less electronegative, so that there is less of an electron "thirst" to be satisfied; thus there is less of an advantage to be gained by losing a hydrogen, and therefore a lower acidity.

Transition metals have a very rich redox chemistry. Some further aspects of it will be discussed in Chapter 20. At this point the text discusses the aqueous chemistry of chromium and of manganese. Cr in the +6 state is a commonly used laboratory oxidizing agent, which exists as CrO_4^{2-}, chromate ion, in basic or neutral solution, and as dichromate, $Cr_2O_7^-$, in acid. Both of these are orange in color. When they are reduced they generally form chromic ion, Cr^{3+}, which is green.

Chromium forms four oxides: CrO (+2), Cr_2O_3 (+3), CrO_2 (+4) and CrO_3 (+6). Of these, CrO is weakly basic (soluble in strong acid) and CrO_3 is acidic (soluble in water to form an acid). Generally speaking, the higher the oxidation state of the central atom in an oxide, the more acidic the oxide. The reason for this is discussed in the **Perspective**. Cr_2O_3 and CrO_2 are *amphoteric*, meaning that they are insoluble in water, but soluble in either strong acid or strong base. Incidentally, CrO_3 finds use as an oxidizing agent in organic chemistry.

The oxides of manganese exhibit the expected trend in acidity: MnO is basic, MnO_2 is amphoteric and Mn_2O_7, a violently reactive red oil, is acidic. The most commonly encountered manganese species is the purple permanganate ion, MnO_4^-, a widely used oxidizing agent. In acid solution it gets reduced to the pink Mn^{2+}, and in basic solution to solid MnO_2.

It is not always easy to rationalize descriptive chemistry in terms of atomic structure; one way or another, however, this is what chemists seek to do. The aqueous chemistry we have been discussing involves a delicate balance among factors peculiar to the atoms themselves (atomic structure), solvation and crystal forces. Certain general trends, correlating electronegativity, acidity, oxidation number and atomic structure are, however, apparent, and are set out in the **Perspective**.

Chapter 15   Oxidation States and Oxidation-Reduction Reactions

Questions

15.1 Definitions of oxidation and reduction

1. For each of the following reactions, write the two balanced half-reactions, and the balanced overall reaction. Tell which half-reaction corresponds to oxidation, and which to reduction.

    (a) $Ca + S \rightarrow CaS$

    (b) $Al + O_2 \rightarrow Al_2O_3$

2. For each of the reaction in Question 1, what is the oxidizing agent, and what is the reducing agent

	Oxidant	*Reductant*
(a)	_____	_____
(b)	_____	_____

3. If A→B is an oxidation, then B→A is a reduction. (True, False)

15.2 The oxidation state of an element

1. Give the oxidation states of all the elements in each of the following compounds

    (a) H_2O

    (b) OH^-

    (c) $Na(s)$

Chapter 15   Oxidation States and Oxidation-Reduction Reactions

    (d) NaOH

    (e) $FeCl_3$

    (f) FeO

    (g) Fe_2O_3

    (h) Fe_3O_4

    (j) H_2O_2

    (k) LiH

    (l) $LiAlH_4$

    (m) ClO^-

    (n) ClO_2^-

    (o) MnO_4^-

    (p) CH_4 (methane)

    (q) C_2H_6 (ethane)

    (r) C_2H_4 (ethylene)

    (s) C_2H_2 (acetylene)

    (t) CH_3COOH (acetic acid)

    (u) C_4H_{10} (butane)

2. (a) Oxidation number depends on the particular Lewis structure used to represent a compound. (True, False)

   (b) Formal charge depends on the particular Lewis structure used to represent a compound (True, False)

3. For each of the following species, give both the oxidation number and the formal charge on the central element. When there is a choice of resonance structures, use one that satisfies the octet rule.

   (a) CH_4

Chapter 15   Oxidation States and Oxidation-Reduction Reactions

    (b) CO_2

    (c) SO_2

    (d) SO_4^{2-}

    (e) ClO_3^-

4. For each of the following groups in the periodic table, tell whether the most commonly observed oxidation state increases, decreases or remains the same as one goes down the periodic chart.

    (a) IA

    (b) IVA

    (c) VIIB (starting in second row)

    (d) VIIA

15.3 Balancing oxidation-reduction equations

1. Balance the following equations, assuming the reaction takes place in acid solution. Also give the balanced half-reactions

    (a) $H_2SO_4 + I^- \rightarrow I_2 + SO_2$

Chapter 15   Oxidation States and Oxidation-Reduction Reactions

(b) $Cr_2O^{2-} + C_2H_4O \rightarrow Cr^{3+} + C_2H_4O_2$

(c) $H_2O \rightarrow H_2 + O_2$

(d) $MnO_4^{2-} \rightarrow MnO_4^- + MnO_2$

(e) $MnO_4^- + H_2O_2 \rightarrow Mn^{2+} + O_2$

2. Balance the following equations, assuming the reaction takes place in basic solution. Also give the balanced half-reactions

(a) $Zn + NO_3^- \rightarrow NH_3 + Zn(OH)_4^{-2}$

(b) $ClO_2(g) \rightarrow ClO_2^- + ClO_3^-$

Chapter 15 — Oxidation States and Oxidation-Reduction Reactions

(c) $H_2O \rightarrow H_2 + O_2$

(d) $Ag_2S + CN^- + O_2 \rightarrow Ag(CN)_2^- + S$

(e) $CrO_4^{2-} + HSnO_2^- \rightarrow HSnO_3^- + CrO_2^-$

15.4 The oxidation states of nitrogen and other group VA elements

1. Write the formulas for the following oxides of nitrogen, and give the oxidation state of nitrogen in each

    (a) nitrogen oxide

    (b) nitrous oxide

    (c) nitrogen dioxide

2. Write the equation for the self-ionization of pure nitric acid. Is this a redox disproportionation?

3. Write the formula for the anhydride of each of the following, and give the oxidation state of nitrogen in each. Which is the stronger acid?

    (a) nitric acid

    (b) nitrous acid

Chapter 15 Oxidation States and Oxidation-Reduction Reactions

4. Write the equation for the decomposition of dinitrogen trioxide. Is this a redox disproportionation?

5. In acid solution, NO_2^- is usually reduced to _____, and ammonium salts are usually oxidized to _____.

6. (a) Complete and balance this reaction

$$Ca_3N_2(s) + H_2O \rightarrow$$

   (b) Is this a redox reaction?

7. Write the formula for the anhydride of each of the following, and give the oxidation state of phosphorus in each. Which is the stronger acid?

   (a) orthophosphoric acid

   (b) phosphorous acid

8. What is the most common oxidation state of

   (a) phosphorus

   (b) antimony

9. Write the formula for the anhydride of each of the following, and give the oxidation state of arsenic in each. Which is the stronger acid?

   (a) arsenic acid

   (b) arsenious acid

15.5 The oxidation states of chlorine and other group VIIA elements

1. For each of the following acids, write the name of the anion, the formula of the anhydride and the oxidation state of chlorine. Which is the strongest acid? Which the weakest?

   (a) perchloric acid

   (b) chloric acid

Chapter 15 Oxidation States and Oxidation-Reduction Reactions

(c) chlorous acid

(d) hypochlorous acid

2. When Cl⁻ is used as a reducing agent, to what form would you guess it is oxidized?

3. In what way does the redox chemistry of fluorine differ from that of the other halogens?

15.6 The oxidation states of the transition metals chromium and manganese

1. Name the following ions or compounds. Give the oxidation state of chromium in each.

   (a) $Cr_2O_7^{2-}$

   (b) CrO_4^{2-}

   (c) Cr^{3+}

   (d) Cr^{2+}

   (e) CrO_3

2. Write balanced equations for the following, and tell whether each is a redox reaction.

   (a) solution of CrO_3 in water

   (b) equilibrium of $Cr_2O_7^{2-}$ and CrO_4^{2-} in water

3. Tell whether each of the following is acidic, basic or amphoteric in water.

   (a) CrO

   (b) Cr_2O_3

Chapter 15   Oxidation States and Oxidation-Reduction Reactions

   (c) CrO_3

4. (a) In acid solution, chromium(VI) exists in the form _____, and is reduced to _____.

   (b) In basic solution, chromium(VI) exists in the form _____, and is reduced to _____.

5. The acid anhydride of $HMnO_4$ is an explosive oil.

   (a) What is its formula?

   (b) What is the name of $HMnO_4$

6. MnO_2 is amphoteric. Would you expect MnO to be acidic, basic or amphoteric?

7. The central element in an acid anhydride (e.g., S in SO_2) always has an oxidation state *lower than* that of the element in the acid.

8. Permanganate ion gets reduced to _____ in acid solution and to _____ in basic solution.

Chapter 15   Oxidation States and Oxidation-Reduction Reactions

Perspective: Descriptive Redox Chemistry of the Elements
and the Acidity of the Oxides

Note: You should have a periodic chart and a table of electronegativities in front of you while you read this section.

Range of Oxidation States

For the representative elements (all except the noble gases, transition metals and rare earths), only the s- and p-orbitals of the valence shell can be partially filled. Since these orbitals have a total capacity of eight electrons, the highest oxidation number available to such an element is the one which would be obtained by stripping the valence shell of electrons, and the lowest is that which would obtain from filling the valence shell to capacity. Thus, for example, nitrogen, with five valence electrons, exhibits oxidation numbers ranging from +5, corresponding to stripping the L-shell, to -3, which would result from filling it.

Nitrogen is unusual in exhibiting the extremes, as well as all intermediate values, of its possible range of oxidation numbers. The most electropositive elements, such as the Group IA and IIA metals (Na and Ca groups), exhibit positive oxidation states (+1 and +2, respectively), since they have low ionization potentials (meaning that they easily lose electrons) and low electron affinities (meaning that they don't add electrons easily). The most electronegative elements — oxygen and the halogens — exhibit mainly negative oxidation states, which are usually associated with ion formation. This, of course, is because they have low ionization potentials, so that they tend not to lose electrons readily, and high electron affinities, meaning that they gain electrons easily. Fluorine, the most electronegative element, exhibits only the -1 oxidation state, but the other halogens, which are much less electronegative, exhibit in addition positive oxidation states, as do sulfur, nitrogen and carbon.

Representative Elements

High positive oxidation states are generally associated with covalent bonding through hybrid orbitals, such as carbon's sp^3 (as in CF_4) and sulfur's d-orbital hybrids (as in SF_6), or with species that have reasonable resonance structures which violate the octet rule (as in ClO_4^-). Low positive oxidation numbers may be associated with either covalence or ionic bonding. Consider MgO and CO, for example. The first is ionic and the second is covalent, yet Mg and C both have oxidation numbers of +2 in these compounds. Recall that we said in the Scope that one of the strengths of the oxidation-state concept is that one doesn't need to know too much about bonding to use it. Conversely, as the MgO/CO analogy shows, one can not infer too much about bonding from oxidation numbers alone. MgO and CO are about as different as compounds can get, yet they are identical from the point of view of oxidation number. Most of this **Perspective** is about correlating oxidation number with chemical behavior, but clearly caution is in order!

The rules given in the text for calculating oxidation numbers really amount to ascribing to the most electronegative and the most electropositive elements in a compound the oxidation states that they exhibit in their ions (e.g., +1 for Na, -1 for F). Thus, it should not be too surprising that an element like carbon, which almost always forms covalent bonds, and is intermediate in electronegativity, can exhibit oxidation states ranging from +4 in CCl_4 to -4 in methane, CH_4, and that in ethane, CH_3CH_3, carbon's oxidation state is -3, and so on, without there being much difference in the actual environment of the carbon among these examples. In other words, the rules make the oxidation numbers of atoms of intermediate electronegativity (and intermediate location on the periodic chart) "float" according to what they are bonded to. This makes some sense, because when a carbon is bound to an electronegative element such as fluorine, we would expect the fluorine to attract electron density to it, polarizing the bond, and becoming the negative end of the bond dipole. Thus in the unstable diatomic molecule CF, F has an oxidation number of -1 and C has an oxidation number of +1.

There is a general rule that for the representative elements on the right-hand side of the periodic chart, as one goes down within a group, the negative and the the most highly positive oxidation states become less prevalent, and the low positive oxidation states become most prevalent. For example, nitrogen and antimony are both in Group VA, but since Sb is metallic it should bind better with highly electronegative elements than does N. This is indeed found; using the criterion of heats of formation, which you will learn about in Chapter 17, NH_3 and $SbCl_3$ are stable, whereas NCl_3 and SbH_3 are highly unstable. This is because the representative elements become less electronegative and more metallic with increasing atomic number, within a group. The negative oxidation states become less prevalent going down because the more metallic elements tend to bond best with electronegative elements such as the halogens, giving rise to positive oxidation states. Thus Sb "prefers" the +3 to the -3 oxidation state, and N "prefers" -3 to +3. Also, the more metallic elements tend to exhibit greater ionic character in their bonding; we said that high positive oxidation states usually imply covalence, so this explains why highly positive oxidation states are less likely to be found as one goes down a group within the representative elements.

The most characteristic positive oxidation states of these elements (remember we're still on the right-hand side of the periodic chart) are the ones corresponding to the loss of all the valence electrons and the ones corresponding to the loss of only the valence p-electrons. Consider lead (Pb) as an example. Its electronic configuration is $(Xe)6s^26p^2$, and it exhibits oxidation states of +2 and +4. According to what we said earlier about high oxidation numbers, the +4 oxidation state should be associated with covalence, and the +2 state with ionicity. In fact, $PbCl_4$ is a liquid at room temperature, whereas $PbCl_2$ is a solid exhibiting considerable ionic character. As one goes down the periodic chart, the increasing electropositivity and metallic nature of the elements increases their tendency to form bonds with at least some ionic character, giving rise to lower, but still positive, oxidation numbers. The reason high positive oxidation numbers usually imply covalence is that highly charged (+4 or greater) positive ions tend not to be formed. The sum of the first four or more ionization energies is so huge that the energy it takes to make the ion cannot be recouped by crystal forces; therefore, when such a high oxidation number exists, it must involve a covalent interaction.

Chapter 15 Oxidation States and Oxidation-Reduction Reactions

Transition Elements

The transition elements are all metallic, and exhibit only positive oxidation states. For the first five columns of the transition elements (Groups IIIB — VIIB), the highest oxidation state obtained is that corresponding to the loss of all the s- and d-electrons (+3 for Sc, +4 for Ti, ..., +7 for Mn). This holds for all rows. Once again, the high oxidation states (e.g., +7 in MnO_4^-) are associated with covalence, and the lower ones with increasingly ionic bonding. (Note that in MnO_4^- we are talking about the Mn-O bond being covalent.) To the right of iron ($3d^6$), only low oxidation numbers (+4 at most) are common, and even to the left oxidation states lower than the maximum are the ones most often found. For the transition metals, it is not always easy to explain the oxidation states which are exhibited. For example, it is easy to explain why Cu, Ag and Au, which are all in Group IB, each exhibit an oxidation state of +1 (loss of a single s-electron, leaving a stable d^{10} configuration), but hard to explain why Cu exhibits in addition a +2 state, Au a +3 state, and Ag no other state.

In contrast with the representative metals, within the transition groups the higher oxidation states become favored as one goes down the periodic chart. This corresponds to a slight increase in electronegativity (slight loss of metallic character) with increasing atomic number. For the representative elements, we had the opposite trend: electronegativity decreased (metallic nature increases) as one went down; in both cases, however, the more electropositive metals tend to exhibit the lower positive oxidation states.

Acidity of the Oxides

Oxidation states are useful for balancing redox equations, as we saw earlier; they also correlate much chemical data. We just saw that oxidation state is a periodic property, and can at least be partly understood on the basis of familiar notions such as metallic nature, electronegativity and electronic configuration. We saw that there is a close relationship between electronegativity and the oxidation states of the elements; we will soon see that there is also a close relationship between these and the acidity of the oxides.

Chapter 15 — Oxidation States and Oxidation-Reduction Reactions

In general, the more electronegative the element, the more acidic its oxide. The oxides of electronegative elements like N and S are all acidic in the Arrhenius sense, meaning that if they are added to water excess hydrogen ion is formed.

$$SO_3 + H_2O \rightarrow \text{sulfuric acid}$$

$$N_2O_5 + 2H_2O \rightarrow \text{nitric acid}$$

In contrast, the oxides of the highly electropositive elements on the left side of the periodic chart are basic:

$$Na_2O + H_2O \rightarrow \text{sodium hydroxide}$$

Since elements of similar electronegativities form covalent bonds together, and elements of greatly differing electronegativities form ionic bonds, and since oxygen is very electronegative, it follows that the strongly basic oxides, in which oxygen is bound to an electropositive element, are ionic, and the strongly acidic oxides are covalent. Thus, SO_3 and N_2O_5 are acidic and covalent, and Na_2O is basic and ionic.

As a further and more subtle example of these trends, consider the tetravalent oxyacids of the elements Si through Cl, which fall within the second period of the representative elements. ("Tetravalent oxyacid" is just a fancy term for an acid whose central element is attached to four oxygens.)

Tetravalent Oxyacids of the Elements Si through Cl

Oxide	Oxy-acid	Electro-negativity	Oxidation Number	Formal Charge*	Acidity	Crystal Structure
SiO_2	$Si(OH)_4$	1.7	+4	0	very weak	network
P_2O_5	$PO(OH)_3$	2.1	+5	+1	weak	molecular
SO_3	$SO_2(OH)_2$	2.4	+6	+2	very strong	molecular
Cl_2O_7	$ClO_3(OH)$	2.8	+7	+3	extremely strong	molecular

*In single-bonded resonance structure.

Note that despite writing the formulas in a funny way, three of the acids are the well-known phosphoric, sulfuric and perchloric acids. The fourth is called silicic acid.

Chapter 15

Oxidation States and Oxidation-Reduction Reactions

Even comparing the oxides of these elements of rather similar electronegativities, it is clear that the oxides of the more electronegative elements are the more acidic. They are also the more covalent, as one would expect from our previous discussion: P_2O_5, SO_3 and Cl_2O_7 are discrete molecules, but SiO_2 forms a covalently-bound network crystal. Such a structure, though still covalent, is an intermediate type between ionic crystals and crystals consisting of discreet molecules, as you will see in Chapter 21 of the Study Guide.

Note also that the more acidic oxides are associated with higher oxidation states on the central element. This is also usually true when a single element exhibits multiple oxidation states; for example, SO_3 (oxidation number +6 on S) is the anhydride of sulfuric acid, H_2SO_4, and SO_2 (oxidation number +4 on S) is the anhydride of sulfurous acid, H_2SO_3, and sulfuric acid is stronger than sulfurous. Why should higher oxidation states be associated with greater acidity? The answer is that for compounds whose bonding is predominently covalent, oxidation state is related to formal charge, which is at least somewhat indicative of the real electron distribution in the compound. This is discussed in the text (Section 15.5) for the oxyacids of chlorine. High formal charge results in the strengthening of the bonds from the central element to the oxygen, and a weakening of the O-H bonds, resulting in greater acidity. Another way of seeing this is to compare the acid and the anion, shown here for perchloric acid, using Lewis structures obeying the octet rule, with formal charges indicated. By losing the proton, an additional non-bonded pair of electrons on oxygen becomes available to be polarized by and thereby help neutralize the formal charge on the chlorine. The greater the formal change on the central atom, the more the species is stabilized by the loss of a proton — in other words, the more acidic it is.

Chapter 15   Oxidation States and Oxidation-Reduction Reactions

Amphoterism

As mentioned above, strongly acidic and strongly basic oxides, respectively, dissolve in water to give excess hydronium and hydroxide ion. A weakly acidic oxide such as SiO_2, silica, is one which is insoluble in water, but at least somewhat soluble in strong base. This is the reason that strong base should not be stored in glass bottles; the glass, which is largely derived from silica, partly dissolves. Similarly, a weakly basic oxide, such as MnO, is one which is insoluble in water, but which dissolves in strong acid. Recall that the more electronegative elements form acidic oxides and that the more electropositive elements form basic oxides. An electronegativity of 1.5 is roughly the dividing line between the elements forming weakly acidic and weakly basic oxides. Close to this value, oxides tend to be amphoteric, meaning that they dissolve both in strong acid and strong base. As the text mentions, high oxidation number and high formal charge tend, in effect, to make the central atom in a species more electronegative. If oxides of more electronegative elements tend to be more covalent and more acidic, then oxides of the same element in higher oxidation states should also be more covalent and more acidic. This is found. We mentioned that MnO (+2 state) is weakly basic; MnO_2 (+4) is amphoteric, and Mn_2O_7 (+7) is an explosive red oil which is the anhydride of permanganic acid, $HMnO_4$. A similar trend is found among the oxides of chromium: CrO is basic, Cr_2O_3 is amphoteric and CrO_3 is acidic.

As an example of amphoterism consider aluminum oxide, Al_2O_3. Its solution in acid and base is according to the following equations:

$$Al_2O_3 + 6H^+ \rightarrow 2Al^{+3} + 3H_2O$$

$$Al_2O_3 + 6OH^- + 3H_2O \rightarrow 2[Al(OH)_6]^{-3}$$

Actually, in strongly acidic solution, the aluminum cation exists in the form $[Al(H_2O)_6]^{+3}$; in strongly basic solution, it exists as $[Al(OH)_6]^{-3}$. $[Al(H_2O)_6]^{+3}$ is an acidic species and will donate a proton ($K_a \approx 10^{-5}$) to give $[Al(OH)(H_2O)_5]^{+2}$; likewise, $[Al(OH)_6]^{-3}$ is basic, and will pick up a proton to form $[Al(H_2O)(OH)_5]^{-2}$. The question then arises: if these ions are all soluble, why is Al_2O_3 not soluble in neutral solution to give species like $[Al(H_2O)_3(OH)_3]$ and $[Al(H_2O)_4(OH)_2]^{-1}$? In other words, so far we have explained why Al_2O_3 is soluble in acid and base, but not why it is insoluble in water. The answer probably lies in the fact that crystalline Al_2O_3 is extremely stable. If there is not enough acid or base in solution, the complex ions formed will be close to neutral. These neutral ions can approach each other and react to precipitate the stable solid. The highly charged species that exist far from acid-base neutrality, however, repel each other so that this approach cannot take place. The pH where the predominent species is neutral is called the *isolectric point*, and those of you who go on to study biochemistry will find that it is very important in protein chemistry. Because neutral protein molecules can approach each other and react to form stable crystals, the solubility of a protein is at a minimum near its isoelectric point. This fact is often used in the purification of proteins.

Chapter 15 — Oxidation States and Oxidation-Reduction Reactions

Answers

15.1 Definitions of oxidation and reduction

1.
    (a) oxidation $Ca \rightarrow Ca^{2+} + 2e^-$
        reduction $S + 2e^- \rightarrow S^{2-}$
        overall $Ca + S \rightarrow CaS$
    (b) oxidation $Al \rightarrow Al^{3+} + 3e^-$
        reduction $O_2 + 4e^- \rightarrow 2O^{2-}$
        overall $4Al + 3O_2 \rightarrow 2Al_2O_3$

2.

	Oxidant	*Reductant*
(a)	S	Ca
(b)	O	Al

3. True

15.2 The oxidation state of an element

1. (a) H=+1, O=-2
   (b) H=+1, O=-2
   (c) Na=0
   (d) Na=+1, O=-2, H=+1
   (e) Fe=+3, Cl=-1
   (f) Fe=+2, O=-2
   (g) Fe=+3, O=-2
   (h) Fe=+8/3, O=-2
   (j) H=+1, O=-1
   (k) Li=+1, H=-1
   (l) Li=+1, Al=+3, H=-1
   (m) Cl=+1, O=-2
   (n) Cl=+3, O=-2
   (o) Mn=+7, O=-2
   (p) H=+1, C=-4
   (q) H=+1, C=-3
   (r) H=+1, C=-2
   (s) H=+1, C=-1
   (t) H=+1, O=-2, C=-2
   (u) H=+1, C=2.5

2. (a) False
   (b) True

3.

	Oxidation Number	*Formal Charge*
(a)	-4	0
(b)	+4	0
(c)	+4	+1

Chapter 15   Oxidation States and Oxidation-Reduction Reactions

(d)   +6   +2
(e)   +5   +2

4. (a) remains the same
    (b) decreases
    (c) increases
    (d) remains the same

15.3 Balancing oxidation-reduction equations

1. (a) oxidation: $2I^- \rightarrow I_2 + 2e^-$
   reduction: $H_2SO_4 + 2H^+ + 2e^- \rightarrow SO_2 + 2H_2O$
   overall: $H_2SO_4 + 2H^+ + 2I^- \rightarrow I_2 + SO_2 + 2H_2O$

   (b) oxidation: $3(C_2H_4O + H_2O \rightarrow C_2H_4O_2 + 2H^+ + 2e^-)$
   reduction: $Cr_2O_7^{2-} + 14H^+ + 6e^- \rightarrow 2Cr^{3+} + 7H_2O$
   overall: $Cr_2O_7^{2-} + 3C_2H_4O + 8H^+ \rightarrow 2Cr^{3+} + 3C_2H_4O_2 + 4H_2O$

   (c) oxidation: $2H_2O \rightarrow O_2 + 4H^+ + 4e^-$
   reduction: $2(2H^+ + 2e^- \rightarrow H_2)$
   overall: $2H_2O \rightarrow O_2 + 2H_2$

   (d) oxidation: $2(MnO_4^{2-} \rightarrow MnO_4^- + e^-)$
   reduction: $MnO_4^{2-} + 4H^+ + 2e^- \rightarrow MnO_2 + 2H_2O$
   overall: $3MnO_4^{2-} + 4H^+ \rightarrow 2MnO_4^- + MnO_2 + 2H_2O$

   (e) oxidation: $5(H_2O_2 \rightarrow O_2 + 2H^+ + 2e^-)$
   reduction: $2(MnO_4^- + 8H^+ + 5e^- \rightarrow Mn^{2+} + 4H_2O)$
   overall: $5H_2O_2 + 2MnO_4^- + 6H^+ \rightarrow 5O_2 + 2M_n^{2+} + 8H_2O$

2. (a) oxidation: $4(Zn + 4OH^- \rightarrow Zn(OH)_4^{2+} + 2e^-)$
   reduction: $NO_3^- + 6H_2O + 8e^- \rightarrow NH_3 + 9OH^-$
   overall: $4Zn + NO_3^- + 6H_2O + 7OH^- \rightarrow 4Zn(OH)_4^{2-} + NH_3$

   (b) oxidation: $ClO_2 + 2OH^- \rightarrow ClO_3^- + H_2O + e^-$
   reduction: $ClO_2 + e^- \rightarrow ClO_2^-$
   overall: $2ClO_2 + 2OH^- \rightarrow ClO_3^- + ClO_2^- + H_2O$

Chapter 15 Oxidation States and Oxidation-Reduction Reactions

(c) oxidation: $4OH^- \rightarrow O_2 + 2H_2O + 4e^-$
reduction: $2(2H_2O + 2e^- \rightarrow H_2 + 2OH^-)$
overall: $2H_2O \rightarrow 2H_2 + O_2$

(d) Note that only S and O change oxidation states.
oxidation: $2(Ag_2S + 4CN^- \rightarrow 2Ag(CN)_2^- + S + 2e^-)$
reduction: $O_2 + 2H_2O + 4e^- \rightarrow 4OH^-$
overall: $2Ag_2S + 8CN^- + O_2 + 2H_2O \rightarrow 4Ag(CN)_2^- + 2S + 4OH^-$

(e) oxidation: $3(HSnO_2^- + 2OH^- \rightarrow HSnO_3^- + H_2O + 2e^-)$
reduction: $2(CrO_4^{2-} + 2H_2O + 3e^- \rightarrow CrO_2^- + 4OH^-)$
overall: $3HSnO_2 + 2CrO_4^{2-} + H_2O \rightarrow 3HSnO_3 + 2CrO_2^- + 2OH^-$

15.4 The oxidation states of nitrogen and other group VA elements

1.

Formula	Oxidation State
NO	+2
N_2O	+1
NO_2	+4

2. $2HNO_3 \rightarrow H_2O + NO_3 + NO_2^+$
Not a redox disproportionation, since N has oxidation number +5 throughout.

3.

Anhydride	Oxidation State
(a) N_2O_5	+5
(b) N_2O_3	+3

HNO_3 is the stronger acid.

4. $N_2O_3 \rightarrow NO + NO_2$
This is a redox disproportionation.

5. NO, N_2

6. (a) $Ca_3N_2 + 6H_2O \rightarrow 2NH_3 + 3Ca(OH)_2$
(b) Not a redox reaction, since all elements retain the same oxidation state (-3 for N).

7.

Anhydride	Oxidation State
(a) P_2O_5	+5
(b) P_2O_3	+3

Chapter 15 Oxidation States and Oxidation-Reduction Reactions

Phosphorous is the stronger acid. (Exception to general rule; see Scope)

8. (a) +5
   (b) +3

9.
	Anhydride	Oxidation State
(a)	As_2O_5	+5
(b)	As_2O_3	+3

Arsenic is the stronger acid

15.5 The oxidation states of chlorine and other group VIIA elements

1.
	Anion Name	Anhydride	Oxidation State
(a)	perchlorate	Cl_2O_7	+7
(b)	chlorate	Cl_2O_5	+5
(c)	chlorite	Cl_2O_3	+3
(d)	hypochlorite	Cl_2O	+1

Perchloric is the strongest, hypochlorous the weakest acid.

2. $Cl_2(g)$ (which can then escape)

3. Fluorine exhibits only the -1 oxidation state.

15.6 The oxidation states of the transition metals chromium and manganese

1.
	Name	Oxidation State
(a)	dichromate	+6
(b)	chromate	+6
(c)	chromic	+3
(d)	chromous	+2
(e)	chromium(VI)oxide	+6

2. (a) $2CrO_3 + H_2O \rightarrow Cr_2O_7^{2-} + 2H^+$ or $CrO_3 + H_2O \rightarrow CrO_4^{2-} + 2H^+$
   Not redox.
   (b) $Cr_2O_7^{2-} + H_2O = 2CrO_4^{2-} + 2H^+$
   Not redox.

3. (a) basic
   (b) amphoteric
   (c) acidic (see answer to 2(a) above!)

4. (a) dichromate ion, chromic ion
   (b) chromate ion, chromite ion $(Cr(OH)_4^-)$

5. (a) Mn_2O_7
   (b) permanganic acid

15-31

Chapter 15   Oxidation States and Oxidation-Reduction Reactions

6. basic (lower oxidation state of Mn)
7. the same as
8. manganous ion, manganese dioxide.

CHAPTER 16: ELECTROCHEMISTRY

Scope

In the last chapter, you saw how redox reactions can be divided into *half-reactions* — an oxidation part and a reduction part. At the time we treated this simply as an artifice to help us balance equations, but in this chapter we shall see that the two half-reactions can actually be carried out separately — for example, in two separate beakers — if a path is provided for electrons to flow from the beaker where oxidation is taking place to the one where reduction is taking place. Such a system is called an *electrochemical cell*, and each beaker is a *half-cell*.

If the overall reaction taking place in the two half-cells is spontaneous, then the flow of electrons can be harnessed to do electrical work, such as lighting a lamp or running a motor, and the cell is called a *galvanic cell*, a *voltaic cell*, or a *battery*. If a non-spontaneous reaction is to be made to go, then electrical work has to be put into the cell, and the cell is then called an *electrolytic cell*.

It turns out that each half-reaction has a number associated with it, called the reduction potential. The reduction potentials for a pair of half-reactions can be used to tell whether a reaction involving the two couples will be spontaneous. (It is important to know that "spontaneous" in this context means that the reaction will *eventually* take place, not that it will take place quickly).

In Chapter 8, you learned that "spontaneous" means $K_{eq} > 1$. In fact, reduction potentials can be used to calculate the equilibrium constant of any redox reaction. Of course, at equilibrium, no net reaction is taking place; a galvanic cell can do work only if it starts out not at equilibrium. On the way to achieving equilibrium, work can be done. These phenomena are described quantitatively by the *Nernst Equation*, and the most important part of the chapter involves using it. Finally, the chemistry of several commercial batteries and electrolytic processes is discussed.

Galvanic Cells

Consider the balanced chemical equation

$$Fe^{3+} + I^- \rightarrow \frac{1}{2}I_2 + Fe^{2+}$$

This reaction is spontaneous in aqueous solution. The two half-reactions are

$$Fe^{3+} + e^- \rightarrow Fe^{2+} \qquad \text{(reduction)}$$

$$I^- \rightarrow \frac{1}{2}I_2 + e^- \qquad \text{(oxidation)}$$

The oxidant, Fe^{3+}, and the reductant, I^-, can be placed in separate beakers, connected by a wire, as shown in Figure 16.1. Of course, they must be in the form of neutral salts, such as $Fe_2(SO_4)_3$ and KI. Since electrons can flow through the wire, it is reasonable to think that electrons should travel from the I^--containing beaker to the Fe^{3+}-containing beaker. Then the iodide, which "wants" to be oxidized, can lose electrons, and the Fe^{3+}, which "wants" to be reduced, can gain electrons.

If electrons were to flow, however, the I^--containing solution would become positively charged, and the Fe^{3+} solution would become negatively charged, since we would be moving negatively charged electrons from the former to the latter. Since charge separation is highly unfavorable energetically, no reaction takes place in the system we have described. Now, however, suppose we add to the system a *salt bridge*. This is a connecting tube containing a strong electrolyte dissolved in some kind of gelatin, or else dissolved in water, but making contact with the solutions in the beaker through porous plugs (Figure 16.2). Now, for each electron that goes through the wire, the net charge can be compensated by the diffusion of ions out of the salt bridge. Let us suppose that the electrolyte in the salt bridge is NH_4NO_3. If two electrons from the I^--containing solution pass through the wire into the Fe^{3+}-containing solution, the following also happens:

— one I_2 molecule is formed in the I^--containing solution

— two NO_3^- ions diffuse from the salt bridge into the I^--containing solution

— two Fe^{3+} ions are reduced to Fe^{2+}

— two NH_4^+ ions diffuse from the salt bridge into the Fe^{3+}-containing solution.

Note that the direction of diffusion of ions from the salt bridge into the solutions is such as to maintain charge neutrality. The overall driving force of the reaction is the spontaneity of the overall redox process itself; this "pushes" electrons through the wire and "sucks" ions out of the salt bridge.

A few more words on the salt bridge. Note that NO_3^- ions diffuse out to the left in Figure 16.2. This should leave an excess of positively charged ions within the salt bridge on the left. In addition, NH_4^+ ions diffuse out on the right, which should leave an excess of negatively charged ions within the salt bridge on the right. Note, however, that the *total* charge within the salt bridge remains balanced. What happens is that as ions diffuse out of the salt bridge, the remaining ions diffuse within the bridge to prevent charge build-up. Clearly, charge cannot be allowed to build up in the opposite ends of the salt bridge, any more that it can within the beakers. In fact, the rate at which the entire redox process can take place is often limited by how quickly the ions within the salt bridge can diffuse to maintain uniform concentration.

Why do we need a salt bridge at all? Why can't we just connect the two beakers with a hollow tube, and let the solutions themselves diffuse into each other? The reason is that if we do, the overall redox reaction will take place where the solutions mix. To the extent that this occurs, we are using up reactants without forcing electrons through the wire. The electrons going through the wire can light a lamp or turn a motor to do work, but electrons transferred directly between the two solutions on mixing cannot be "tapped" for their energy. The object is to make as much of the electron flow as possible go through the wire. Although there is a small amount of diffusion of reactants into the salt bridge, where they can react directly, this can be made insignificant by proper design of the cell. Thus the salt bridge performs a positive and a negative function: it facilitates electroneutrality in the reacting solutions, and it presents the mixing of these solutions.

Electrodes

Now, some terminology. The surfaces at which reaction takes place — where the I^- loses electrons and the Fe^{2+} gains them — are called *electrodes*. Platinum is often used as an *inert* or *passive electrode* because it itself is not easily oxidized or reduced. In many industrial processes graphite is used. You will see later that there is also such a thing as an *active electrode*, which actually takes part in the reaction. The *anode* is defined as the electrode where oxidation takes place, and the *cathode* is the electrode where reduction takes place. (One way to remember which is which is that "anode" and "oxidation" both begin with vowels, and "cathode" and "reduction" both begin with consonants.)

Chapter 16   Electrochemistry

The Faraday

From the balanced chemical equation that we started with, it should be clear that by the time Avogadro's number, N_A, of electrons have passed through the wire, one mole of Fe^{3+} will have been converted to one mole of Fe^{2+}, and one mole of I^- will have been converted to one-half mole of I_2. The total amount of charge contained in a mole of electrons is called the *faraday*, and is equal to about 96,500 coulombs. It is usually symbolized by a script F, but since my word processor does not have a script font, I will use a bold-face **F**. Incidentally, Michael Faraday, the early 19th century English physicist for whom the faraday is named, performed many experiments with electrochemical cells. As we commented in Chapter 12, it is these experiments which gave the first hints that the forces holding matter together must be electrical in nature.

Active Electrodes

For historical reasons, certain electrochemical cells have special names. One is the *Daniell cell*, in which the following half-reactions take place.

$$Zn(s) \rightarrow Zn^{2+} + 2e^- \quad \text{(oxidation)}$$
$$Cu^{2+} + 2e^- \rightarrow Cu(s) \quad \text{(reduction)}$$

The cell is illustrated in Figure 16.3. The anode is solid zinc, and the cathode is solid copper. These are *active electrodes*; the anode dissolves and the cathode increases in mass as reaction proceeds. As the text mentions, if we used, say, platinum for the cathode, it would become copper plated.

Note that in the Daniell cell one of the reactants, Zn, is in the solid phase. Therefore even if the solutions containing Zn^{2+} and Cu^{2+} mix, the reaction cannot take place; however, the zinc electrode cannot be permitted to come in contact with Cu^{2+} ions, or else the cell reaction will take place on the surface, and we will have copper-plated zinc. Thus we do not need as stringent separation of the two half-cells as a salt bridge provides; a porous glass plug, as shown in Figure 16.3, will suffice.

Chapter 16 — Electrochemistry

Current

Electrical current is defined as the rate at which charge is transferred; for example, one *ampere* is defined as one coulomb per second; 1A = 1C/s. Consider the Daniell cell. Note from the half-reactions we wrote down that it takes 2F, or 193,000C, to reduce one mole of Cu, which weighs 63.5g. Suppose in a Daniell cell a mole of copper is plated out in one hour (3600 seconds). Then the average current must have been 193,000C/3600s=53.6A. Example 16.1 provides a more complicated problem based on the same idea.

Incidentally, 53.6A is a lot of current; a 100watt light bulb draws about 1A, though at a much higher voltage. (We will talk about the voltage across an electrochemical cell very shortly.) Certain metals, such as aluminum, are extracted from their ores by electrochemical processes, and their cost is directly related to the cost of electricity. In fact, it is usually more economical to situate aluminum plants near sources of electricity, and ship ore to them, than to situate the plants near the ore, and transmit power to them.

Cell Voltage

The *electrical potential difference* between two points, measured in volts, V, is a measure of the work done on or by a charge in moving from one point to the other. In the SI system of units, if a charge of one coulomb (1C) moves through a voltage difference of +1V, then 1J of work was done on it.

$$1 \text{ volt} = 1 \frac{\text{joule}}{\text{coulomb}}$$

This could result from moving the charge from a place where the potential is 0V to a place where it is 1V, or from a place where the potential is +1000V to one where it is +1001V; as the text makes clear, it is potential differences, not absolute values, that count. The zero of any potential is arbitrarily defined.

If a mole of electrons is moved through a potential of +1V — that is, if these negative particles are allowed to flow from a negatively charged to a positively charged electrode one volt apart in potential — then 96,500J of energy (1F×1V) is released. If the flow is taking place in a wire with some resistance, then the energy release is in the form of heat; if the wire is intercepted by a small motor, then work can be done. Similarly, if a faraday of electrons is forced through a potential difference of -1V — that is, if an external source of energy is used to move these negative electrons from a positive to a negative electrode 1V apart in potential — then the energy it takes to do this is 96,500J.

It is found experimentally that three factors influence the measured *EMF* (*electromotive force*, or *cell voltage*) of an electrochemical cell. These are

— the chemical nature of the cell ingredients

— their concentrations

— the temperature.

In this chapter we will discuss the first two; the third will be considered in Chapter 18.

First of all, we wish to isolate the effects of chemical composition on cell voltage; that means that we want to be able to compare, say, the two cells we have discussed so far in this chapter and be able to say which has a higher cell voltage. Recall that the two overall reactions were

$$I^- + Fe^{3+} \rightarrow \frac{1}{2}I_2 + Fe^{2+}$$

$$Zn(s) + Cu^{2+} \rightarrow Zn^{2+} + Cu(s)$$

Since concentrations and temperature, as well as chemical nature, have effects, clearly we're going to have to compare the voltages of the two cells under the same conditions of concentration and temperature. You will see a bit later that we do this by making all our comparisons at 25°C and at concentrations of 1M for dissolved substances. It turns out that when the two cells are set up in this way, the Fe^{3+}/I^- cell always exhibits an EMF of 0.24V, and the Daniell cell always exhibits a potential of 1.10V.

Another experimental finding is that if we have two redox reactions such as the following

$$A + B \rightarrow C + D \qquad (I)$$

$$C + D \rightarrow E + F \qquad (II)$$

and if we measure the voltages across the cells in which they occur (for now, call these voltages V_I and V_{II}), then their sum, $V_{III} = V_I + V_{II}$ is equal to the V that we would measure for the overall process.

Chapter 16 — Electrochemistry

$$A + B \rightarrow E + F \tag{III}$$

In other words, when chemical reactions (I and II) add to give an overall process, (III), then the cell EMF's add to give the cell EMF of the overall process ($V_{III} = V_I + V_{II}$). This is true just as long as temperature and concentrations are the same for all three cells.

To give a specific example, the following two reactions can be run in electrochemical cells.

$$Cu^{2+} + H_2(g) \rightarrow Cu(s) + 2H^+ \tag{I}$$

$$Zn(s) + 2H^+ \rightarrow Zn^{2+} + H_2(g) \tag{II}$$

It is found experimentally that $V_I = 0.34V$ and $V_{II} = 0.76V$. The reactions add to give the overall process for the Daniell cell,

$$Cu^{2+} + Zn(s) \rightarrow Cu(s) + Zn^{2+} \tag{III}$$

and we already mentioned that $V_{III} = 1.10V$, which you can see is equal to $V_I + V_{II}$. This sort of thing always happens, just as long as the cells are all run under the same conditions of temperature and concentration.

Standard Conditions

There is a set of standard conditions that are used to tabulate cell potentials. These are

— temperature of 25°C

— concentration of 1M for dissolved substances, 1atm for gases, the pure state for pure solids and liquids, such as H_2O, $Cu(s)$ and $Zn(s)$

-.3i

A substance exhibiting the concentration described above — such as 1M NaCl(aq), $H_2(g)$ at 1atm or $Cu(s)$ — is said to be in its *standard state*. Note that the term "standard state" specifies the pressure and the concentration, but not the temperature. A substance can be in its standard state at any temperature; however, data on materials in their standard states are generally given for T=298K (25°C).

Let us discuss how a half-cell can be constructed to perform the reaction

$$H_2(g) \rightarrow 2H^+ + 2e^-$$

Chapter 16   Electrochemistry

under standard conditions; this will be called the *standard hydrogen electrode* (*SHE*). Such an electrode is illustrated in Figure 16.4. Hydrogen gas, maintained at 1atm, is bubbled over an inert electrode immersed in a solution containing 1M H^+. The electrode is usually made of platinum, but doesn't have to be. What makes the electrode standard are the pressure of the H_2 and the concentration of the H^+.

Cell Conventions

Since it is cumbersome to draw a picture for each cell we talk about, we use a short-hand notation. For the oxidation of H_2 at the SHE we write

$$Pt(s)|H_2(1atm)|H^+(1M)$$

In this notation, vertical bars denote a phase barrier; thus the platinum electrode, H_2(1atm) and H^+(1M) are in the solid, gaseous and aqueous phases respectively. The electrode is always written all the way on the left or the right (see below), and the ingredients are often given in the order in which they react; thus, for the half-cell, as written, $H_2(g)$ is reacting to form $H^+(aq)$. This is an oxidation, so this half-cell must be the anode.

For the reduction of copper under standard conditions we write the half-cell as follows.

$$Cu^{2+}(1M)|Cu(s)$$

Note that in this half-cell $Cu^{2+}(aq)$ forms $Cu(s)$, left-to-right. This is a reduction, hence this half-cell is a cathode. Putting this together with the SHE anode gives

$$Pt(s)|H_2(1atm)|H^+(1M)||Cu^{2+}(1)|Cu(s) \qquad (I)$$

The double bar represents a salt bridge or other barrier.

Note that this cell carries out reaction (I), given earlier. The voltage, which we called V_I, and which we said was equal to 0.34V, is actually the EMF of this cell measured under standard conditions; this is given the symbol $\Delta E°$; ΔE is used, rather than V, for a cell potential, and the ° means standard conditions. The text uses a script E, but we will use bold-face instead.

The cell that carries out reaction (II) can be written

Chapter 16   Electrochemistry

$$Zn(s)|Zn^{2+}(1M)||H^{+}(1M)|H_2(1atm)|Pt(s) \qquad \text{(II)}$$

Note that here Zn(s) is the anode, and the SHE is the cathode. In the normal cell shorthand, the anode is always on the left. Recall that for cell (II) we said $\Delta E = 0.76V$. For the Daniell cell we can write

$$Zn(s)|Zn^{2+}(1M)||Cu^{2+}(1M)|Cu(s) \qquad \text{(III)}$$

Again, the anode is on the left. Clearly, from what we have said so far, $\Delta E° = 1.10V$ for the Daniell cell.

Half-Cell Potentials

Now, suppose I have the following arbitrary oxidation and reduction half-reactions.

$$A \rightarrow B + e^- \qquad \text{(a)}$$

$$X + e^- \rightarrow Y \qquad \text{(b)}$$

If I know $\Delta E°$ for the two cells

$$Pt(s)|A(1M), B(1M)||H^+(1M)|H_2(g)|Pt(s) \qquad \text{(a)}$$

$$Pt(s)|H_2(g)|H^+(1M)||X(1M), Y(1M)|Pt(s) \qquad \text{(b)}$$

Then, as for the Daniell cell, for the overall process

$$A + X \rightarrow B + Y \qquad \text{(c)}$$

we have

$$Pt(s)|A(1M), B(1M)||X(1M), Y(1M)|Pt(s) \qquad \text{(c)}$$
$$\Delta E°_a + \Delta E°_b = \Delta E°_c$$

In other words, if I tabulate $\Delta E°$'s for a bunch of half-reactions versus the SHE, then I know the $\Delta E°$ for any pair of the half-reactions taking place together. Such tabulated values against the SHE are denoted $E°$, rather than $\Delta E°$, and are called *standard half-cell potentials*.

We still have to discuss the sign of the half-cell potential. For reaction (III), the Daniell cell, illustrated in Figure 16.3, electrons flow through the wire from the anode to the cathode. Zn^{2+} ions come off the Zn anode into solution, leaving excess electrons behind. These electrons flow through the wire and combine with Cu^{2+} ions on the surface of the Cu(s) cathode. Since the direction of spontaneous electron flow must be from a relatively negative to a relatively positive potential, the cathode must be positively charged with respect to the anode. This is true for any galvanic cell.

Chapter 16 — Electrochemistry

Recall that the cathode is where reduction takes place. In cell (I) above, Cu^{2+} is reduced at the cathode, and the SHE is the anode; therefore the Cu(s) electrode will be positive. For a similar reason, in cell (II) the Zn(s) anode must be negative. Note that the standard half-cell potentials in a tabulation such as that in Table 16.1 give $E°=+0.337V$ for the copper half-reaction and $E°=-0.763V$ for the zinc half-reaction.

The particular kind of standard half-cell potentials tabulated in the text — and in most widespread chemical use today — are called *standard reduction potentials*. A standard reduction potential for a redox couple is positive if the couple is spontaneously reduced in a cell for which the other electrode is the SHE, and negative if the spontaneous process taking place against the SHE is oxidation. Thus if a standard Zn^{2+},Zn half-cell is combined with the SHE, then the fact that $E_{Zn^{2+},Zn}$ is negative tells us that the zinc electrode will be the anode and that, under standard conditions, Zn tends to be oxidized spontaneously against the SHE; that is, the reaction

$$Zn(s) + 2H^+(1M) \rightarrow H_2(1\,atm) + Zn^{2+}(1M)$$

will be spontaneous. For any two couples, $\Delta E°$, the voltage across the cell made up of the two corresponding half-cells is given by

$$\Delta E° = E°_{red} - E°_{ox}$$

that is, the standard cell potential is equal to the standard reduction potential of the couple undergoing reduction, minus that of the couple undergoing reduction; thus, for the Daniell cell we have

$$\Delta E°_{Daniell} = E°_{Cu^{2+},Cu} - E°_{Zn^{2+},Zn}$$
$$= 0.337 - (-0.763)$$
$$= +1.100V$$

If a standard cell potential is positive, then the reaction is spontaneous as written. Thus, for the *reverse* of the Daniell cell reaction, we have

$$Zn^{2+}(aq) + Cu(s) \rightarrow Cu^{2+} + Zn(s)$$
$$\Delta E° = E°_{Zn^{2+},Zn} - E°_{Cu^{2+},Cu}$$
$$= -0.763 - (+0.337)$$

$$= -1.100\text{V}$$

This reveals that the reaction is not spontaneous under standard conditions; in fact, it reveals that the reverse reaction *is* spontaneous under standard conditions, something we already knew.

If a standard reduction potential is highly positive, it means that the oxidized member of the couple has a great tendency to be reduced; for example, $E°_{F_2,F^-} = 2.87\text{V}$, a highly positive value. This corresponds to the half-reaction

$$F_2(g) + 2e^- \to 2F^-(aq)$$

The fact that $E°$ is so large means that fluorine has a huge tendency to be reduced; in other words, it is an extremely strong oxidizing agent. By the same token, $F^-(aq)$ is a very poor reducing agent. On the other hand, for the reaction

$$Na^+ + e^- \to Na(s)$$

we have $E°_{Na^+,Na} = -2.71$. This highly negative number means that the *reverse* of the written half-reaction is likely to occur; that is to say, sodium has a great tendency to be oxidized — is a strong reducing agent — and Na^+ is a very poor oxidizing agent. Thus, the positions of the various couples in a table of standard reduction potentials give the relative strength of oxidizing and reducing agents.

As an example, suppose we wish to find an oxidizing agent strong enough to oxidize $Pb(s)$ to Pb^{2+}. From Table 16.1, $E°_{Pb^{2+},Pb} = -0.126$. We want to find another couple so that the overall $\Delta E° > 0$, implying spontaneity; that is, we need

$$\Delta E° = E°_? - E°_{Pb^{2+},Pb} > 0$$

Thus we need

$$E°_? > E°_{Pb^{2+},Pb}$$

This says that any couple with a standard reduction potential more positive than -0.126V will suffice. Let us use the Cu^{2+},Cu couple as an example. $E°_{Cu^{2+},Cu} = +0.337\text{V}$, so that $\Delta E° = 0.337\text{V} - (-0.126\text{V}) = +0.463\text{V}$. Thus the reaction

$$Cu^{2+} + Pb(s) \to Cu(s) + Pb^{2+}$$

will be spontaneous under standard conditions (when Cu^{2+} and Pb^{2+} are both 1M); in other words,

Chapter 16　　　　　　　　　　　　　　　　　　　　　　　　　　　　　　　　Electrochemistry

Cu^{2+} is a strong enough oxidizing agent to oxidize $Pb(s)$ to Pb^{2+}, again under standard conditions.

Let us repeat something you should already know about standard conditions. Standard conditions refer to concentrations, not to temperature. A substance can be in its standard state at any temperature. Tables of $E°$'s are usually tabulated at 25°C, however, and most of the predictions made from these tables are good only at the temperature of tabulation. Also, standard state depends on the phase the substance is in; thus $HCl(g)$ is in its standard state at 1atm pressure, whereas $HCl(aq)$ is in its standard state at 1M concentration. Standard states will be used extensively in the discussions of thermodynamics in Chapters 17 and 18, so it is not a bad idea to make sure you know what they mean now.

Nernst

What we have shown so far is that a table of standard reduction potentials enables one to predict the direction of any redox reaction under standard conditions; in fact, we were able to predict more than just direction — we were actually able to predict the cell voltage of a cell operating under standard conditions. We saw that there is an intimate relationship between the standard reduction potential and the strength of reducing and oxidizing agents, and between the sign of a cell potential for a full cell and the direction of spontaneous reaction. On the other hand, all these results pertained to systems operating under standard conditions. Recall that cell potentials depend on concentration and temperature as well as on chemical composition. We will show that concentrations can be altered so as to change the magnitude and even the sign of the overall cell potential, so that by altering concentrations of components even the direction of spontaneous reaction can be altered. As we said earlier, we will defer consideration of temperature effects until Chapter 18.

First of all, let us consider what happens in a cell such as the Fe^{3+}/I^- cell described earlier when concentrations are altered. Recall that the two half-reactions are

$$I^- \rightarrow \frac{1}{2}I_2 + e^- \quad \text{(oxidation)}$$

$$Fe^{3+} + e^- \rightarrow Fe^{2+} \quad \text{(reduction)}$$

From the table of standard reduction potentials given in Appendix F, we have

$$\Delta E° = E°_{Fe^{3+},Fe^{2+}} - E°_{I_2,I^-}$$

$$= +0.771V - (+0.535V)$$

$$= +0.236V$$

The fact that $\Delta E° > 0$ means that the overall reaction

$$I^- + Fe^{3+} \rightarrow \frac{1}{2}I_2 + Fe^{2+}$$

will be spontaneous, under standard conditions. Now, it is found experimentally that the equilibrium constant for this reaction is equal to about 10^4. Recall that K_{eq} is equal to Q, the equilibrium quotient, when all the concentrations are equilibrium concentrations, and that Q is given, for this reaction, by

$$Q = \frac{[I_2]^{1/2}[Fe^{2+}]}{[I^-][Fe^{3+}]}$$

Under standard conditions, all the concentrations are 1M, so that Q=1. But recall that when $Q < K_{eq}$ then a net forward reaction will occur spontaneously; this is the case here, since $1 < 10^4$. Thus the fact that $E° > 0$ and the fact that $Q < K_{eq}$ are telling us the same thing.

What would we have to do to make the reverse reaction be the spontaneous one? Recall that this will occur if $Q > K_{eq}$, which can be accomplished by making the concentrations of the products, as the reaction is written, much greater than those of the reactants; then LeChatelier's principle will "push" the reaction to the left. For example, let us pick $[I_2]=0.01M$, $[Fe^{2+}]=1.0M$ and $[I^-]=[Fe^{3+}]=0.001M$. This gives $Q=10^5$, which is about ten times greater than K_{eq}. What would happen if we were to set up the experimental apparatus shown in Figure 16.2 with these concentrations? I_2 would be reduced to I^-, and Fe^{2+} would be oxidized to Fe^{3+}; electrons would flow in the opposite direction to their flow under standard conditions, and the electrodes would switch polarity; that is, the iodine half-cell would no longer be the negatively charged anode, but would become the positively charged cathode, and the ion half-cell would become the anode. The observed cell voltage would switch sign as the direction of spontaneity reversed.

We use the symbol ΔE to represent the observed cell voltage; there is no superscript ° because we no longer assume standard conditions. If we want to be able to predict how concentration will alter the

Chapter 16 — Electrochemistry

observed ΔE, then we need an equation that starts out with $\Delta E°$ for a given chemical system, plugs in concentrations, and comes out with ΔE. The equation that does this is the *Nernst equation*, which was first derived by the German chemist Walther Nernst in the 19th century. We will derive it in Chapter 18, but for now you will have to take our word for its validity. Even at this point, however, you will be able to see how its implications accord with what you already know about chemical equilibrium.

The Nernst equation can be written

$$\Delta E = \Delta E° - \frac{0.0592}{n}\log_{10}Q$$

Here Q is the familiar equilibrium quotient, and n is the number of electrons transferred from the oxidizing agent to the reducing agent in the balanced chemical equation as written. Thus, for the Fe^{3+}/I^- reaction as we wrote it, n=1 and $Q=[I_2]^{1/2}[Fe^{2+}]/[I^-][Fe^{3+}]$; however, if we had written the equation

$$2I^- + 2Fe^{3+} \rightarrow I_2 + Fe^{2+}$$

then we would have had n=2 and $Q = \frac{[I_2][Fe^{2+}]^2}{[I^-]^2[Fe^{3+}]^2}$. Let us stick to our original form for now, and examine what the Nernst equation says.

There are two particularly important sets of experimental conditions to examine. The first is standard conditions, where all concentrations, and hence Q, are equal to unity, as shown earlier. Recall that for our reaction $\Delta E°=0.236V$. Under standard conditions, the Nernst equation then gives

$$\Delta E = 0.236V + 0.0592\log_{10}(1)$$

But $\log_{10}(1) = 0$, so that under standard conditions

$$\Delta E = 0.236V = \Delta E°$$

Well, this should be comforting. We already knew that under standard conditions the cell voltage is equal to the standard cell potential, and all this calculation does is to prove that the Nernst equation reflects this.

The other particularly important set of conditions where we want to apply the Nernst equation is where the system is at equilibrium. Recall that if the forward reaction is spontaneous $\Delta E > 0$, and that if

Chapter 16 — Electrochemistry

the reverse reaction is spontaneous $\Delta E < 0$. If neither direction is spontaneous, we must have $\Delta E = 0$, and of course this must correspond to chemical equilibrium. ΔE really represents the driving force of the reaction under the cell conditions, and if ΔE goes to zero, there is no driving force in either direction; again, equilibrium. Finally, recall that if $\Delta E > 0$ then there is a voltage difference between the electrodes, and electrons can flow through the wire and be used to heat a heater, light a lamp or turn a motor. As reaction proceeds, however, concentrations approach their equilibrium concentrations (that is, reactants get used up and products get formed), and ΔE diminishes. At equilibrium, there is no net reaction taking place, so there is no electron flow through the wire, and no voltage difference across the cell; the cell is completely discharged, and no work can be done.

Thus, at equilibrium, $\Delta E = 0$. Also, we know that at equilibrium $Q = K_{eq}$, so that

$$\Delta E = \Delta E° - \frac{0.0592}{n} \log_{10} Q$$

$$0 = \Delta E° - \frac{0.0592}{n} \log_{10} K_{eq}$$

$$\Delta E° = \frac{0.0592}{n} \log_{10} K_{eq}$$

$$\log_{10} K_{eq} = \frac{n \Delta E°}{0.0592}$$

$$K_{eq} = 10^{\frac{n \Delta E°}{0.0592}}$$

What is remarkable is that the value of $\Delta E°$ that we can calculate from a table of standard reduction potentials completely determines K_{eq} for a given reaction — at 25°, of course, since that's the temperature for which the data are tabulated. This is a profound result, but perhaps not an entirely unexpected one. After all, we said earlier that a positive $\Delta E°$ implies a spontaneous forward reaction under standard conditions, and we *almost* said in Chapter 8 that $K_{eq} > 1$ implies a spontaneous forward reaction under standard conditions. I say "almost" because we didn't have the notion of standard conditions then; we did, however, have the notion of Q, and we said that $Q < K_{eq}$ implies forward spontaneity. Well, since $Q = 1$ under standard conditions, $K_{eq} > 1$ means spontaneity under standard conditions.

Note from the last equation above that if $\Delta E° > 0$ then K_{eq} is equal to ten to a positive power, which is a number greater than one; if $\Delta E° < 0$ then K_{eq} is ten to a negative power, which is less than one, and if $\Delta E° = 0$ then K_{eq} is ten to the zero, or one. Thus $K_{eq} > 1$ and $\Delta E° > 0$ are entirely equivalent criteria of forward spontaneity under standard conditions.

For our reaction as written, we have

$$K_{eq} = 10^{\frac{0.236}{0.0592}} = 10^{3.99} = 9.69 \times 10^3$$

If we had written the reaction the second way, with n=2, we would have had

$$K'_{eq} = 10^{\frac{(2)(0.236)}{0.0592}} = 10^{7.97} = 9.40 \times 10^7 = (K_{eq})^2$$

But recall from Chapter 8 that when you double the reaction you square the equilibrium constant; the Nernst equation reflects this, too, as it must.

Note that the Nernst equation predicts that $\Delta E \neq 0$ even if $\Delta E° = 0$, provided that $Q \neq 1$. Such a situation is provided in a *concentration cell*, Example 16.7. The cell is

$$Zn(s) | Zn^{2+}(0.024M) || Zn^{2+}(0.480M) | Zn(s)$$

If this cell is constructed, the Zn^{2+} in the left-hand compartment, as written, will become more concentrated, and the zinc anode will shrink, while on the right the reverse will occur. The standard reduction potential is given by

$$\Delta E° = E°_{Zn^{2+},Zn} - E°_{Zn^{2+},Zn} = 0$$

yet the cell will exhibit a measurable potential! What is the driving force of the cell reaction? Clearly, if we mixed the two Zn^{2+} solutions of differing molarities they would diffuse into each other and form a solution of uniform composition. For the same reason, in a concentration cell the spontaneous reaction is the one which causes the two solutions to become more similar in concentration. The ultimate explanation is that entropy increases when this occurs.

In the example, n=2 and Q=(0.024/0.480)=0.05, giving

$$\Delta E = 0 - \frac{0.0592}{n} \log_{10} Q$$

$$= -\frac{0.0592}{2} \log_{10}(0.05)$$

Chapter 16　　　　　　　　　　　　　　　　　　　　　　　　　　　Electrochemistry

$$= -(0.0296)(-1.30)$$

$$= +0.0385 \text{V}$$

Note that we had to use n=2 because we used

$$Q = \frac{[Zn^{2+}]_{low}}{[Zn^{2+}]_{high}}$$

At the anode, the reaction was

$$Zn(s) \rightarrow Zn^{2+} + 2e^-$$

and the reverse took place at the cathode. The overall process was

$$Zn(\text{high concentration}) \rightarrow Zn(\text{low concentration})$$

For each net transfer of a zinc ion from the region of high to the region of low concentration, two electrons had to move through the wire.

It is interesting that, as in this example, the Nernst equation can sometimes be used even when the overall equation does not look like a redox equation; you can use the Nernst equation whenever you can divide the reaction up into an oxidation and a reduction half-reaction, no matter what the overall reaction looks like. Another example is provided by the use of electrochemical measurements to calculate K_{sp}'s. The in-text example in Section 16.8 involves AgBr. I will solve it in a slightly different way — without reference to the spontaneous cell direction.

The overall reaction for which K_{eq} is $K_{sp}(AgBr)$ is

$$AgBr(s) = Ag^+(aq) + Br^-(aq) \qquad \text{(overall)}$$

We need to have two half-reactions which can continue to give this. Fortunately, Appendix F provides

$$AgBr(s) + e^- \rightarrow Ag(s) + Br^- \qquad \text{(I)}$$

$$Ag^+ + e^- \rightarrow Ag(s) \qquad \text{(II)}$$

By taking (I–II) we get the overall reaction; therefore

$$\Delta E°_{overall} = E°_I - E°_{II} = 0.0713\text{V} - 0.799\text{V} = -0.728\text{V}$$

This is a one-electron process, so that

$$K_{sp} = K_{eq} = 10^{\frac{-0.728}{0.0592}} = 5.1 \times 10^{-13}$$

Lead Storage Cell

Section 16.10 describes a number of commercial batteries. I will not redescribe them here, except for the commercial lead storage battery, which is interesting for several reasons. The cell diagram is

$$Pb(s)|PbSO_4(s)|H_2SO_4|PbO_2(s),PbSO_4(s)|Pb(s)$$

At the anode, metallic lead is oxidized to lead sulfate; at the cathode, lead oxide is reduced to lead sulfate. At both electrodes, both the reactant and the product are in the solid phase. For this reason, no salt bridge or other barrier is needed; the solids stick to the electrodes and do not migrate. Lead sulfate is formed at both anode and cathode during discharge, so sulfate ion must be supplied from somewhere; "somewhere" is the strong H_2SO_4 solution in which the electrodes are immersed. Obviously, as the cell discharges the H_2SO_4 becomes more dilute; as this occurs the specific gravity drops. The easiest way to test whether the battery is charged or discharged is to measure the specific gravity of the electrolyte solution. This is what the gas station attendant is doing when he sticks that thing that looks like a gigantic eye-dropper into your battery.

Actually, a single lead-acid cell develops a potential of about 2V, though at solutions of H_2SO_4 as concentrated as 35% our version of the Nernst equation cannot be trusted, since it relies on the law of mass action, which fails at such high electrolyte concentrations. A 12V car battery consists of six such cells hooked together in series. When such a battery is charged, the $PbSO_4$ sticking to the electrodes gets converted back to Pb and to PbO_2. What causes these batteries to eventually go bad is both the change in $PbSO_4$ crystal structure alluded to in the text, and the fact that some of the $PbSO_4$ falls off the electrodes and down to the bottom of the casing, and becomes unavailable for regeneration. The detailed design of the electrode grids is intended to minimize this problem.

Electrolysis

Recall that we said that if electrons are allowed to flow from a low to a high potential, the system can do work on the outside world. All the cells we have discussed so far operate in this manner. The cell reactions were spontaneous, and as the reactions proceeded electrons flowed spontaneously through the wire, from the negative anode to the positive cathode. A cell which is discharging spontaneously in this manner is called a *galvanic* cell. We also said that is is possible to make electrons flow from a positive electrode to a negative electrode, by doing work on the system. For example, consider the Daniell cell again (Figure 16.3). If an external power supply is inserted in the wire with its negative end toward the zinc electrode, then any potential developed by the power supply will oppose the cell potential.

We said (and later rationalized, using standard reduction potentials) that the Daniell cell operating under standard conditions exhibits an EMF of 1.10V, and that electrons flow through the wire from the zinc anode to the copper cathode. Suppose the power supply, inserted as just described, is set up to contribute its own potential of 1.10V. This will exactly counteract the tendency of the electrons to flow; the zinc electrode and the copper electrodes will be connected to a device which maintains them at precisely their respective voltages, so that the electrons "see" no potential difference. If the power supply's voltage is set to greater than 1.10V, then the negative pole, which is attached to the zinc electrode, "forces electrons out" through the zinc electrode. This is accompanied by the following solution process.

$$Zn^{2+} + 2e^- \rightarrow Zn(s)$$

Since this is a reduction, the zinc half-cell is now the cathode! Similarly, the pole connected to the copper electrode is positive, so it "sucks electrons up" through the copper electrode, giving rise to the following process.

$$Cu \rightarrow Cu^{2+} + 2e^-$$

Since this is an oxidation, copper is now the anode.

The overall cell reaction is given by

$$Cu(s) + Zn^{2+} \rightarrow Cu^{2+} + Zn(s)$$

This is the reverse of the process that is spontaneous, if the cell is operating under standard conditions, and a cell can operate this way only if an external power supply performs work on the system, work which pushes the electrons back through the wire in opposition to their natural tendency. A cell operating in this mode is called an *electrolytic* cell. This corresponds to the charging of a battery; for example, after connecting up a Daniell cell to a power supply for a while and letting it run, Zn(s) will build up on the left (in Figure 16.3) and Cu^{2+} will increase in concentration on the right. Then, after removing the power supply, a large reservoir of reactants will have been formed, and the cell can be used to do work.

Note that when we went from galvanic to electrolytic operation the zinc electrode continued to be negative with respect to the copper electrode; what changed is that the zinc electrode went from being the anode to being the cathode. In a galvanic cell the anode is negative, and in an electrolytic cell the anode is positive, but the anode is always where oxidation takes place.

Electrolysis is a commonly used technique for producing pure or unstable substances; for example, water can be electrolysed to give H_2 and O_2 gases. In acid solution the half-reactions are

$$[2H^+ + 2e^- \rightarrow H_2(g)] \times 2 \qquad \text{(cathode)}$$

$$2H_2O(l) \rightarrow O_2(g) + 4H^+ + 4e^- \qquad \text{(anode)}$$

$$2H_2O(l) \rightarrow O_2(g) + 2H_2(g) \qquad \text{(overall)}$$

Non-aqueous systems are widely used in industrial electrolytic processes. For example, molten NaCl is electrolysed in the *Downs cell* (Figure 16.9) to produce pure sodium and chlorine gas.

$$[Na^+ + e^- \rightarrow Na(l)] \times 2 \qquad \text{(cathode)}$$

$$2Cl^- \rightarrow Cl_2(g) + 2e^- \qquad \text{(anode)}$$

$$2Na^+ + 2Cl^- \rightarrow 2Na(l) + Cl_2(g) \qquad \text{(overall)}$$

Elemental sodium and chlorine are so reactive that they are not found in nature except as ions. Note that under standard conditions the electrolysis of sodium chloride would exhibit $\Delta E^\circ = E^\circ_{Na^+,Na} - E^\circ_{Cl_2,Cl^-} = -2.71 - (+1.36) = -4.07V$; this is a very large ΔE°, indicating that the reaction is highly unfavored. (Of course, standard conditions in our table refer to 1M concentrations in aqueous solution, and molten NaCl is far from this, but the conclusion still holds.) Even so, all it takes is a sufficiently high voltage to cause the reaction to occur.

Similarly, Al_2O_3, *bauxite*, dissolved in molten *cryolite*, Na_3AlF_6, is reduced to aluminum in the *Hall process*. In this solution, the Al_2O_3 exists as Al^{3+} and O^{2-} ions. The cell reactions are

$$[Al^{3+} + 3e^- \rightarrow Al(l)] \times 4 \quad \text{(cathode)}$$

$$[C(s) + 2O^{2-} \rightarrow CO_2(g) + 4e^-] \times 3 \quad \text{(anode)}$$

$$4Al^{3+} + 3C(s) + 6O^{2-} \rightarrow 4Al(l) + 3CO_2(g) \quad \text{(overall)}$$

The carbon is supplied by a carbon electrode, often made from byproducts of petroleum refining. Since carbon has a strong tendency to be oxidized to CO_2, its use facilitates the process; if instead oxygen were oxidized to O_2 at the anode, much more work would have to be performed (much more electricity consumed) per pound of aluminum produced. The availability of a cheap source of carbon allows great savings in carrying out the process.

Let us now return to aqueous systems. If pure water is electrolysed, H_2 is produced at the cathode and O_2 is produced at the anode. In general, if a salt is dissolved in water, and the resulting solution electrolysed, the cation will be reduced in preference to hydrogen if it is a better oxidizing agent than H_3O^+ or H_2O; for example, Cu^{2+} is more readily reduced than H^+, so Cu, not H_2 will be formed at the cathode in the electrolysis of a solution of a cupric salt, under standard conditions. Similarly, Br^- is more readily oxidized than H_2O, so that Br_2, not O_2, is formed at the anode in the electrolysis of a bromide under standard conditions.

According to the table of standard reduction potentials, it is harder to oxidize Cl^- ($E°_{Cl_2,Cl^-} = +1.36V$) than H_2O ($E°_{H_2O,O_2} = 1.23V$), so that one would predict that solutions of chloride salts under standard conditions should form O_2 at the anode when electrolysed; surprisingly, Cl_2 is formed. Note that the $E°$'s are close together; if the power supply were carefully maintained at a voltage between $+1.23V$ and $+1.36V$ (assuming H_2 formed at the cathode), then presumably only O_2 could form. If the voltage were to exceed 1.36V then, in principle, either process could occur. Presumably Cl_2 is formed not because this reaction is the more spontaneous, as we have been using the term, but rather because is faster. As you will see in forthcoming chapters, ΔE's and $\Delta E°$'s really measure thermodynamic spontaneity — the relative stabilities of reactants and products — and this may or may not be correlated with reaction rate — how *quickly* a product is formed — in a given system.

Chapter 16　　　　　　　　　　　　　　　　　　　　　　　　　　Electrochemistry

Questions

16.1 The use of oxidation-reduction reactions to produce electricity: the galvanic cell

1. What function does a salt bridge provide?

2. What process takes place

   (a) at the anode?

   (b) at the cathode?

3. A *faraday* is the amount of _____ contained in a mole of _____.

4. For the reaction:

$$Zn(s) + Cu^{2+} \rightarrow Cu(s) + Zn^{2+}$$

   write the half-cell reactions taking place at

   (a) the anode:

   (b) the cathode:

5. The ampere is the SI unit of _____, and 1 amp = 1_____ per _____.

6. How many amperes is 1F/s?

Chapter 16   Electrochemistry

16.2 The EMF of a galvanic cell

1. Suppose I define sea level to be the zero of gravitational potential energy.

    (a) What is the sign of the potential energy (+, -, or 0) at each of these locations:

    _____(i) top of Mt. Everest (~10,000 meters above sea level)

    _____(ii) Long Beach, Long Island (sea level)

    _____(iii) bottom of Marianas Trench (~10,000 meters below sea level).

    (b) What is the sign of the change in potential energy on moving the rock

    _____(i) from Long Beach to the top of Everest

    _____(ii) from Long Beach to the bottom of the Marianas Trench

2. Now, suppose I define the zero of gravitational potential energy to be the top of Mt. Everest.

    (a) What is the sign of the potential energy (+, -, or 0) at each of these locations:

    _____(i) top of Mt. Everest (~10,000 meters above sea level)

    _____(ii) Long Beach, Long Island (sea level)

    _____(iii) bottom of Marianas Trench (~10,000 meters below sea level).

    (b) What is the sign of the change in potential energy on moving the rock

    _____(i) from Long Beach to the top of Everest

    _____(ii) from Long Beach to the bottom of the Marianas Trench

3. What do the answers to Questions 1 and 2 demonstrate?

16.3 Cell conventions and half-cell potentials

1. $E°_{Zn^{2+}|Zn} = -0.763$ and $E°_{Ag^+|Ag} = +0.799$, based on the choice of $E°$ _____ as the zero of electrochemical potential.

2. Suppose I define a new electrochemical potential scale, called $E°'$, instead of $E°$, such that $E°'_{Ag^+|Ag} = 0$. Then

16-24

Chapter 16　　　　　　　　　　　　　　　　　　　　　　　　Electrochemistry

(a) $E^{\circ\prime}_{H^+|H_2|Pt} = $ _____.

(b) $E^{\circ}_{Zn^{+2}|Zn} = $ _____.

3. (a) Write the balanced chemical reaction for the oxidation of zinc with silver ion:

   (b) $\Delta E^{\circ\prime}$ (based on Question 2) for this reaction is _____.

   (c) ΔE° for this reaction is _____.

   (d) What do the answers to (b) and (c) demonstrate?

4. If $[Ag^+] = 1.0$ M and $[Zn^{+2}] = 2.5$ M, write the shorthand notation for a cell to carry out the reaction in Question 3.

16.4 Significance of the sign of the cell potential

1. For the reaction in Question 3 of the last section, the sign of ΔE° is _____. This means the reaction will proceed (to the left, to the right) as written, under _____ conditions.

2. (a) For the reverse of the same reaction, the sign of ΔE° is _____.

   (b) Write the cell notation for this reverse reaction:

16.5 The table of Standard Reduction Potentials

1. The more negative the reduction potential, the _____ the reducing agent, and the _____ the oxidizing agent in the redox couple.

2. Assuming standard conditions, write a balanced chemical equation and give ΔE° for a spontaneous redox reaction involving each pair of constituents, based on table 16.1 in the text. If no reaction is

16-25

Chapter 16   Electrochemistry

possible, say so.

(a) Cu^{2+} and $Zn(s)$

(b) F_2 and $Pb(s)$

(c) $Cu(s)$ and Br^-

(d) Mn^{2+} and F_2

(e) H^+ and $Zn(s)$

(f) H^+ and $Cu(s)$

(g) H^+ and Cu^{2+}

(h) H_2 and Cu^{2+}

3. What is meant by the statement that an electrochemical cell is being operated under "standard conditions"?

Chapter 16　　　　　　　　　　　　　　　　　　　　　　　　　　Electrochemistry

16.6 Concentration dependence of the cell EMF

1. What is the difference between ΔE and $\Delta E°$?

2. If you know $\Delta E°$ for a reaction, what else do you need in order to calculate ΔE?

3. For the oxidation of zinc with silver ion,

    (a) Write the Nernst equation.

    (b) What is ΔE if $[Zn^{2+}]=0.01M$ and $[Ag^+]=0.1M$?

    (c) What is ΔE if $[Zn^{2+}] = [Ag^+] = 1.0M$?

    (d) What is ΔE is the system is at equilibrium?

    (e) Which of (b), (c) and (d) correspond(s) to standard conditions?

16.7 Proof that the cell potential is independent of the form of the net reaction

1. Simplify the following expressions involving logarithms.

    (a) $\log(x)^2 =$

    (b) $\log(x^{-1}) =$

    (c) $\log(x^{1/2}) =$

    (d) $\log(x^n) =$

    (e) $\log_{10}(10^x) =$

Chapter 16   Electrochemistry

(f) $10^{\log_{10}(x)} =$

(g) $\ln(e^x) =$

(h) $e^{\ln(x)} =$

(j) $\log\left(\dfrac{x}{y}\right)^n =$ (in terms of $\dfrac{x}{y}$)

(k) $\log\left(\dfrac{x}{y}\right)^n =$ (in terms of x and y)

2.   (a) Write the Nernst Equation for:

       (i) $Zn(s) + 2Ag^+ \rightarrow Zn^{2+} + 2Ag(s)$

       (ii) $5Zn(s) + 10Ag^+ \rightarrow 5Zn^{2+} + 10Ag(s)$

(b) Show that both equations predict the same value of ΔE once the values of $[Zn^{2+}]$ and $[Ag^+]$ are defined.

Chapter 16   Electrochemistry

16.8 Use of the Nernst equation to determine the equilibrium constant

1. When a redox reaction is at equilibrium,

    (a) $\Delta E =$ _____

    (b) $Q =$ _____

    (c) The direction the reaction goes spontaneously is _____

2. For the oxidation of zinc with silver ion,

    _____(a) Write the Nernst equation for the reaction at equilibrium.

    _____(b) Calculate K_{eq}.

3. For the salt $Ni(OH)_2$

    (a) Write the reaction whose K_{eq} is K_{sp} for the salt.

    (b) In Appendix F of the text, find and write two half-reactions which can be combined into this equation.

    (c) Calculate $\Delta E°$ for the K_{sp} reaction, using the $E°$'s of the reactions in part (b).

(d) Write the Nernst equation for this system.

(e) Calculate K_{sp}.

(f) Check your answer in Appendix E.

16.9 The relation between two criteria for predicting the direction of a spontaneous reaction

1. (a) Write the general form of the Nernst Equation.

    (b) Rewrite the equation, substituting for ΔE and Q their special values when equilibrium obtains.

    (c) If $Q > K_{eq}$

        (i) Which is more positive, $\Delta E°$ or $\left(\dfrac{0.0592}{n} \log Q\right)$?

        (ii) In which direction will the reaction tend spontaneously?

        (iii) What will be the sign of ΔE?

    (c) If $Q < K_{eq}$

        (i) Which is more positive, $\Delta E°$ or $\left(\dfrac{0.0592}{n} \log Q\right)$?

Chapter 16　　　　　　　　　　　　　　　　　　　　　　　　　　　　　Electrochemistry

　　　(ii) In which direction will the reaction tend spontaneously?

　　　(iii) What will be the sign of **ΔE**?

16.10 Commercial batteries

1. *The Lead Storage Cell*

    (a) Write the half reactions and the overall chemical reaction for this cell.

    (b) Calculate **ΔE°**.

    (c) Why is no salt bridge needed?

2. *The Dry Cell*

    Write the half reactions and the overall chemical reaction for this all.

Chapter 16 Electrochemistry

3. *The Mercury Cell*

    Write the half reactions and the overall chemical reaction for this cell.

4. *The Nickel-Cadmium Cell*

    (a) Write the half reactions and the overall chemical reaction for this cell.

    (b) Write the equilibrium quotient expression of the overall reaction.

    (c) Why does the voltage remain virtually constant over the life of the cell? Explain in terms of your answer to (b).

Chapter 16

Electrochemistry

16.11 Electrolysis

1. (a) For a galvanic cell, ΔE is (positive, negative).

   (b) For an electrolytic cell, ΔE is (positive, negative).

2. (a) In a galvanic cell, the positive electrode is the (anode, cathode).

   (b) In an electrolytic cell, the positive electrode is the (anode, cathode).

3. (a) In a galvanic cell, reduction takes place at the _____.

   (b) In an electrolytic cell, reduction takes place at the _____.

4. (a) For an electrolytic cell operating under standard conditions, how big does the externally imposed voltage have to be?

   (b) Why should the voltage not be allowed to become much bigger than the answer to (a)?

5. What would you expect to form at the anode and cathode during the electrolysis of aqueous solutions of the following salts?

	Salt	Anode	Cathode
(a)	CaI_2		
(b)	CuF		
(c)	$Fe_2(SO_4)_3$		
(d)	$FeSO_4$		

Chapter 16 — Electrochemistry

Answers

16.1 The use of oxidation-reduction reactions to produce electricity: the galvanic cell

1. Separates reacting solutions and maintains charge neutrality.
2. (a) oxidation
   (b) reduction
3. charge, electrodes
4. (a) $Zn(s) \rightarrow Zn^{2+} + 2e^-$
   (b) $Cu^{2+} + 2e^- \rightarrow Cu(s)$
5. current, 1 amp = 1 coulomb/second
6. $\approx 96,500$ amp

16.2 The EMF of a galvanic cell

1. (a)
   - (i) +
   - (ii) 0
   - (iii) —

   (b)
   - (i) +
   - (ii) —

2. (a)
   - (i) 0
   - (ii) —
   - (iii) —

   (b)
   - (i) +
   - (ii) —

3. Potential energy differences, not absolute values of the potential energy, have physical significance.

16.3 Cell conventions and half-cell potentials

1. $E°_{H^+|H_2}$
2. (a) —0.799V
   (b) —1.562V
3. (a) $Zn(s) + 2Ag^+ \rightarrow 2Ag(s) + Zn^{2+}$
   (b) 1.562V
   (c) 1.562V
   (d) $\Delta E°$'s, not $E°$'s, have physical significance.
4. $Zn(s)|Zn^{2+}(2.5M)||Ag^+(1.0M)|Ag(s)$

16.4 Significance of the sign of the cell potential

1. positive, to the right, standard
2. (a) negative
   (b) $Ag(s)|Ag^+(1.0M)||Zn^+(2.5M)|Zn(s)$

16.5 The table of Standard Reduction Potentials

1. stronger, weaker
2. (a) $Cu^{2+} + Zn(s) = Cu(s) + Zn^{2+}$ $\Delta E° = 1.100V$
   (b) $F_2 + Pb(s) = Pb^{2+} + 2F^-$ $\Delta E° = 300V$
   (c) no reaction possible

Chapter 16 — Electrochemistry

(d) $5F_2 + 2Mn^{2+} + 8H_2O = 10F^- + 2MnO_4^- + 16H^+ \quad \Delta E° = 1.36V$
(e) $Zn(s) + 2H^+ = H_2 + Zn^{2+} \quad \Delta E° = 0.763V$
(f) reaction not spontaneous, because $Cu(s) + 2H^+ = Cu^{2+} + H_2$ has $\Delta E = -0.377V$
(g) no reaction possible
(h) $H_2 + Cu(s) = 2H^+ + Cu(s) \quad \Delta E = 0.337V$

3. All reactants and products are in their standard states: 1M concentration for substances in solution, 1atm partial pressure for gases, pure phases for liquids and solids.

16.6 Concentration dependence of the cell EMF

1. $\Delta E°$ is the voltage measured across a cell operating under standard conditions; ΔE is the voltage measured under at arbitrary concentrations.

2. Concentrations of all components.

3. (a) $\Delta E = \Delta E° - \dfrac{0.059}{2}\log\dfrac{[Zn^{2+}]}{[Ag^+]^2}$
   (b) $\Delta E = \Delta E° = +1.562V$
   (c) $+1.562V$
   (d) 0
   (e) (c) only, even though (b) gives the same ΔE.

16.7 Proof that the cell potential is independent of the form of net reaction

1. (a) $2\log(x)$
   (b) $-\log(x)$
   (c) $(1/2)\log(x)$
   (d) $n\log(x)$
   (e) x
   (f) x
   (g) x
   (h) x
   (j) $n\log\left(\dfrac{x}{y}\right)$

   (k) $n(\log(x) - \log(y))$

2. (a)

   (i) $\Delta E = \Delta E° - \dfrac{0.0592}{2}\log\dfrac{[Zn^{2+}]}{[Ag^+]^2}$

   (ii) $\Delta E = \Delta E° - \dfrac{0.0592}{10}\log\dfrac{[Zn^{2+}]^5}{[Ag^+]^{10}}$

   (b) $\dfrac{0.0592}{10}\log\dfrac{[Zn^{2+}]^5}{[Ag^+]^{10}} = \dfrac{0.0592}{10}\log\left(\dfrac{[Zn^{2+}]}{[Ag^+]^2}\right)^5$

   $= \dfrac{5(0.0592)}{10}\log\dfrac{[Zn^{2+}]}{[Ag^+]^2}$

   $= \dfrac{0.0592}{2}\log\dfrac{[Zn^{2+}]}{[Ag^+]^2}$

   so the two expressions are equal.

16.8 Use of the Nernst equation to determine the equilibrium constant

1. (a) 0
   (b) K_{eq}
   (c) (neither direction)

2. (a) $0 = \Delta E° - \dfrac{.0592}{2} \log_{10} \dfrac{[Zn^{2+}]}{[Ag^+]^2}$

   $+1.562V = \dfrac{.0592}{2} \log_{10} K_{eq}$

   (b) $K_{eq} = 10^{\frac{2(1.562)}{0.0592}} = 5.89 \times 10^{52}$

3. (a) $Ni(OH)_2(s) = Ni^{2+} + 2OH^-$
   (b) $Ni(OH_2)(s) + 2e^- \rightarrow Ni(s) + 2OH^-$ $E° = -0.72V$

   $Ni^{2+} + 2e^- \rightarrow Ni(s)$ $E° = -0.25V$

   (c) $\Delta E° = -0.72V - (-0.25V) = -0.47V$
   (d) $\Delta E = 0 = -0.47 - \dfrac{0.0592}{2} \log K_{sp}$
   (e) $K_{sp} = 10^{\frac{-2(0.47)}{0.0592}} = 7.56 \times 10^{-15}$
   (f) From Appendix E2, $K_{sp}(Ni(OH)_2) = 2 \times 10^{-15}$

   Note that $\Delta E° = -0.45$ would give $K_{sp} = 1.59 \times 10^{-15}$, so the calculated values of K_{sp} are very sensitive to $\Delta E°$. Values such as this are hard to measure, and our value is probably within experimental error.

16.9 The relation between two criteria for predicting the direction of a spontaneous reaction

1. (a) $\Delta E = \Delta E° - \dfrac{0.0592}{n} \log Q$

   (b) $0 = \Delta E° - \dfrac{0.0592}{n} \log K_{eq}$

   (c)
       (i) $\dfrac{0.0592}{n} \log Q$
       (ii) left
       (iii) negative

   (d)
       (i) $\Delta E°$
       (ii) right
       (iii) positive

16.10 Commercial batteries

1. (a)

   $Pb(s) + SO_4^{2-} \rightarrow PbSO_4(s) + 2e^-$ (anode)

   $PbO_2(s) + SO_4^{2-} + 4H^+ + 2e^- \rightarrow PbSO_4(s) + 2H_2O$ (cathode)

   $Pb(s) + PbO_2(s) + 4H^+ + 4SO_4^{2-} \rightarrow 2PbSO_4(s) + 2H_2O$ (overall)

   (b) $\Delta E° = 1.685V - (-0.356V) = +2.04V$
   (c) Both anode and cathode reactions require solids, so diffusion of ions in in solution will not

Chapter 16   Electrochemistry

short-circuit cell.

2.

$$Zn(s) \rightarrow Zn^{2+} + 2e^-$$ (anode)

$$2MnO_2(s) + 2NH_4^+ + 2e^- \rightarrow Mn_2O_3(s) + 2NH_3 + H_2O$$ (cathode)

$$Zn(s) + 2MnO_2(s) + 2NH_4^+ \rightarrow Zn^{2+} + Mn_2O_3(s) + 2NH_3$$ (overall)

3.

$$Zn(amalgam) + 2OH^- \rightarrow ZnO(s) + H_2O$$ (anode)

$$HgO(s) + H_2O + 2e^- \rightarrow Hg(l) + 2OH^-$$ (cathode)

$$Zn(amalgam) + HgO \rightarrow ZnO(s) + Hg(liq)$$ (overall)

4. (a)

$$Cd(s) + 2OH^- \rightarrow Cd(OH)_2(s) + 2e^-$$ (anode)

$$NiO_2(s) + 2H_2O + 2e^- \rightarrow Ni(OH)_2 + 2OH^-$$ (cathode)

$$Cd(s) + NiO_2(s) + 2H_2O \rightarrow Cd(OH)_2(s) + Ni(OH)_2(s)$$ (overall)

(b) Q=1 (there are no concentration terms, since everything is either a solid or a solvent!)
(c) Q=1=K_{eq}, so Q doesn't change during the reaction. $\Delta E = \Delta E°$ throughout.

16.11 Electrolysis

1. (a) positive
   (b) negative

2. (a) cathode
   (b) anode

3. (a) anode
   (b) anode

4. (a) Greater than $\Delta E°$.
   (b) Other reactions than the desired one might take place.

5.

	Salt	Anode	Cathode
(a)	CaI$_2$	I$_2$(s)	H$_2$
(b)	CuF	O$_2$	Cu(s)
(c)	Fe$_2$(SO$_4$)$_3$	O$_2$	Fe^{2+}
(d)	FeSO$_4$	O$_2$	H$_2$

CHAPTER 17: ENERGY, ENTHALPY AND THERMOCHEMISTRY
Scope

Thermodynamics

The *Concise Oxford Dictionary* defines thermodynamics as the "science of relations between heat and other (mechanical, electrical, etc.) forms of energy." We have alluded many times to the fact that chemical and physical systems evolve toward lower energy and/or higher entropy; however we have not defined these terms precisely. In this chapter we introduce a rigorous definition of the *energy* and a related function, called the *enthalpy*. Many of the energetic transformations of matter, for example the heat given off when a chemical reaction, such as combustion, takes place in a calorimeter, can be understood using these ideas.

In Chapter 18 we will define the *entropy* and a related function, called the *Gibbs free energy*. These predict whether a process can occur or not, and are intimately related to other criteria of chemical spontaneity, such as the values of K_{eq} and of $E°$ for a process. The major weakness of thermodynamics is that it has absolutely nothing to say about reaction rates. A mixture of H_2 and O_2 has an enormous tendency to form H_2O, but a flask containing this mixture at room temperature is stable indefinitely, unless a spark is applied or a catalyst is added. In Chapter 19 we will discuss the factors influencing reaction rates.

State Functions

Recall that in Chapter 3 we said that PV=nRT is a *state function*; we said that the *state* of a system is specified by all the variables that uniquely define the properties of a given sample; for example, when any three of (P, V, n and T) are specified for an ideal gas, then its state is defined, and all the macroscopic properties of the sample are guaranteed to be the same as those of any other sample of the same substance with the same values of the chosen parameters. Thus, if P, V and n are fixed, T can have only one value.

Chapter 17   **Energy, Enthalpy and Thermochemistry**

Any parameter that is fixed when the state is fixed is called a *state function*; thus, P, V, n and T are state functions. Note that if a system goes from one state to another, then the *change* in any state function has only one possible value; for example, if an ideal gas goes from state 1 to state 2, then obviously $\Delta P = P_2 - P_1$, $\Delta T = T_2 - T_1$, and so on, and if the two states are defined there can only be one ΔP and one ΔT. This may seem too obvious to mention, but you will soon see that many state functions are not as intuitively understandable as P, V and T, and that many important variables in chemical processes are not state functions. When we discuss them, we will need rules such as the one we just mentioned. Another statement of the same rule is that when a system goes from, say, state 1 to state 2, the change in any state function is *independent of path*; for example, if $V_1 = 1L$ and $V_2 = 3L$, then $\Delta V = 2L$, even if the system first expanded up to 4L before contracting down to its final value. Note that for any function, X, whether a state function or not, we always define ΔX to mean $X_{final} - X_{initial}$.

Heat, Work and the First Law*

You may have noticed that the different units we use for energy are seemingly unrelated. The SI unit, the joule, has the fundamental value, $1 kg \cdot m^2/s^2$, and is actually a unit of work:

$$\begin{aligned} \text{Work} &= (\text{force})(\text{distance}) \\ &= (\text{mass})(\text{acceleration})(\text{distance}) \\ &= (kg)\left(\frac{m}{s^2}\right)(m) \\ &= \frac{kg \cdot m^2}{s^2} = \text{joules} \end{aligned}$$

On the other hand, chemists have for a long time used the calorie to measure energy. The calorie was originally defined as the amount of heat it takes to raise 1g of water by 1°C. This original definition had nothing to do with work — it had to do with heat. Yet the joule and the calorie are both valid units of energy. The reason for this is that it became clear in the early part of the nineteenth century that work could be converted into heat. Friction is one common means of doing this: it is what causes your hands to burn if you slide down a rope too fast, or a drill-bit to heat up in operation.

---
* We discuss these topics in a somewhat different sequence than the text.

Chapter 17

Benjamin Thompson (Count Rumford) was the first to demonstrate a quantitative relationship between heat and work. When cannons were bored, tremendous quantities of heat were released, and therefore the operation was performed under water. Thompson showed in 1798 that the work of a given horse boring a cannon over a given time period always caused a given mass of water to rise in temperature by the same amount. Later and better experiments of this sort — where both the work done and the temperature rise of a mass of water could be measured precisely — gave the result that to raise 1g of water by 1°C it takes a work input of 4.184J.

Nowadays, we say that work and heat are different forms of energy. Thermodynamics is the science which describes how heat and work — and other forms of energy, such as the potential energy stored in chemical bonds — can be interconverted. When an electric battery is connected to a motor raising a weight, it does work. When both ends are connected to a short piece of wire, it produces heat. In both cases some of the potential energy in the chemical bonds of the reactants (for example, MnO_2, Zn and NH_4^+, in the dry cell you learned about in Section 16.10) is converted into work (motor) or heat (wire); the rest is converted into the potential energy of bonds in the products (Mn_2O_3, Zn^{2+}, NH_3 and H_2O). The total energy of the universe remains constant; the energy of the battery decreases, and this decrease is just matched by the heat produced and/or work done.

If I give you a discharged battery, you have no way of knowing how much heat and how much work it has produced, since you don't know whether it was connected to a motor or a wire. You do know, however, that a certain amount of chemical energy has been transformed into some *combination* of heat and work. This leads to the equation:

$$\Delta E = q + w$$

which expresses the *first law of thermodynamics*. ΔE, q and w all refer to the same system. For the battery, all are negative for the processes we have discussed; for the surroundings, all are positive: within the surroundings, heat has been released, work has been done, and the energy has gone up. Unless we explicitly state otherwise, we will be talking about changes taking place within a system; thus this equation can be read, "the increase (or decrease) in the energy of a system is equal to the heat added to (or taken from) the system, plus the work done on (or done by) the system." Note in particular that if work is done on the system then w>0 and this makes a positive contribution to ΔE. (Some older texts define w in the opposite sense.)

Energy, E, is a state function, since (still using the same example) taking the battery from one state to another — say, charged to discharged — always changes the energy by the same amount. Heat and work are not state functions, since discharging the battery does not necessarily cause the same amount of heat to be released or work done. But heat plus work, or energy, *is* a state function, and this is in fact another way of expressing the first law.

PΔV Work and Enthalpy

Work involves a force acting through a distance. One example of this is a pressure changing a volume:

$$\begin{aligned} P\Delta V &= \text{(pressure)(volume)} \\ &= \left(\frac{\text{force}}{\text{area}}\right)(\text{volume}) \\ &= \frac{\text{(force)}}{\text{(distance)}^2}(\text{distance})^3 \\ &= \text{(force)(distance)} \\ &= \text{work} \end{aligned}$$

Note that if the volume of a system increases, then it must be doing work on its surroundings; for example, if a compressed gas in a cylinder equipped with a weighted piston is allowed to expand, then the gas does work by lifting the weight. If $\Delta V>0$, then $P\Delta V>0$, and w, as we have defined it, must be negative. Similarly, if a larger weight is then added to the piston, re-compressing the gas, then $\Delta V<0$, and w>0; that is, work is done on the system. Thus, if this so-called "pressure-volume" or "PV" work is the only kind we have, then

$$w = -P\Delta V$$

If a process takes place in a sealed, rigid volume (a *bomb*), then it cannot expand or contract, so that it cannot do work on, or have work done on it by, its surroundings. Therefore, at constant volume

Chapter 17 — Energy, Enthalpy and Thermochemistry

$$\Delta E_v = q_v + w_v = q_v + 0 = q_v$$

In other words, any decrease in the energy of a constant-volume system must be accompanied by the transfer of an equal amount of heat to its surroundings, and any increase in the energy at constant volume must be accompanied by the absorbtion of heat from the surroundings.

In the laboratory, it is more common to do an experiment in a beaker (constant P) than in a bomb (constant V). Then:

$$\Delta E_p = q_p + w_p = q_p - P\Delta V$$

$$q_p = \Delta E_p + P\Delta V = \Delta H_p$$

Since ΔE_p, P and ΔV are state functions, so is $\Delta E_p + P\Delta V$, and this quantity is called ΔH_p, the *enthalpy change* at constant pressure, which is numerically equal to the heat absorbed at constant pressure.

If you run a beaker experiment, and discover that heat is given off (that is, the reaction is exothermic), then the enthalpy of the products must be less than that of the reactants. (ΔH must be negative; the heat content goes down by the same amount as the heat given off.) If you want to find ΔE, the energy change, then you must know ΔV, and use the equation:

$$\Delta E_p = \Delta H_p - P\Delta V$$

For reactions in liquid solution, ΔV is usually small, so that $P\Delta V$ is insignificant and $\Delta H \simeq \Delta E$. But in the gas phase, a reaction can have a large ΔV; therefore, $P\Delta V$ can be large, so that $\Delta H \neq \Delta E$. For example, in the reaction

$$N_2O_4(g) \rightarrow 2NO_2(g)$$

the volume doubles. For 1 mole N_2O_4 reacting at standard temperature and pressure

$$\Delta V = V_{final} - V_{initial}$$

$$= 2(22.4L) - 22.4L$$

$$= 22.4L$$

$$P\Delta V = \text{work done on surroundings}$$

$$= (1\text{atm})(22.4L)$$

$$= (22.4 L\cdot atm)\left(\frac{101 J}{L\cdot atm}\right)$$

$$= 2.26 \frac{kJ}{\text{mole } N_2O_4}$$

$$\Delta E_p = \Delta H_p - 2.26 \frac{kJ}{\text{mole } N_2O_4}$$

Thus, ΔE for this reaction is 2.26kJ less than ΔH. The way to understand this result is to realize that if the reaction takes place at constant pressure, some of the heat absorbed ($q_p = \Delta H_p$) goes into expanding the system; that is, into doing work on the surroundings; therefore the heat that is left to increase the energy of the system is somewhat less than the total.

So far all we have defined is the enthalpy for a constant pressure process, ΔH_p. A more general definition of the enthalpy is given by

$$H = E + PV$$

It follows that if a system goes from state 1 to state 2 then

$$\Delta H = H_2 - H_1 = (E_2 + P_2V_2) - (E_1 + P_1V_1)$$

$$= (E_2 - E_1) + P_2V_2 - P_1V_1$$

$$= \Delta E + P_2V_2 - P_1V_1$$

If the initial and final pressures are the same, then we recover our earlier expression.

$$\Delta H = \Delta E + P(V_2 - V_1)$$

$$= \Delta E + P\Delta V$$

Hess' Law

Since most chemical reactions are run at constant pressure (for example, in a beaker), there are a lot of enthalpy data tabulated for chemical use. Much of this data was collected simply by measuring the heat given off or absorbed in chemical reactions at constant pressure. It is found experimentally that when two reactions add to give an overall reaction, their ΔH's add to give the overall ΔH. For example, if we have the system

$$A \rightarrow 2B \qquad \text{(I)}$$

$$B + C \rightarrow D \qquad \text{(II)}$$

Chapter 17 — Energy, Enthalpy and Thermochemistry

$$A + 2C \rightarrow 2D \qquad (III)$$

then

$$\Delta H_{III} = \Delta H_I + 2\Delta H_{II}$$

This law of enthalpy summation is known as *Hess' law*. Note that ΔH's relate to the stoichiometric coefficients of the chemical equation as written. Thus if a reaction is doubled, ΔH must be doubled. This makes sense, because clearly when two moles of a substance react at constant P twice as much heat is absorbed or released as when one mole reacts.

Hess' law is really a consequence of the fact that ΔH is a state function. If we start with A+2C and wind up with 2D, then ΔH cannot depend on the path we took to get there, and all Hess' law states is that any combination of reactions (any path) from the same initial reactants to the same final products is accompanied by the same overall ΔH.

Standard Heats of Formation

There are several factors, besides the chemical nature of the reactants and products, that affect ΔH for a chemical process. One is the temperature although, it has only a small effect; another is the phase (gas, liquid or solid) that the substance is in. Thus, to apply Hess' law to calculating ΔH for some overall process, we must be careful to specify the states of the reactants and products. Given this, suppose we rewrite reaction (I) above, specifying such conditions for A and B.

$$A(25°C, l, 1atm) \rightarrow 2B(25°C, g, 1atm) \qquad (I)$$

Now suppose we had a table where we could look up $H_A(25°C, l, 1atm)$ and $H_B(25°C, g, 1atm)$. Then we would know that

$$\Delta H_I = 2H_B(25°C, g, 1atm) - H_A(25°C, l, 1atm)$$

We have a good understanding of what a ΔH for process is — it is the heat absorbed at constant pressure — but what could H for a substance possibly mean? How could it be measured? The answer is that it doesn't matter how we choose our H's (i.e., it doesn't matter how we define the state H=0), just as long as all the ΔH's we get by combining them come out right. Pursuing the same example, suppose $\Delta H_I = 2$ kJ/mol. Then it doesn't matter if our table of H's looks like

Chapter 17 — Energy, Enthalpy and Thermochemistry

H's at 25°C

Substance	H(kJ/mol)
A(l)	6
B(g)	4

or if it looks like

H's at 25°C

Substance	H(kJ/mol)
A(l)	-4.86
B(g)	-1.43

In either case $\Delta H_I = 2H_B - H_A = 2.00$ kJ/mol. In fact, there are two sorts of tabulations that serve the function that these entirely fictitious H tables serve in this example. One is a tabulation of *standard heats of formation*, which are commonly used for inorganic compounds, and the other is a tabulation of *standard heats of combustion*, which are commonly used for organic compounds containing only C, H and O.

The *standard heat of formation* of a substance, $\Delta H°_f$, is defined as the enthalpy change on forming one mole of the substance in its *standard state* from the elements in their *standard states*. The standard state of an element is the most stable form at 1atm pressure and whatever temperature is chosen for the tabulation — generally 25°C. The standard state of anything else is the pure liquid or solid, or 1atm pressure if a gas, or 1M concentration if in solution. This is the same standard state that we used for tabulating $\Delta E°$'s in Chapter 16.

To show how this works, it is found experimentally that at 1atm and 25°C, the reaction

$$C(\text{graphite}) + \frac{1}{2}O_2(g) \rightarrow CO(g)$$

releases 110.52kJ/molCO. Since graphite is the most stable form of C and the gaseous state is the most stable form of O_2 at 25°C and 1atm, these elements are in their standard states. Furthermore, 1atm is the standard state for CO(g). Therefore, at 25°C

$$\Delta H°_f(CO) = -110.52 \text{kJ/mol}$$

It is also found experimentally that at 1atm and 298K the reaction

$$C(\text{graphite}) + O_2(g) \rightarrow CO_2(g)$$

exhibits $\Delta H = 393.51 \text{kJ/mol}$; therefore

$$\Delta H°_f(CO_2) = -393.51 \text{kJ/mol}$$

It follows, by subtracting the first from the second of these reactions and applying Hess' law, that for the reaction

$$CO(g) + \frac{1}{2}O_2(g) \rightarrow CO_2(g)$$

under the same conditions of T and P,

$$\Delta H° = \Delta H°_f(CO_2) - \Delta H°_f(CO)$$

$$= -3.93.51 \text{kJ/mol} - (-110.52 \text{kJ/mol})$$

$$= -282.99 \text{kJ/mol}$$

This $\Delta H°$ is the standard enthalpy change at 298K for the combustion of CO to CO_2. Again, recall that it pertains to the process in which one mole of CO at 1atm pressure is burned (in O_2 at 1atm pressure) to form 1mole of CO_2 at 1atm pressure.

For an arbitrary reaction

$$\alpha A + \beta B \rightarrow \gamma C + \delta D$$

we have

$$\Delta H = \delta \Delta H°_f(D) + \gamma \Delta H°_f(C) - \beta \Delta H°_f(B) - \alpha \Delta H°_f(A)$$

There are a few tricky aspects of the definition and use of $\Delta H°_f$'s. One is that what must appear on the left-hand side of the equation defining a $\Delta H°_f$ is elements in their standard states, and the standard state of an element is the most stable state at 1atm pressure and the temperature of the tabulation (usually 25°C). This implies that these elements are always in specified pure phases. The substances whose $\Delta H°_f$ is tabulated are not under quite this severe a restriction; for example, even though H_2O is most stable as a liquid at 25°C and 1atm pressure, the standard heat of formation of $H_2O(g)$ as well as $H_2O(l)$ is found

Chapter 17   Energy, Enthalpy and Thermochemistry

in tables; the difference between them is ΔH_{vap} for H_2O at the temperature of tabulation. Similarly, HCOOH(aq) as well as HCOOH(l) is found. The values differ by ΔH_{soln} of HCOOH. Both the pure liquid and the solution are in their standard states; one is in the standard state for a pure liquid, and one is in the standard state for a solution (1M concentration).

In effect, standard heats of formation "act like" absolute enthalpies, like H_A and H_B in our earlier example. The arbitrary zero of the $\Delta H°_f$ scale is the standard state of the elements. It may not be obvious why we can arbitrarily set $\Delta H°_f = 0$ simultaneously for all the elements. The answer involves the fact that what we are really trying to get at is ΔH's for *processes*; these are the physically measurable and meaningful numbers. Whenever there is a process that takes A to B, a ΔH is definable, and can be measured, or calculated using $\Delta H°_f(A)$ and $\Delta H°_f(B)$; however, if A and B are elements there is no such process, at least in the context of ordinary chemical interactions. Therefore, as long as we do not leave the realm of ordinary chemical interactions, we can pick H's for the elements arbitrarily; the definition of standard heats of formation amounts to the convenient, though arbitrary, choices that, first, the elements in their standard states will all be given the same value of H, and, second, that that value will be zero.

Standard Heats of Combustion

The decision that H for any element in its standard state is to be zero is not the only possible choice for the zero of the enthalpy scale; tables of *standard heats of combustion*, $\Delta H°_{comb}$, in effect make the choice that the enthalpy is to be taken as zero for $H_2O(l)$, $CO_2(g)$ and $O_2(g)$, all at 1atm pressure. Again tabulation is usually done at 25°C. The standard heat of combustion for a substance is ΔH for the reaction in which one mole of substance is burned in $O_2(g)$ to form $CO_2(g)$ and $H_2O(l)$, all at 1atm pressure. Recall that $\Delta H°_{comb}$'s are used only for compounds containing nothing but C, H and O.

We can see how this works using the same reaction we used for $\Delta H°_f$.

$$CO(g) + \frac{1}{2}O_2(g) \rightarrow CO_2(g)$$

First of all, by definition, $\Delta H°$ of this reaction must be $\Delta H°_{comb}(CO)$. Its value is not tabulated in the

Chapter 17   Energy, Enthalpy and Thermochemistry

text,

but it must be equal to the $\Delta H°$ we calculated earlier for this reaction, namely -282.99kJ/mol. Considering this reaction from the point of view of an arbitrary absolute enthalpy scale, we have

$$\Delta H° = H°(CO_2) - H°(CO) - \frac{1}{2}H°(O_2)$$

We said that the definition of $\Delta H°_{comb}$ implies that $H°(CO_2)=H°(O_2)=0$; thus $\Delta H°=H°(CO)=\Delta H°_{comb}(CO)$.

Now recall that for the reaction

$$C(graphite) + O_2(g) \rightarrow CO_2(g)$$

we had $\Delta H°=\Delta H°_f(CO_2)=-393.51kJ/mol$; it is clear from the definition of $\Delta H°_{comb}$ that also $\Delta H°=\Delta H°_{comb}(C(graphite))$, so that $\Delta H°_{comb}(C(graphite))=-393.51kJ/mol$. Table 17.2 reveals that this is indeed the case. Thus $\Delta H°_f$ and $\Delta H°_{comb}$ will in general differ for any substance (exception: $O_2(g)$), but for any process the calculated $\Delta H°$ will be the same, regardless of which convention is used.

Bond Energies

Despite their names, *bond energies* and *bond dissociation energies* are defined as *enthalpies* of specific processes. The bond dissociation energy, often symbolized D, is the enthalpy change of breaking a *specific* bond; for example, for the reaction

$$H_2(g) \rightarrow 2H(g)$$

ΔH is equal to +435.9kJ/mol, and this is equal to $D(H_2)$, the bond dissociation energy of H_2. The bond energy, on the other hand, refers not to a specfic bond, but rather to a type of bond; for example, most C—H bonds have roughly the same dissociation energy, approximately 416kJ/mole, and this average value is the C—H bond energy. Similarly, most C—C single bonds have a dissociation energy near 348kJ/mole, and this value is taken as the C—C bond energy.

A bond energy, or a bond dissociation energy, is the enthalpy of *breaking* a bond. The sum of the bond dissociation energies of a substance is thus the enthalpy of breaking the substance up into isolated

Chapter 17 — Energy, Enthalpy and Thermochemistry

atoms; if bond energies are used, then the same statement holds, but the result will not be as accurate, since bond energies are only approximate. Using methane, CH_4, as an example, the sum of the bond energies ($4 \times$ bond energy of C—H = (4)(413 kJ/mole) = 1652 kJ/mole) is $\Delta H°$ for the reaction

$$CH_4(g) \rightarrow C(g) + 4H(g)$$

Thus to *form* $CH_4(g)$ from atomic fragments, $\Delta H°$ is approximately *minus* the sum of the bond energies. In general, then, $\Delta H°$ for a reaction is the sum of the bond energies of the reactants minus the similar sum for the products. This is in contrast to the use of $\Delta H°_f$ and $\Delta H°_{comb}$, for which $\Delta H°$ was the sum of the tabulated values for the products minus the sum for the reactants.

Example 17.10 shows how to calculate $\Delta H°$ for the isomerization of dimethyl ether to ethyl alcohol.

$$CH_3-O-CH_3(g) \rightarrow CH_3CH_2OH(g)$$

These compounds contain the following types and numbers of bonds.

Number of Bonds, by Type

	CH_3-O-CH_3	CH_3CH_2OH	Net Change	Bond Energy (kJ/mol)
C-H	6	5	-1	413
C-C	0	1	+1	348
C-O	2	1	-1	351
O-H	0	1	+1	463

Thus $\Delta H°$ should be given by the following sum of bond energies

$$\Delta H = 1(C-H) - 1(C-C) + 1(C-O) - 1(O-H)$$
$$= (413 - 348 + 351 - 463) \text{kJ/mol}$$
$$= -47 \text{kJ/mol}$$

As the text, shows, the exact value, using $\Delta H°_f$'s, is -51.1 kJ/mol; the 8% difference is a rough measure of how good an approximation bond energies are.

Chapter 17 — Energy, Enthalpy and Thermochemistry

Heat Capacity

The *heat capacity* is the amount of heat that has to be added to a system to raise the temperature by 1K (or 1°C). The heat capacity of a gram of a substance is called its *specific heat*, and the heat capacity of a mole of substance is called the *molar heat capacity*; converting between them is trivial, as Example 17.11 demonstrates.

Recall that the heat put into a system at constant volume is equal to ΔE, whereas the heat put in at constant pressure is equal to ΔH. Suppose one wishes to raise the temperature of a system by 1K; if this is done at constant V, then q_v, the amount of heat it takes to do this is equal to ΔE, and if this is done at constant P, then q_p, the amount of heat it takes to do this is ΔH. Since $\Delta E \neq \Delta H$, in general, then $q_p \neq q_v$. Thus the heat capacities at constant P and at constant V must differ; we denote them C_p and C_v, and based on what we have said the following relationships hold:

$$q_v = \Delta E = C_v \Delta T$$

$$q_p = \Delta H = C_p \Delta T$$

As the text notes, C_p and C_v do vary somewhat with temperature, but for the purposes of this text they can be assumed constant.

We can use the definition of the enthalpy to calculate the difference between C_p and C_v for an ideal gas.

$$H = E + PV$$

$$\Delta H = \Delta E + (P_2 V_2 - P_1 V_1)$$

$$P_1 V_1 = nRT_1$$

$$P_2 V_2 = nRT_2$$

$$P_2 V_2 - P_1 V_1 = nRT(T_2 - T_1) = nR\Delta T$$

$$\Delta H = \Delta E + nR\Delta T$$

For one mole of ideal gas

$$\Delta H = C_p \Delta T = \Delta E + R\Delta T$$

$$= C_v \Delta T + R\Delta T$$

Chapter 17 — Energy, Enthalpy and Thermochemistry

$$C_p = C_v + R$$

Thus, for an ideal gas, C_p is bigger than C_v. The reason is that the constant pressure system expands and does work as it is heated. The heat that is added to raise the temperature also must be sufficient to do this work of expansion; thus, more heat is needed at constant P than at constant V, where no work can be done. Since liquids and solids do not expand much on heating, little work is done on raising the temperature even at constant pressure, and $C_p \approx C_v$. In algebraic terms, at constant pressure $\Delta(PV) = P(\Delta V)$, and for liquids and solids ΔV upon raising the temperature is almost zero.

Recall (Chapter 4) that for a mole of an ideal gas

$$E_{trans} = \frac{3}{2}RT$$

We also said that for an ideal gas kinetic energy is the only energy there is; since there are no intermolecular attractions or repulsions, there is no potential energy. Thus, for a process involving an ideal gas, we have

$$\Delta E = \Delta E_{trans} = \frac{3}{2}R\Delta T$$

$$C_v = \frac{q_v}{\Delta T} = \frac{\Delta E}{\Delta T} = \frac{3}{2}R = 12.5 \frac{J}{mol \cdot K}$$

$$C_p = C_v + R = \frac{5}{2}R = 20.8 \frac{J}{mol \cdot K}$$

Table 7.5 shows that this calculated value for C_p holds only for helium, among the gases listed. It also holds for Ne and Ar, and other monatomic gases. For polyatomic gases C_p's are higher; for these, E_{trans}, the translational kinetic energy, remains equal to $(3/2)RT$, but polyatomic molecules have other forms of kinetic energy as well, most notably those associated with rotation and vibration. Some of the heat added to a gas consisting of polyatomic molecules goes into increasing this type of motion, thereby increasing C_p (and C_v).

Chapter 17   Energy, Enthalpy and Thermochemistry

Calorimetry

It is easier to measure work than heat. For example, I can easily calculate the work needed to carry a rock of mass m up a flight of stairs of height h: w = mgh, where g is the acceleration due to gravity (g = 3.0×10^8 m/s^2). On the other hand, if I am heating a pot on the stove, how can I measure the amount of heat going into the water? Because this is hard to do, the calorie was initially *defined* as the amount of heat it takes to raise 1g of water 1°C. Thus, if the pot contains 900 g of water, and it rises in temperature from 20°C to 90°C, then the heat put in must have been, by definition

$$q = (900g)\left(1\frac{cal}{g°C}\right)(90-20)°C$$

$$= 63,000 cal$$

Note that this definition makes the specific heat of water equal to 1 cal/g°C. This can then be used to calculate the specific heat of other things. For example, let me take the pot of water I have just heated, place it in a well-insulated vessel (a *calorimeter*), and add to it 2000 g (about 10 moles) of lead, initially at 20°C. I observe that this causes the temperature to drop from 90°C to 85.5°C. The first law says that the heat gained by the lead must equal the heat lost by the water

$$(10 \text{ moles Pb})(C_{Pb})(85.5-20) = (900 H_2O)(1\frac{cal}{mole°C})(90-85.5)°C$$

$$C_{Pb} = \text{molar heat capacity Pb} \approx 6\frac{cal}{mole°C}$$

Thus, *however* much energy a calorie represents, calorimetry allows scientists to make measurements involving heat.

Nowadays we would say C_{Pb} = (6cal/mole·K)(4.184J/cal) = 25J/mole·K. This value is exhibited by most solids, and that fact is called the *Law of Dulong and Petit*, after the French scientists who discovered it in 1819, by doing experiments of just this kind.

ΔH's are commonly measured in a calorimeter. This is a well insulated chamber in which processes can be carried out. By measuring the change in temperature, and knowing the heat capacity of the system, the enthalpy change of a process can be calculated. The caloric content of a food, for example, is

Chapter 17 — Energy, Enthalpy and Thermochemistry

the amount of heat it gives off when burned in a calorimeter. In your body, foodstuffs are partly burned to produce CO_2, H_2O, the heat that keeps your body at 98.6° F and the work that you do. The unburned remainder is excreted.

Example 17.13 sets out an example of a calorimetric calculation. In a *bomb calorimeter* a reaction, such as combustion, takes place in a sealed, constant volume, so that any observed rise in temperature must be accounted for by C_v; this means that we end up calculating a ΔE for the process. (Sometimes, however, calorimetry is done at constant P, in which case C_p applies, and what we get is a ΔH for the process.)

It is important to realize that even an empty calorimeter has a C_v; that is, it takes heat to raise its temperature. In Example 17.13 a sucrose sample of known heat of combustion (at constant V; that makes it an energy, not an enthalpy of combustion!) is burned in a calorimeter. Since the sample gives off 33.04kJ, and the temperature rises by 2.966K, we have

$$C_v(\text{calorimeter}) = \frac{33.04\,\text{kJ}}{2.966\,\text{K}} = 11.14\,\frac{\text{kJ}}{\text{K}}$$

When 1.015×10^{-2} mol of of benzene is subsequently burned in the same calorimeter, the temperature rises by 3.008K; therefore the molar ΔE of combustion of benzene is given by

$$\Delta E = -\frac{(11.14\,\text{kJ/K})(3.008\,\text{K})}{1.015 \times 10^{-2}\,\text{mol}} = -3301\,\frac{\text{kJ}}{\text{mol}}$$

The negative sighn comes in because the process is exothermic.

Note that in this example C_v(calorimeter) was the only important C_v. The heat capacity of the contents of the calorimeter was neglected in the calculation. Clearly, however, some of the heat generated in the reaction must have been used to raise the temperature of the contents. For bomb calorimetry we can justify this neglect because the bomb is massive, and the reaction being examined — generally combustion — is highly exothermic, so that the contents do not need to weigh much to produce a lot of heat. There are other set-ups, however, where both the heat capacity of the calorimeter and that of the contents have to be included in the calculation.

Chapter 17   Energy, Enthalpy and Thermochemistry

Questions

17.1 State functions

1. Define a state function.

2. If a system expands in volume, what is the sign of ΔV? __+ve__

17.2 Heat and work

1. If a system expands, what is the sign of w? __−ve__

2. If a system expands against a constant pressure, P, then

    (a) $P\Delta V$ is the work done __by__ the system

    (b) $-P\Delta V$ is the work done __on__ the system

    (c) w = __$-P\Delta V$__ (in terms of P and ΔV)

3. If heat is added to a system, what is the sign of q? __+ve__

4. If a system heats its surroundings, what is the sign of q? __−ve__

17.3 The first law of thermodynamics

1. Which of these are state functions? (q, w, E)   __E__

2. If a system in a bomb heats its surroundings, what is the sign of ΔE? __−ve__

3. If a system expands isothermally against a constant pressure P,

    (a) what is the sign of w? __−ve__

    (b) what is the sign of ΔE? __−ve__

    (c) q = __$+P\Delta V$__ (in terms of P and V)

Chapter 17 — Energy, Enthalpy and Thermochemistry

4. Give an equation that expresses the first law of thermodynamics:

    $\Delta E = q + w$

5. State the first law in words.

    At constant mass, energy can neither be created or destroyed but its different forms can be converted.

17.4 Energy and enthalpy

1. The symbol for heat added to a system at constant volume is __q_v__.

2. Why can't a reaction carried out in a work?

    Since work = force × distance and there is no ΔV or distance $w = 0$.

3. The heat absorbed at constant volume is a measure of Δ __E__.

4. The heat absorbed at constant pressure is a measure of Δ __H__.

5. By definition, H = __$E + PV$__.

6. At constant external pressure, P

    (a) $\Delta H =$ __$\Delta E + P\Delta V$__ (in terms of E, P and V)

    (b) $\Delta H =$ __q_p__ (in terms of E, q and w)

    (c) $\Delta E =$ __$\Delta H - P\Delta V$__ (in terms of H, P and V)

    (d) $\Delta E =$ __$\Delta H + w$__ (in terms of H, q, and w)

17.5 Thermochemistry

1. State Hess' Law, in your own words:

2. Suppose these two reactions have known enthalpies:

    $$A + 2B \to 2C \quad \Delta H_1$$

17-18

$$C \rightarrow D \quad \Delta H_2$$

(a) Write a balanced chemical equation for the reaction of A and B to form D.

(b) What is ΔH for this reaction, in terms of ΔH_1 and ΔH_2 ?

3. (a) Any substance *must be* at 298 K to be in its standard state.

(b) Any substance *must be* at 1 atm pressure to be in its standard state.

(c) In the notation $\Delta H°_f$ the ° means $0°C$ and the f stands for *formation*.

(d) If an element has more than one allotrope, the standard state is the *least stable* of them.

(e) In order to calculate correct values of $\Delta H°_f$ *all elements and all products* must be in their standard states.

(f) $\Delta H°_f$ for graphite is *positive*, for diamond is *negative* and for amorphous carton is *zero*.

4. Using only data in Table 17.2, calculate the heat of formation of $CH_4(g)$

Chapter 17 Energy, Enthalpy and Thermochemistry

17.6 Bond energies

1. What is the difference between a bond energy and a bond dissociation energy?

2. "The reason bond energies are useful is that all bonds between the same pairs of atoms -- such as C-C or C-H or N-H -- have similar energies." (True, False)

3. Suppose the bond dissociation energy of the molecule AB is 200 KJ/mole. What is ΔH for the reaction:

$$A(g) + B(g) \rightarrow AB(g)$$

17.7 Heat capacity

1. "For any normal chemical substance, the molar heat capacity is greater than the specific heat". (True, False)

2. "For an ideal gas, $C_p > C_v$". (True, False)

3. "For any gas, $C_p > C_v$". (True, False)

4. "For polyatomic gases, $C_p < (5/2)R$". (True, False)

5. A solid is found to have a specific heat near room temperature of about 0.25 J/gK. About what is its molecular weight?

Chapter 17   Energy, Enthalpy and Thermochemistry

17.8 Calorimetry

1. A reaction is performed in a calorimeter, and as a result the temperature decreases. The sign of ΔH for the reaction must be _____ .

2. Less than 100% of the sugar you eat gets converted to CO_2 and water; some derived products get excreted. Therefore, the heat of combustion of sucrose as determined in a calorimeter provides an (overestimate, underestimate) of the useful energy you get from it.

Chapter 17 — Energy, Enthalpy and Thermochemistry

Answers

17.1 State functions

1. A state function is any variable which has the same value whenever the system is in the same state.
2. +

17.2 Heat and work

1. −
2. (a) by
   (b) on
   (c) $-P\Delta V$
3. +
4. −

17.3 The first law of thermodynamics

1. E
2. −
3. (a) −
   (b) −
   (c) $+P\Delta V$
4. $\Delta E = q + w$
5. Energy is not created or destroyed, but its different forms can be interconverted.

17.4 Energy and enthalpy

1. q_v
2. Work involves motion; the walls of a bomb cannot move.
3. E
4. H
5. $E + PV$
6. (a) $\Delta E + P\Delta V$
   (b) q
   (c) $\Delta H - P\Delta V$
   (d) $\Delta H + w$

17.5 Thermochemistry

1. When chemical reactions are added algebraically, their ΔH's are added arithmetically. (Also true for ΔE and other *extensive* state functions.)
2. (a) $A + 2B \rightarrow 2D$
   (b) $\Delta H_1 + 2\Delta H_2$
3. (a) may be
   (b) (true)
   (c) standard, (True)
   (d) most stable
   (e) True
   (f) zero, positive, positive

Chapter 17 — Energy, Enthalpy and Thermochemistry

4. (Use data for CH_4, H_2 and C)

17.6 Bond energies

1. A bond energy is a typical value; a bond dissociation energy is an average value.
2. True
3. -200 kJ/mole

17.7 Heat capacity

1. True (since mol. wt. > 1)
2. True
3. True
4. False
5. From Dulong and Petit, $\dfrac{25 \text{ J/mole·K}}{.25 \text{ J/g·K}} = 100 \text{ g/mole}$

17.8 Calorimetry

1. +
2. underestimate

CHAPTER 18: ENTROPY, FREE ENERGY, AND EQUILIBRIUM

Scope

When a rock falls under the influence of gravity, its potential energy decreases. Physical systems (of which chemical systems are one kind) tend to move "downhill" energetically. For any physical process which takes place spontaneously, there is some kind of energy which is decreasing. The history of thermodynamics is partly the history of finding out what kinds of energy characterize which processes in this manner. This chapter leads up to the definition of the *free energy*, G, which is guaranteed to decrease whenever a process at constant temperature and pressure takes place spontaneously. In order to develop the concept of free energy, we first have to introduce the *entropy*, S, a new state function which measures how much "freedom," or "disorder," or "randomness" a system has.

At one time it was believed that all spontaneous processes were exothermic, and certainly many are, but physical systems also have a tendency to move toward a state in which they have more freedom, if such a state is available; for example, an ideal gas initially confined to a small vessel will spontaneously fill a larger volume if a valve leading to another empty vessel is opened. This occurs spontaneously even though ΔE, q and w for the process are all zero! As another example, the dissolution of urea in water in spontaneous, yet it is endothermic: the solution cools as the urea dissolves, and thus it tends to absorb heat from its surroundings. Since $\Delta H \approx \Delta E$ for a solution process, and since endothermicity at constant P implies $\Delta H > 0$, we conclude that for this spontaneous process also $\Delta E > 0$. As we discussed in Chapter 6, for a case like this the unfavorable increase the energy is more than offset by the favorable increase in entropy: the additional freedom the solute molecules obtain when they dissolve. It is not possible to answer the question *why* the tendency to freedom or randomness is favored, any more than it is possible to say why energy is conserved; we can say only that it appears to be a fundamental law of the universe.

Some Sample Processes

It is very useful when defining thermodynamic functions, such as the entropy, to be able to refer to simple physical processes as examples. Processes involving ideal gases are especially useful conceptually, because the equation of state, PV=nRT, allows calculations to be done readily, and thus the physical

Chapter 18
Entropy, Free Energy, and Equilibrium

principles involved are not hidden in a morass of calculations. You should realize that the principles themselves are subtle, and you should make every effort to understand the examples, so that you can refer back to them for guidance when you are having difficulty with more complicated applications of thermodynamics.

First we will take up two processes that were discussed in Examples 17.1 and 17.2, and are discussed again in Section 18.1. These are both *isothermal* (occuring at constant temperature) expansions of an ideal gas. Process 1 is expansion into a vacuum, and Process 2 is expansion against a constant external pressure. After this we will discuss a third process, a *reversible* isothermal expansion. These are the same three processes that are summarized in Table 18.1 in the text.

For all the processes the net change is

ideal gas (8atm, 1L, 298K, 0.327mol) → ideal gas (2atm, 4L, 298K, 0.327mol)

The thing to remember throughout is that for an ideal gas E, the energy, depends only on temperature, since from the kinetic theory of gases,

$$E = E_{trans} = \frac{3}{2}RT$$

As we recalled in Chapter 17, all the energy of an ideal gas is kinetic, since there are no intermolecular forces. Thus, $\Delta E=0$ for all three processes, since $\Delta T=0$.

First consider Process 1 (Figure 18.1), in which we allow the gas, initially confined to a 1L bulb at a pressure of 8atm, to expand through a stopcock into a 3L bulb, initially empty. At the end of the experiment the gas fills both bulbs. Let us consider these bulbs, together, to constitute a *system* and the rest of the universe to constitute the *surroundings*. Note that the entire system (which starts out as one full and one empty bulb) is a constant volume system; when the gas flows through the stopcock the system doesn't change size, but rather the gas simply redistributes itself within it. Recalling that $w_{sys}=-P\Delta V$, it is clear that since $\Delta V=0$, w=0. We are told that the temperature does not change, and therefore $\Delta E=0$, and since $\Delta E=q+w$, we must have q=0 as well.

A few comments on this (not in text). First of all, the result is different if the gas is not ideal. If the same experiment were to be performed on a non-ideal gas at a pressure where attractive forces dominate, then separating the molecules would raise the energy; thus, $\Delta E > 0$ for the expansion. We would still have $w=0$, since this is defined by how the system is built ($\Delta V_{sys}=0$), so $\Delta E = q+w$ would mean $q>0$. Since q refers to q_{sys}, heat would have been transferred from the surroundings to the system to keep the temperature constant. If the experiment had been run on a non-ideal gas at very high pressures, where repulsive forces dominate, then separating molecules would have lowered the energy, so that $\Delta E < 0$; again $w=0$, and the first law gives $q<0$, so that heat would have been transferred from the system to the surroundings.

To further understand Process 1, consider what would happen if it were to be allowed to take place in a well-insulated vessel, where $q=0$; that is, by insulating the vessel, we ensure that no heat can be transferred, in the same way that we ensured that $w=0$ by making the system rigid. Now we have $\Delta E = q+w = 0$, for either an ideal or a non-ideal gas. For the ideal gas everything is as it was before; since there never was a tendency for heat to flow, making sure that it doesn't does not change anything. For the non-ideal gas in the realm of net attractive interactions we said that separating the molecules raises the potential energy; since ΔE must be 0, the only way this can happen is for the kinetic energy to decrease by the same amount as the potential energy increases. But since temperature is proportional to kinetic energy this means that the process must be accompanied by a temperature decrease ($\Delta T < 0$). Likewise, if inside a rigid, insulated container a non-ideal gas is allowed to expand within a pressure range where repulsive forces dominate, $\Delta T > 0$.

Still another point about Process 1: note that the q and w which we have found to be equal to zero pertain to heat and work transferred to or from the system from the surroundings. In particular, even though within the system the gas expands, the ΔV that counts when calculating w is the ΔV of the entire system, which is zero. So to summarize our understanding of Process 1, we know that $\Delta V=0$, since the gas is ideal and we were told that $\Delta T=0$. We know that $w=0$ since the overall system does not do work on the surroundings; it follows from the first law that $q=0$. Furthermore, even if we had not been told that $\Delta T=0$ we would have been able to figure it out from the molecular picture of an ideal gas: expansion is nothing more than separating the molecules, and for an ideal gas this involves no ΔE;

Chapter 18   Entropy, Free Energy, and Equilibrium

nothing happens to the molecular velocities, so $\Delta T=0$.

In Process 2 (Figure 18.2) the system is a cylinder with a movable piston. Inside the cylinder the gas occupies a volume of 1L and is at an internal pressure, P_{int}, of 8atm. This system sits in a room where the external pressure, P_{ext}, is 2atm. At some time a pin holding the piston down is released, and the piston rises until the internal and external pressures are equal. At this point V=4L, and $P_{int}=P_{ext}=2$atm. The temperature of the room is controlled by a thermostat, so that heat is transferred to or from the cylinder so as to maintain a constant temperature.

Here, $\Delta V=3$L; that is, the system really does expand, and the work done on the surroundings is $P_{ext}\Delta V = (2\text{atm})(3\text{L}) = (6\text{L·atm})(101.3\text{J/L·atm}) = 608\text{J}$. For the system, then, w=−608J, since $w = -P_{ext}\Delta V$ (work is taken *out of* the system). As in Process 1, $\Delta T=0$, and therefore $\Delta E=0$. Then

$$\Delta E = 0$$
$$= q + w$$
$$= q - 608\text{J}$$

so that q = +608J. In other words, the system does work, which tends to lower the energy of the system; however, we said the energy stays constant. By the first law, heat must be added to the system to compensate.

Again, it is instructive to consider what would have happened if the process had been run in an insulated cylinder, for which q=0 (also not in text). Here, as expansion took place and work was done, we would have had $\Delta E<0$, since w<0 and q=0. Since the gas is ideal, so that all the energy is kinetic, the kinetic energy, and hence the temperature, would have gone down. The process would still have proceeded until $P_{int}=P_{ext}=2$atm, but we would have had $\Delta T<0$ instead of $\Delta T=0$, and naturally the final volume would not have been as large as in Process 2, because of the lower temperature; thus ΔV would have been smaller than for Process 2, and the work done, $w_{surr} = P_{ext}\Delta V$, would have been less as well. The process would not have been isothermal. There is a term for a process where q=0; it is called an *adiabatic* process.

Chapter 18

Entropy, Free Energy, and Equilibrium

For Process 3 the system is the same as the system described for Process 2, except that instead of holding the piston down initially with a pin which will be suddenly released we use a weight just heavy enough to cause P_{ext} to be equal to 8atm, the initial value of P_{int}. Then we imagine removing the weight a very little bit at a time — as if it consisted of a heap of sand, and we could remove it grain by grain. As each grain is removed the system expands a little bit, maintaining $P_{int} = P_{ext}$, as these both decrease. Note that at the end of the process $P_{ext} = 2$atm as before. Such a process is called *reversible* because by making only a tiny reversal in the external conditions the process will go backwards; for example, if we add back the grain of sand we just removed, the system will contract just back to the volume it had before we removed it. The system is always in equilibrium, in that at any point in the process we can stop, and nothing will happen. In contrast, Processes 1 and 2 were *spontaneous*: the initial states were not at equilibrium (for example, $P_{int} > P_{ext}$ in Process 2), and the processes proceeded until an equilibrium state was achieved. Of course, a reversible process is an idealized abstraction, since removing each grain of sand causes P_{ext} to decrease to less than P_{int} momentarily, before the system can adjust, but by making the steps small enough we can keep the difference between P_{int} and P_{ext} as small as we like. Conceptually, the extrapolation to infinitessimally small steps corresponds to the reversible process.

Now let us consider what q and w are for Process 3. Note first of all that just as for Process 2, $\Delta E = 0$, since $\Delta T = 0$, so again we are going to have $q = -w$. The work done on the surroundings is going to be greater than the corresponding work for Process 2, for the following reason. The initial expansion is done against an external pressure of 8atm, and the final expansion against an external pressure of 2atm. In Process 2 all the work was done against an external pressure of 2atm. Since $w_{surr} = P_{ext} \Delta V$, the increments of ΔV that occurred against higher pressures than 2atm give greater contributions to the work done than they do in Process 2.

If we let δV ("δ" is a lower case "Δ") represent an incremental volume charge, then

$$w = -\sum_i P_{ext,i} (\delta V)_i$$

For Process 2 all the $P_{ext,i}$ are equal to 2atm; for Process 3 all except the last one are greater than 2atm. Thus w is more negative (and w_{surr} is more positive) for Process 3 than for Process 2. In calculus notation

Chapter 18   Entropy, Free Energy, and Equilibrium

this goes over to

$$w = -\int_{V_1}^{V_2} P_{ext} dV$$

For Process 2, $P_{ext} = 2$ atm throughout, so that

$$w = -(2\text{atm})(V_2-V_1) = -6\text{L·atm}$$

as we had before. For Process 3, $P_{ext} = P_{int}$, but P_{int} is given by the ideal gas law.

$$P_{int} = \frac{nRT}{V}$$

$$w = -\int_{V_1}^{V_2} \frac{nRT}{V} dV = -nRT\ln\left(\frac{V_2}{V_1}\right)$$

This equation applies to the work of the system for a reversible, isothermal expansion of an ideal gas. Even if your course does not assume calculus, this formula is important, and you should know it. For the particular conditions we have been describing, w for Process 3 comes out to −11.1 L·atm, or −1120 J. Therefore, since $\Delta E = q+w = 0$, we have q=1120 J. We summarize our calculations so far in the following table, which essentially duplicates Table 18.1 in the text.

Note that even though ΔE is zero for all three processes, w and q vary. This should not be surprising; the initial and final conditions are the same for all three, and ΔE is a state function, so that ΔE must be the same for all. q and w are path dependent, however, so they can, and do, differ. Note that w is most negative for Process 3. In fact, for *all* possible processes between these initial and final states, Process 3 exhibits the most negative value of w. The reasoning is as follows. The work done is maximized when P_{ext} is maximized. But for a reversible process P_{ext} is as great as it can possibly be, at any point in the process. If P_{ext} were any greater, the process would reverse and work would be done *on* the system.

Chapter 18 Entropy, Free Energy, and Equilibrium

ΔE, q and w for Three Processes

	Process 1	Process 2	Process 3
n	0.327 mol	0.327 mol	0.327 mol
P_1	8 atm	8 atm	8 atm
P_2	2 atm	2 atm	2 atm
V_1	1 L	1 L	1 L
V_2	4 L	4 L	4 L
T_1	298 K	298 K	298 K
T_2	298 K	298 K	298 K
P_{ext}	—	2 atm	nRT/V
w	0	−608 J	−1120 J
q	0	608 J	1120 J
ΔE	0	0	0

Since w is as negative as it can be between these initial and final states, q is as positive as it can be. Note that these values, q_{rev} and w_{rev}, must be uniquely defined; there is only one maximum value. This allows us to define a new state function, the entropy, S. For any *isothermal* process, the entropy change is defined by

$$\Delta S = \frac{q_{rev}}{T}$$

where q_{rev} is the amount of heat it would be necessary to add to the system on the way from the initial to the final state by a *reversible* path. (If the initial and final states are not at the same temperature, one can imagine adding heat and changing the temperature bit by bit along a reversible path, then adding the q_{rev}/T terms for each bit, giving:

$$\Delta S = \sum_i \frac{q_{rev,i}}{T}$$

In calculus notation, this goes to

$$\Delta S = \int \frac{dq_{rev}}{T}$$

The path must still be reversible for this to define the entropy.)

Note that ΔS is uniquely defined by the initial and final states of the system. Thus ΔS is the same for Processes 1, 2 and 3 that we have been discussing. This ΔS is equal to q/T for Process 3.

$$\Delta S = \frac{q_{rev}}{T} = \frac{1120 J}{298 K} = 3.76 J/K$$

ΔS for Phase Changes

In general, to calculate ΔS, the entropy difference, between any two states, one devises a reversible path between the two states, and calculates ΔS using the above formula; again, S is a state function, so that the entropy change is the same even if the actual process of interest is not reversible. Three outstanding questions concerning the entropy are, first, how do we define the zero of entropy, so that we can look up numbers for compounds and calculate ΔS for processes, much as we used tables of $\Delta H°_f$ or $\Delta H°_{comb}$ to calculate $\Delta H°$ for processes; second, why is entropy defined the way it is, and how is it related to spontaneity of a process and to molecular freedom, or disorder, as we indicated it is; and third, what do you do with ΔS once you know it; that is, what good is it? We will answer all these questions in this chapter, but first we will discuss the entropy changes associated with reversible phase transitions.

Recall that a reversible process is one which takes place under equilibrium conditions. Water and ice in their standard states are in equilibrium at the melting point of water, so the melting of ice at 0°C is a reversible process; so is the vaporization of water at 100°C and 1atm pressure. Since both these processes are isothermal as well as reversible, ΔS is given simply by q/T. Furthermore, since both processes also take place at constant pressure, we know that $q=\Delta H$. Strictly speaking, both these processes refer to phase changes under standard conditions: the water and ice are in their pure states at 1atm, and the vapor is at 1atm pressure; thus we have

$$\Delta S°_{fus} = \frac{\Delta H°_{fus}}{T_{mp}}$$

and

Chapter 18 — Entropy, Free Energy, and Equilibrium

$$\Delta S°_{vap} = \frac{\Delta H°_{vap}}{T_{bp}}$$

where T_{mp} is the melting point and T_{bp} is the boiling point.

Trouton

For most liquids, it is found that $\Delta S°_{vap}$ is in the range 88 ± 5 J/K. The major exception is hydrogen-bonded liquids, for which $\Delta S°_{vap}$ is about 20J/K higher. Some examples are given in Table 18.2; for now, consider benzene, a "normal" liquid. For it, $\Delta S°_{vap} = \Delta H°_{vap}/T_{bp} = (30800\text{J/mol})/353\text{K} = 87.2$ J/K. For water, on the other hand, which is hydrogen-bonded, $\Delta S°_{vap} = (40700\text{J/mol})/273\text{K} = 109$ J/K. Note that $\Delta S°_{vap}$ will always be positive, since $\Delta H°_{vap}$ and T are always positive. We have alluded earlier in the text to the idea that an entropy increase corresponds to an increase in molecular freedom or disorder, and so it makes sense that $\Delta S°_{vap}$ is positive, since gas molecules have greater freedom than molecules in a liquid. The fact that non-hydrogen-bonded liquids all gain about the same amount of entropy on vaporization means that they gain about the same amount of freedom on vaporization. The vapors both of these and of hydrogen-bonded liquids are very close to ideal at 1atm pressure, and the main type of freedom gained by molecules on vaporizing is freedom of motion: the ability to move so as to fill the entire container, and to be unrestricted by interaction with surrounding molecules to rotate and to vibrate. In an ideal gas, this freedom is unrestricted. Thus the fact that ΔS_{vap} is close 88KJ/mol for all normal liquids means that in these liquids these motions must be restricted to about the same extent. The fact that for hydrogen-bonded liquids ΔS_{vap} is greater means that these motions are more severely restricted in them; as an example, water molecules in the liquid, by being hydrogen-bonded to about four neighbors, have their rotational motion much more severly restricted than do benzene molecules in liquid benzene. This additional restriction, which is relieved in the vapor, contributes to elevating ΔS_{vap} above what it is for benzene.

As Example 18.1 shows, Trouton's rule allows ΔH_{vap} to be estimated if T_{bp} is known. For nitrobenzene, for example, $T_{bp} = 484$K. Trouton's rule gives

$$\Delta H°_{vap} = T_{bp} \Delta S°_{vap}$$

Chapter 18 — Entropy, Free Energy, and Equilibrium

$$= (484K)(88 \pm 5 J/mol \cdot K)$$

$$= 42600 \pm 2400 kJ/mol$$

The experimentally determined value is 40.4kJ/mol, within our error margin.

$\Delta S°_{fus}$ is always less than $\Delta S°_{vap}$ for a given substance, indicating that a molecule gains less freedom going from the solid to the liquid state than it does going from the liquid to the vapor state. This is not surprising: strong intermolecular forces dominate in both the liquid and the solid, and are largely absent in the vapor.

Second Law

For any process taking place in an *isolated* system, the entropy always increases. This is known as the *second law of thermodynamics*. Since the universe is an isolated system (what could be on the other side of its boundary?), the entropy of the universe must always be increasing — another statement of the second law! Let us see how this accords with our calculations for the three processes we considered above (not in text). For all three processes, ΔS_{sys} was given by q_{rev}/T for Process 3; that is, for all of them

$$\Delta S_{sys} = \left(\frac{q_{rev}}{T}\right)_{Process\ 3}$$

$$= \frac{1}{T}\int_{V_1}^{V_2} P_{ext} dV = \frac{1}{T}\int_{V_1}^{V_2} P_{int} dV$$

$$= \frac{1}{T}(nRT)\int_{V_1}^{V_2} \frac{dV}{V}$$

$$= nR\ln\left(\frac{V_2}{V_1}\right)$$

It turned out that $\Delta S_{sys} = 3.77 J/K$ for the conditions employed. For all of these processes, ΔS_{surr} is equal to the actual heat transferred divided by the temperature of transfer.

$$\Delta S_{surr} = \frac{q_{surr}}{T}$$

For Process 1, $q_{surr}=0$, so that $\Delta S_{surr}(Process 1)=0$. For Process 2, $q_{surr}=-608J$ (negative since the surroundings lose heat), so that ΔS_{surr} (Process 2) = –608J/298K = –204kJ. For Process 3, q_{surr} = –1120kJ, so that ΔS_{surr} = –1120J/298K = –3.77J/K. For the entire universe,

$$\Delta S_{universe} = \Delta S_{sys} + \Delta S_{surr}$$

For Process 1

$$\Delta S_{universe} = (3.77-0)J/K = 3.77 J/K$$

For Process 2

$$\Delta S_{universe} = (3.77-2.04)J = 1.73 J/K$$

For Process 3

$$\Delta S_{universe} = (3.77-3.77)J/K = 0 J/K$$

Thus for Process 1 and Process 2, which are spontaneous, the total entropy of the universe increases, whereas for Process 3, which is reversible, there is no total entropy charge.

This same approach can be used to explain Example 18.2. When water freezes, $\Delta S°$ is given by $\Delta H°_{fus}/T_{mp}$. Thus, for freezing one mole of water, at *any* temperature*

$$\Delta S_{sys} = \frac{\Delta H°_{fus}}{T_{mp}} = \frac{6010 kJ/mol}{273 K}$$

$$= 22.0 J/K$$

T_{mp} is used because ΔS_{sys} is always calculated along a reversible path, and the only place melting is reversible is at T_{mp}; here the liquid and solid are at equilibrium. For the surroundings, on the other hand,

$$\Delta S_{surr} = \frac{q}{T}$$

where q is the heat transferred and T is the *actual* temperature of transfer. We have

$$q_{surr} = -\Delta H°_{fus}$$

$$\Delta S_{surr} = \frac{-\Delta H°_{fus}}{T}$$

$$\Delta S_{universe} = \Delta S_{sys} + \Delta S_{surr}$$

$$= \frac{\Delta H°_{fus}}{T_{mp}} - \frac{\Delta H°_{fus}}{T}$$

---

* This is almost true; $\Delta S°$ does change somewhat with temperature, but, as for $\Delta H°$, this variation is slight and can be ignored for the purposes of this course.

$$= \Delta H°_{fus}\left(\frac{1}{T_{mp}} - \frac{1}{T}\right)$$

Clearly, when $T=T_{mp}$, $\Delta S_{universe} = 0$; here, as we mentioned, the process is reversible. On the other hand, if $T<T_{mp}$ then this formula shows that $\Delta S_{universe}>0$; in other words, below 0°C, the freezing of water increases the entropy of the universe. Here, of course, the process is spontaneous, so this is in accord with the second law. Finally, if $T>T_{mp}$, $\Delta S_{universe}<0$. This means that the *reverse* process is spontaneous, and, in fact, the melting of ice is spontaneous at temperatures greater than 0°C.

The Gibbs Free Energy

The second law gives a criterion for the spontaneity of a process. If you want to know if a process will be spontaneous, calculate the $\Delta S_{universe}$ that would result from it. If this is positive, then the process will be spontaneous. It turns out that, surprisingly, this is not difficult to do, despite the fact that there are a lot of surroundings out there! Consider a process taking place at constant T and P. Because of constant P, we can write

$$q_{sys} = \Delta H_{sys}$$
$$q_{surr} = \Delta H_{surr} = -\Delta H_{sys}$$

Thus we can write

$$\Delta S_{universe} = \Delta S_{sys} - \frac{\Delta H_{sys}}{T}$$

If this is positive, recall that the process will be spontaneous. Multiplying through by $-T$ gives

$$-T\Delta S_{universe} = \Delta H_{sys} - T\Delta S_{sys} = \Delta G_{sys}$$

This relationship defines ΔG_{sys}, the change in the *Gibbs free energy* of the system. Note that when $\Delta S_{universe}>0$, $\Delta G_{sys}<0$, provided that the system is at constant P and T. Thus, for any process at constant P and T, we know $\Delta S_{universe}$ if we know ΔG_{sys}! The following two criteria of spontaneity are thus identical, again for a process taking place at constant P and T

$$\Delta S_{universe}>0$$
$$\Delta G_{sys}<0$$

When ΔG is negative, the process will be spontaneous; when ΔG is positive, the reverse process will be spontaneous; when ΔG is zero, the process will be at equilibrium.

Standard Entropies and Free Energies

Recall that we defined the standard enthalpy, ΔH° for a process as the enthalpy change accompanying the process at 1atm pressure, with the reactants in standard states, which meant the pure substance for a liquid or solid, and to which we must add 1M concentration for substances in solution. Standard state data, such as standard enthalpies of formation, are usually tabulated at 298K. Recall also that we could calculate ΔH° from tabulated data on ΔH_f°, standard heats of formation, defined as the enthalpy of forming a mole of a compound in its standard state from the elements in their standard states, where an element in its standard state, must, in addition to the above criteria, also be in its most stable form.

We define the standard free energy, ΔG°, for a process as the free energy change accompanying the process when all the substances are in their standard states.

$$\Delta G^\circ = \Delta H^\circ - T\Delta S^\circ$$

Clearly, all we need to do to calculate ΔG° is to first calculate ΔH°, which we already know how to do, then calculate ΔS°, then combine them as prescribed in this equation.

Absolute Entropies

It turns out that there is a natural zero for the entropy; the *third law of thermodynamics* says that the entropy of a perfect crystal at 0K is zero. By integrating q_{rev}/T from absolute zero to any desired temperature, the *absolute entropy* of any substance can be determined. These are usually tabulated as S° at T=298K. Note that even pure elements in their standard states have non-zero values of S°.

The physical interpretation of the third law is that the amount of freedom or disorder is the same in any perfect crystal at 0K; in fact, this makes sense, because there is only one distinguishable arrangement of atoms in a perfect crystal, no matter what the substance is. Contrast this with a liquid or a gas, in which there are many, many such distinguishable arrangements. Even for a perfect crystal

Chapter 18 Entropy, Free Energy, and Equilibrium

above 0K, thermal motion affects the positions of the atoms, so that there are many possible sets of coordinates for the atoms. Only at 0K does the number of distinguishable arrangements go to zero, and then only for a perfect crystal.

There are some trends visible in tables of $S°$ at 298K; for the most part they are understandable in terms of greater $S°$ implying greater freedom or disorder. In general, $S°$ becomes greater as one goes from solids to liquids to gases. $S°$ also becomes greater as one goes from simpler to more complex molecules. More complex molecules have more ways to vibrate and rotate than simple molecules, and this adds to their freedom, or entropy. Among solid elements, $S°$ tends to increase with increasing atomic number; greater atomic number means more electrons, which means more possible electronic excited states, which adds to the total entropy. All of these comparisons take place at constant temperature. In addition, you should know that when the temperature of a substance is raised, its entropy always increases, since more motion is imported to the molecules. For a chemical reaction of the form

$$\alpha A + \beta B \rightarrow \gamma C + \delta D$$

you already know that

$$\Delta H° = \gamma \Delta H_f°(C) + \delta \Delta H_f°(D) - \alpha \Delta H_f°(A) - \beta \Delta H_f°(B)$$

For $\Delta S°$, we have

$$\Delta S° = \gamma S°(C) + \delta S°(D) - \alpha S°(A) - \beta S°(B)$$

Values of $S°$, like values of $\Delta H°_f$, are usually tabulated at 298K. Such a table, together with this equation, can be used to calculate $\Delta S°$ for a process. Note that for the formation of a compound in its standard state from the elements in their standard states, $\Delta H°_f$ is given directly from the table. For the same reaction, however, $\Delta S°_f$ has to be calculated.

The Standard Free Energy

Naturally, given $\Delta H°$ and $\Delta S°$, we can calculate $\Delta G°$, using

$$\Delta G° = \Delta H° - T\Delta S°$$

This is a very important equation. Recall that if $\Delta S_{universe} > 0$ for a process, then $\Delta G_{sys} < 0$ for the process, provided that the initial and final states are at the same P and T. But if $\Delta S_{universe} > 0$ it means

Chapter 18 — Entropy, Free Energy, and Equilibrium

the process is spontaneous. $\Delta G°$ is the ΔG for a process which takes reactants in their standard states to products in their standard states. If $\Delta G°$ is negative, then this process will be spontaneous. If $\Delta G°$ is zero, the process will be reversible (at equilibrium). If $\Delta G°$ is positive, then the reverse process will be spontaneous.

In Example 18.5b, the text shows how to calculate $\Delta G°$ for the reaction

$$CaCO_3(s) = CaO(s) + CO_2(g)$$

By substracting $\Delta H°_f$ for the reactants from those of the products, we get $\Delta H°_f = +177.9 kJ$. By doing the same for $S°$ we get $\Delta S° = +160.4 J/K$. Therefore

$$\Delta G° = 177900 J - T(160.4 J/K)$$

at T=298K this comes out to +130.1kJ, so that the reaction will not proceed at 298K, under standard conditions.

Temperature Variation of $\Delta G°$

As we have mentioned earlier, $\Delta H°$ and $\Delta S°$ are, in general, temperature-dependent, but this dependence is small, and can be ignored for our purposes. In contrast, $\Delta G°$ depends strongly on the temperature. In fact, the last equation can be used to calculate $\Delta G°$ for the decomposition of $CaCO_3$ at *any* temperature. As an example, let us use it to find the temperature at which the process will be in equilibrium; here $\Delta G°=0$, so that

$$T = \frac{177900 J}{160.4 J/K} = 1109K = 836°C$$

At this temperature, $CO_2(g)$ at 1atm pressure will be in equilibrum with calcium oxide and calcium carbonate; above 836°, $\Delta G°$ will be negative and the reaction will be spontaneous as written. Thus we have the important result that starting with a table of $\Delta H°_f$ and $S°$ at 298K, we can calculate $\Delta G°$ at any temperature! Incidentally, this is the first time we have encountered standard states at other than 298K. Recall again that what defines standard states is pressure and concentration, not temperature. The equilibrium we just calculated pertains to all the ingredients in their standard states, even though the temperature is 836°C. (Actually, the fact that $\Delta H°$ and $\Delta S°$ are not completely temperature-independent will cause the true equilibrium temperature to be somewhat different from our calculated value.)

$\Delta G°$ and K_{eq}

We just noted that when $\Delta G°<0$ it means that a reaction will be spontaneous, provided that all ingredients are in their standard states. Earlier (Chapter 16) we noted that when $K_{eq}>1$ the same conclusion can be drawn; if everything is in its standard state then Q, the equilibrium quotient, is equal to unity, and if $K_{eq}>1$ then $Q<K_{eq}$ and the forward reaction will be spontaneous. Thus $\Delta G°$ must be related to K_{eq}.

In fact, whether a given reaction with the ingredients *not* in standard states will be spontaneous depends not only on K_{eq} but also on the concentrations. For example, suppose we have the reaction

$$A = B$$

and suppose $K_{eq}=2$. If $[B]=0.1M$ and $[A]=0.01M$, then $Q=[B]/[A]=10$ and $Q>K_{eq}$, so that the reverse reaction will be spontaneous. Thus, for a general chemical process not at standard conditions, ΔG, which predicts the spontaneity, must depend on concentrations as well as on $\Delta G°$. There are several ways of expressing this relationship. One useful form is

$$\Delta G = \Delta G° + RT \ln Q$$

This expression is derived in Appendix I of the text, and, in somewhat greater detail, in the **Perspective** at the end of this chapter. For now, let us see how it is used, and what its implications are.

First of all, let us suppose the system is in equilibrium; then $\Delta G=0$ and $Q=K_{eq}$, so that

$$\Delta G° = -RT \ln K_{eq}$$

$$K_{eq} = e^{\frac{-\Delta G°}{RT}}$$

We see from this equation that when $\Delta G°<0$, implying a spontaneous forward process under standard conditions, $K_{eq}>0$; likewise, when $\Delta G°=0$ (equilibrium under standard conditions) $K_{eq}=1$, and of course when $\Delta G°>0$ (reverse reaction spontaneous under standard conditions) $K_{eq}<1$. This is just as it should be, which is comforting.

Chapter 18   Entropy, Free Energy, and Equilibrium

Now let us consider the equation under standard conditions. Here, Q=1, so that Q=0, and $\Delta G = \Delta G°$, which is also what we expect.

Note that since we learned how to calculate $\Delta G°$ at any temperature, we can calculate K_{eq} at any temperature, using $K_{eq} = e^{-\Delta G°/RT}$. In fact, given simply a table of $\Delta H°_f$ and $S°$ for common substances at 298K, we can calculate K_{eq} for any process involving them, at any temperature! Recall that we calculated that for the decomposition of limestone at 298K, $\Delta G° = +130.2 \text{ kJ/mol}$. We can calculate K_{eq} as follows:

$$K_{eq} = e^{-\Delta G°/RT}$$

$$= e^{-\frac{130200 \text{ J/mol}}{(8.314 \text{ J/mol·K})(298\text{K})}}$$

$$= e^{-52.55}$$

$$= 1.50 \times 10^{-23}$$

$\Delta G°$ and the Nernst Equation

You may have noted that the equation for ΔG

$$\Delta G = \Delta G° - RT\ln Q$$

is similar to the Nernst equation

$$\Delta E = \Delta E° + \frac{0.0592}{n}\ln Q$$

and that K_{eq} is related to $\Delta G°$ in a similar manner to the way it is related to $\Delta E°$

$$K_{eq} = e^{-\Delta G°/RT}$$

$$K_{eq} = 10^{-n\Delta E°/0.592}$$

This is because the Nernst equation is an immediate consequence of our equation for ΔG.

To show this, we start by writing out the definition for ΔG at constant P and T and expanding out the terms.

$$\Delta G = \Delta H - T\Delta S$$

$$= \Delta E + P\Delta V - T\Delta S$$

$$= q + w + P\Delta V - T\Delta S$$

Now, let us suppose the process we are considering is reversible; we will return to this assumption later. For a reversible process,

$$q = q_{rev}$$

$$\Delta S = \frac{q_{rev}}{T}$$

$$q_{rev} = T\Delta S$$

so that our equation simplifies to

$$\Delta G = w + P\Delta V$$

Throughout this and the last chapter, we have been assuming that all work is PV work; now let us assume that some other kind is also possible (for an electrochemical cell it will of course turn out to be electrical). Then we have

$$w = w_{PV} + w_{other}$$

$$\Delta G = w_{PV} + w_{other} + P\Delta V$$

but, as we pointed out in Chapter 17,

$$w_{PV} = -P\Delta V$$

so that

$$\Delta G = w_{other}$$

This equation says that for a reversible process at constant P and T, the decrease in ΔG_{sys} is equal to the amount of non-PV work done on the surroundings.

We already noted in Chapter 16 that when n moles of electrons move through a potential ΔE, then the work done on the surroundings is $nF\,\Delta E$, where F is the faraday. Then the w_{sys} is $-nF\Delta E$. Equating this to w_{other} gives

$$\Delta G = -nF\,\Delta E$$

Just one or two points of clarification before succeeding. First of all, the way one does electrical work reversibly is analogous to the way one does PV work reversibly. Recall that for PV work (Process 3 at the beginning of this chapter) we ensured reversibility by maintaining $P_{int}=P_{ext}$, and we noted that since the system was always expanding against the maximum possible pressure, w_{PV} was a maximum for the

Chapter 18 — Entropy, Free Energy, and Equilibrium

process. The way one does electrical work reversibly is to maintain $\Delta E_{ext} = \Delta E_{int}$; that is, to always keep the cell in equilibrium with an externally imposed voltage. Just as in the PV case, no work can actually be done unless the equilibrium is disturbed — by removing the weight bit by bit or lowering the voltage bit by bit — but we can disturb it as little as we like, and thereby approach the truly reversible process as closely we like. As in the PV case, this process ensures the maximum possible amount of work done for a given change of state. In fact, the way electrochemical potentials are measured in the laboratory is to see how much voltage is necessary to just stop the flow of current; the cell is thus being observed in a true equilibrium state, and the equation

$$\Delta G = -n\mathbf{F}\,\Delta E$$

gives ΔG for the cell reaction.

Let us now elaborate this into the familiar form of the Nernst equation.

$$-n\mathbf{F}\,\Delta E = \Delta G = \Delta G^\circ + RT \ln Q$$

$$\Delta E = \frac{-\Delta G^\circ}{n\mathbf{F}} - \frac{RT}{n\mathbf{F}} \ln Q$$

We define the constant $-\Delta G^\circ / n\mathbf{F}$ to be ΔE°, and applying the factor 2.303 to $\ln Q$ to give $\log_{10} Q$ results in

$$\Delta E = \Delta E^\circ \frac{-2.303 RT}{n\mathbf{F}} \log_{10} Q$$

$$= \Delta E^\circ - \frac{0.0592}{n} \log_{10} Q$$

The factor 0.0592 is thus equal to $2.303 RT/\mathbf{F}$, at T=298K.

Now we see why it turned out that ΔE° predicted K_{eq}: ΔE° is just ΔG° times a constant, and ΔG° predicts K_{eq}. Likewise, ΔE predicts spontaneity only because ΔG predicts spontaneity, and ΔE is ΔG times a constant. Recall that ΔG_{sys} at constant P and T is a measure of $\Delta S_{universe}$, so that all these criteria of spontaneity derive ultimately from the second law of thermodynamics, from the idea that the entropy of the universe increases for any spontaneous process.

Chapter 18 — Entropy, Free Energy, and Equilibrium

Temperature Dependence of K_{eq}

We have already alluded to the fact that $\Delta H°$ and $\Delta S°$ do not change much with temperature, but that $\Delta G° = \Delta H° - T\Delta S°$ does. Since $\Delta G°$ is directly related to K_{eq}, K_{eq} also changes significantly with temperature. We can put this on a quantitative basis as follows.

$$\Delta G° = -RT\ln K_{eq} = \Delta H° - T\Delta S°$$

$$\ln K_{eq} = \frac{-\Delta H°}{RT} + \frac{\Delta S°}{R}$$

This says that if K_{eq} is measured at a number of temperatures, and if $\ln K_{eq}$ is plotted against $1/T$, the result will be a straight line with slope $-\Delta H°/R$, and the y-intercept will be $\Delta S°/R$. Note that if $\Delta H° > 0$ (endothermic reaction) the slope will be negative; this means that $\ln K_{eq}$ — and hence K_{eq} — gets smaller as $1/T$ gets bigger — that is, as T gets smaller. Thus a positive $\Delta H°$ (endothermic reaction) means that K_{eq} gets larger with higher temperature, a fact we have noted before. Similarly, a negative $\Delta H°$ (exothermic reaction) means that K_{eq} gets smaller with increasing temperature.

This equation provides a new method (besides calorimetry) of determining $\Delta H°$ for a process. If K_{eq} is measured at two temperatures, then

$$\ln K_1 = \frac{-\Delta H°}{RT_1} + \frac{\Delta S°}{R}$$

$$\ln K_2 = \frac{-\Delta H°}{RT_2} + \frac{\Delta S°}{R}$$

$$\ln K_1 - \ln K_2 = \ln(K_1/K_2) = \frac{-\Delta H°}{R}\left(\frac{1}{T_1} - \frac{1}{T_2}\right)$$

This is called the van't Hoff equation, and Example 18.15 illustrates its use. $K_1 = 0.157$ at $T_1 = 900K$, and $K_2 = 0.513$ at $K_2 = 1000K$. Thus

$$\ln(0.157/0.513) = \frac{\Delta H°}{8.314 \text{J/mol·K}}\left(\frac{1}{900} - \frac{1}{1000}\right)$$

$$\Delta H° = 88.6 \text{kJ/mol}$$

As the example further shows, $\Delta H°$ does in fact change somewhat with temperature, but this change is small compared to the change in $\Delta G°$.

Chapter 18   Entropy, Free Energy, and Equilibrium

One more thing is worth noting. Recall the expression

$$\ln K_{eq} = \frac{-\Delta H^\circ}{RT} + \frac{\Delta S^\circ}{R}$$

At very high T

$$\ln K_{eq} = \frac{\Delta S^\circ}{R}$$

since there the ΔH° term becomes small. If ΔS° is positive, then $K_{eq} > 1$ at high temperatures, and the reaction will be spontaneous there under standard conditions. As the temperature is lowered, the ΔH term becomes more important. If ΔH° is negative, then this term raises K_{eq} still further as the temperature is lowered, and the reaction will continue to be spontaneous. If, on the other hand, ΔH° is positive, then lowering the temperature decreases K_{eq}, and at low enough temperatures it may go through unity, and the reaction become non-spontaneous under standard conditions. Similarly, if ΔS° is negative the reaction is non-spontaneous at high temperatures under standard conditions, but if ΔH° is also negative this favors reaction at low temperatures, and may lead to reversal of the direction of spontaneity if the temperature is lowered.

Postcript

Many aspects of your study of chemistry so far have culminated in this chapter. The idea of equilibrium, so central to chemistry, was related to the second law of thermodynamics: a system is in equilibrium when the process taking place doesn't cause the entropy of the universe to change. We then saw that at constant temperature and pressure, ΔG_{sys} can be used to predict $\Delta S_{universe}$. This is useful because we can calculate ΔG°_{sys} from tabulated values of ΔH° and ΔS°; this gives the equilibrium constant, $\Delta G^\circ = -RT\ln K_{eq}$. With K_{eq} and concentrations, we can make quantitative predictions about how far a reaction will go in what direction, as we did in Chapters 8 through 11. We also saw why the sign of ΔH° determines the effect of temperature on K_{eq}, and put this on a quantitative basis with the van't Hoff equation.

It is well to consider what we have not done. First of all, we have done little to explain why different compounds exhibit the heats of formation and absolute entropies that they do; that is, we have not, in this chapter, tried to rationalize any descriptive chemistry. The basis for this was laid in Chapters 12 through 14, and is taken up again starting with Chapter 20. Also, remember that just because thermodynamics says a process is spontaneous, it doesn't mean it will occur in a reasonable amount of time. Factors responsible for reaction rates are at least as important as thermodynamic factors in determining what actually does happen; in a sense, thermodynamics only tells what *could* happen. *Chemical kinetics*, which treats reaction rates, is taken up in the next chapter.

Chapter 18 — Entropy, Free Energy, and Equilibrium

Questions

18.1 Energy and spontaneity: the need for a second law

1. In an endothermic process, the surroundings get (warmer, cooler).

2. When an ideal gas expands isothermally, its energy (increases, decreases, neither increases nor decreases, may either increase or decrease).

3. When a real gas expands isothermally, its energy (increases, decreases, neither increases nor decreases, may either increase or decrease).

4. When an ideal gas expands isothermally into a vacuum,

    (a) (w = 0, w > 0, w < 0)

    (b) (q = 0, q > 0, q < 0)

    (c) ($\Delta E = 0$, $\Delta E > 0$, $\Delta E < 0$)

5. When an ideal gas expands isothermally against a constant external pressure,

    (a) (w=0, w>0, w<0)

    (b) (q=0, q>0, q<0)

    (c) ($\Delta E=0$, $\Delta E>0$, $\Delta E<0$)

18.2 Reversible processes

1. A process in which a system is always in equilibrium with its surroundings is called a _____ process.

2. The work done in a reversible process is the _____ work possible with the same initial and final states.

3. No reversible process can be spontaneous. (True, False)

18.3 Derivation of the reversible work of expansion of an ideal gas and
18.4 The maximum work that can be done by an isothermal expansion of an ideal gas

1. For the isothermal expansion of an ideal gas,

    (a) the process is reversible if ($P_{int}=P_{ext}$, $P_{int}<P_{ext}$, $P_{int}>P_{ext}$)

Chapter 18 — Entropy, Free Energy, and Equilibrium

(b) the process is spontaneous if ($P_{int}=P_{ext}$, $P_{int}<P_{ext}$, $P_{int}>P_{ext}$)

(c) the reverse process (contraction) is spontaneous if ($P_{int} = P_{ext}$, $P_{int}<P_{ext}$, $P_{int}>P_{ext}$)

(d) The sign of w is (+, −, 0)

2. For the reversible isothermal expansion of an ideal gas, (give algebraic expressions)

   (a) if the volume doubles, w = _____

   (b) if the volume increases by a factor x, w = _____

   (c) if the volume *decreases* by a factor x (contraction), w = _____

   (d) if the volume increases by the factor e (the base of natural logarithms), w = _____

   (e) ΔE = _____

   (f) q = _____ (in terms of w)

18.5 The entropy change for an isothermal process

1. For an isothermal, reversible process, if the entropy increases, heat must have been (added to, taken from) the system.

2. If some process takes a system from State 1 to State 2, ΔS differs depending on whether the process

   (a) is reversible (True, False)

   (b) is isothermal (True, False)

   (c) involves work only, heat only or work and heat both (True, False)

   (d) has positive, negative or zero ΔE (True, False)

   (e) is spontaneous (True, False)

3. If at some temperature, T, vaporization of a liquid is reversible, T is called the _____.

4. If at some temperature, T, a solid melts reversibly, T is called the _____.

5. Consider the following two processes:

   (I) water evaporating from an open jar at room temperature

   (II) water evaporating from a tea kettle at 100°C

   ΔS_{vap} is (greater for (I), greater for (II), about the same for both).

Chapter 18 — Entropy, Free Energy, and Equilibrium

6. According to Trouton's rule, the higher the boiling point of a normal liquid, the (bigger, smaller) its enthalpy of vaporization.

7. Hydrogen bonding in the liquid tends to (increase, decrease) ΔS_{vap}.

18.6 The second law of thermodynamics

1. Which of these systems is isolated?

    (a) an insulated calorimeter

    (b) an insulated cylinder with a piston

    (c) a tea kettle

    (d) Albania

    (e) the Earth

    (f) the universe

2. If the temperature of an isolated system spontaneously decreases, its entropy must have (increased, decreased, can't say).

3. If the temperature of an isolated system spontaneously increases, its entropy must have (increased, decreased, can't say).

4. If the temperature of a non-isolated system spontaneously decreases, its entropy must have (increased, decreased, can't say).

5. According to the second law of thermodynamics, any spontaneous process is accompanied by the entropy increase *of the system*.

18.7 The Gibbs Free Energy Function

1. For a spontaneous process at constant T and P,

    (a) ΔS_{sys} (is positive, is negative, is zero, can't say)

    (b) ΔS_{total} (is positive, is negative, is zero, can't say)

    (c) $\Delta H_{sys} = -\Delta H_{surr}$ (True, False)

    (d) $\Delta S_{surr} = -\Delta H_{sys}/T$ (True, False)

(e) $T\Delta S_{sys} - \Delta H_{sys}$ (is positive, is negative, is zero, can't say)

(f) $\Delta G = $ _____ (in terms of ΔS and ΔH)

(g) ΔG (is positive, is negative, is zero, can't say)

2. For *any* process at constant T and P, $\Delta G_{sys} = -T\Delta S_{total}$, so that if we know T and ΔG for the process, we know ΔS for the universe. (True, False)

18.8 Calculation of ΔG

1. Suppose for some reaction $\Delta G° < 0$. What must be true of the reactants and products to ensure that the reaction will proceed spontaneously at constant T and P?

2. (a) $H° = 0$ for a perfect crystal at $T = 0K$. (True, False)

   (b) $S° = 0$ for a perfect crystal at $T = 0K$. (True, False)

   (c) $\Delta H°_f = 0$ for an element in its standard state. (True, False)

   (d) $\Delta S°_f = 0$ for an element in its standard state. (True, False)

3. State the third law of thermodynamics

4. If $\Delta G°_f < 0$ at 298K for some substance X, then under what conditions will X tend to form spontaneously from the elements at this temperature?

5. $S°$ tends to increase as

   (a) atomic number (increases, decreases) for an element

Chapter 18 Entropy, Free Energy, and Equilibrium

(b) T (increases, decreases)

(c) chemical complexity (increases, decreases)

6. Which of the following usually changes the most as T changes? ($\Delta S°$, $\Delta H°$, $\Delta G°$)

18.9 $\Delta G°$ and the equilibrium constant

1. If I know K_{eq} for a process, I can immediately calculate ΔG. (True, False)

2. Acetic acid, abbreviated HAc, ionizes in water according to

$$HAc = H_3O^+ + Ac^-$$

for which $K_{eq} = 1.8 \times 10^{-5}$.

(a) Give an expression for $\Delta G°$ at 300 K

(b) If all reactants and products are in their standard states,

(i) what is Q equal to?

(ii) give a formula for ΔG

(c) At equilibrium

(i) what is Q equal to?

Chapter 18 — Entropy, Free Energy, and Equilibrium

(ii) give an expression for ΔG

3. If $Q < K_{eq}$, then $RT \ln Q$ (>, <, =) $RT \ln K_{eq}$ and ΔG (<, >, =) 0 and the (forward, reverse) reaction will take place spontaneously.

4. Which of the following can be used to predict whether a reaction will take place spontaneously at constant T and P?

    (a) ΔS_{system}

    (b) ΔG_{system}

    (c) ΔH_{system}

    (d) ΔS_{total}

    (e) K_{eq}

5. Suppose a process takes a system from State 1 to State 2, and these states are not at the same T and P. If the process was spontaneous, must it be true that $\Delta G < 0$? (Yes, No)

6. The more positive $\Delta G°$, the (bigger, smaller, can't say) K_{eq}.

18.10 $\Delta G°$ and the Nernst equation

1. According to Equation 18-28 in the text,

$$\Delta G = -nF\Delta E_{cell}$$

here, ΔG is equal to the work done (by, on) the cell in a (spontaneous, reversible) process at constant _____.

2. When is $\Delta E° = \Delta E$?

Chapter 18   Entropy, Free Energy, and Equilibrium

3. When conditions are such that $\Delta E = \Delta E°$, it must also be true that $\Delta G = \Delta G°$ (True, False).

4. Which of these can be used to calculate K_{eq}

   (a) $\Delta E°$

   (b) ΔE

   (c) ΔE together with Q

   (d) ΔG

   (e) ΔG together with Q

18.11 The temperature dependence of ΔH, ΔS, ΔG and K_{eq}

1. For each of the following situations, tell what happens to $\Delta G°$ and to K_{eq} for a process at constant T and P as T increases:

	$\Delta G°$	K_{eq}	$\Delta H°$	$\Delta S°$
(a)	_____	_____	positive	positive
(b)	_____	_____	positive	negative
(c)	_____	_____	negative	positive
(d)	_____	_____	negative	negative

2. For each of the following processes discussed in the text, what are the signs of ΔH and of ΔS for the system?

   (a) alcohol spontaneously evaporating from your arm (Section 18.1)

   (b) ammonium nitrate spontaneously dissolving in water (Section 18.2)

   (c) an ideal gas expanding isothermally and spontaneously against a constant external pressure

(Section 18.1)

(d) an ideal gas expanding isothermally and reversibly (Section 18.2)

(e) steam condensing (Section 18.2)

(f) the exothermic reaction $2H_2O_2\ (l) \rightarrow 2H_2O\ (l) + O_2(g)$

Chapter 18   Entropy, Free Energy, and Equilibrium

Perspective: Dependence of the Free Energy on Concentration, and the Derivation of the Equilibrium Constant Expression

We will derive a result for an ideal gas, then "wave our hands," and simply say that it applies to other things as well. Note that the molarity of an ideal gas is given by

$$\frac{n}{V} = \frac{P}{RT}$$

thus the concentration of an ideal gas is proportional to the pressure. We will derive an expression for how ΔG depends on pressure for an ideal gas, then use a similar expression for molarities of substances in solution.

First we will need a preliminary result; namely, that $\Delta H=0$ for the isothermal expansion of an ideal gas. Recall that the enthalpy is defined by

$$H = E + PV$$

Thus the enthalpy change for a process is given by

$$\Delta H = \Delta E + (P_2 V_2 - P_1 V_1)$$

Of course, when $P_1 = P_2$ a simpler expression results.

$$\Delta H_p = \Delta E_p + P\Delta V$$

In any case, for an ideal gas, Boyle's law says that at constant temperature $P_1 V_1 = P_2 V_2$, so that $\Delta H = \Delta E$. But if there is no temperature change (and we will be assuming constant-temperature processes), $\Delta E = 0$ for an ideal gas, so that $\Delta H = 0$.

Now we are going to consider the free-energy change accompanying the isothermal expansion of an ideal gas. We use the result obtained earlier for the entropy change accompanying such a process.

$$\Delta G = \Delta H - T\Delta S$$

$$= 0 - T\Delta S$$

$$= -nRT\ln\frac{V_2}{V_1} = nRT\ln\frac{P_2}{P_1}$$

The last relationship comes from the fact that for an ideal gas $V_2/V_1 = P_1/P_2$, together with the fact that $\log(1/x) = -\log(x)$.

Chapter 18 — Entropy, Free Energy, and Equilibrium

Note that if P is measured in atmospheres, then ΔG to take a mole of an ideal gas from its standard state to pressure P is simply $RT\ln P$, since here $n=1$, $P_2=P$ and $P_1=1$. In other words, if G° is the free energy of a gas in its standard state, however this is defined, then G^P, the free energy at pressure P, is given by

$$G^P = G^\circ + RT\ln P$$

if P is in atmospheres; we will return later to a consideration of what happens if other units are used.

It may be confusing that we are using symbols like G° and G^P for the first time; however, all the last equation says is that when a mole of ideal gas is taken from 1 atm to pressure P, then

$$\Delta G = RT\ln P = G_{final} - G_{initial} = G^P - G^\circ$$

In the end it will be the ΔG's that count.

For our archetypal reaction

$$\alpha A + \beta B \rightarrow \gamma C + \delta D$$

taking place at pressure P we can write

$$\Delta G^P = \gamma G^P(C) + \delta G^P(D) - \alpha G^P(A) - \beta G^P(B)$$

Now, for each G^P we substitute the expression we derived above.

$$\Delta G^P = \gamma G^\circ(C) + \delta G^\circ(D) - \alpha G^\circ(A) - \beta G^\circ(B)$$
$$+ \gamma RT\ln P_C + \delta RT\ln P_D - \alpha RT\ln P_A - \beta RT\ln P_B$$
$$= \Delta G^\circ + RT\ln P_C^\gamma + RT\ln P_D^\delta - RT\ln P_A^\alpha - RT\ln P_B^\beta$$
$$= \Delta G^\circ + RT\ln \frac{P_C^\gamma P_D^\delta}{P_A^\alpha P_B^\beta}$$
$$= \Delta G^\circ + RT\ln Q$$

where we have used the properties of logarithms to compress the expression. Of course, we are still assuming that P is in atmospheres. When A, B, C and D are substances in solution, rather than gases, then

$$Q = \frac{[C]^\gamma [D]^\delta}{[A]^\alpha [B]^\beta}$$

There are two very important cases of this equation. One is when all reactants and products are in their standard states. Then Q=1, since when a substance is in its standard state P=1atm or concentration=1M. If Q=1, then $\Delta G=\Delta G°$, since lnQ=ln1=0. This is reassuring; at least our equation is correct under standard conditions! The other important case is when the system is in equilibrium, so that $\Delta G=0$. Then

$$\Delta G° = -RT\ln Q$$

Obviously, under these conditions, Q has a value that is uniquely determined by $\Delta G°$ for the reaction; Q *at equilibrium* is a constant for the reaction, called the *equilibrium constant*, and usually symbolized K_{eq}, or simply K, so that the last equation reads

$$\Delta G° = -RT\ln K_{eq}$$

Expressing this slightly differently

$$K_{eq} = e^{-\Delta G°/RT}$$

The equilibrium constant is one of the most important concepts in chemistry. We already noted in the Scope that if $\Delta G°<0$, then the reaction will be spontaneous, provided that all the reactants and products are in their standard states. Earlier in the text we alluded to the fact that $K_{eq}>1$ ensures the same thing. Now we have proven the latter, given the former.

Units of Q and K_{eq}

Recall, as we have mentioned before, that standard state specifies concentration and pressure, not temperature. What we have really arrived at is a way of calculating the free energy change that accompanies taking a substance from its standard state to some other concentration, and used that to calculate ΔG for a process. For a gas, we used P for the new pressure, and said that it had to be expressed in atmospheres; this is just a way of saying that what P really represents is P/P°, where P° represents the pressure that we have defined as the standard state pressure, namely 1atm, expressed in the same units as P. Thus, if we were expressing pressure in kilopascals, we would use, for each P in the above equations, the value P/101.2kPa, since 101.2kPa=1atm, the concentration which defines the standard state for a gas. Likewise, when we write concentrations for substances in solution, such as [A], what we really mean is $[A]/[A]°$, the concentration divided by the concentration which defines the standard state of the substance. This concentration is 1M, so that if we express concentrations in molarity, we are using the correct numerical values in the expression for Q or for K_{eq}. If instead we expressed our concentration in other units, then we would use, instead of $[A]$ in the above equations, $c_A/c_A°$, where $c_A°$ is the standard state concentration, 1M, expressed in the other units; for example, if we used millimolarity (millimoles per liter) as c, then c_o would be 1000mmol/L. Finally, for a pure liquid or solid, the standard state is whatever the concentration of the pure substance is; for example, pure liquid water has a concentration of 55.6mol/L, and this is the standard state, so pure liquid water would exhibit a factor in the Q or K_{eq} expression equal to $\frac{55.6 \text{mol/L}}{55.6 \text{mol/L}}$, or unity.

Note that if this is done, each concentration factor in Q or in K_{eq} becomes unitless; that is, $P/P°$ and $c/c°$ are both unitless. This means that from this point of view K_{eq} and Q are really unitless; in fact, they must be unitless, strictly speaking, because one can only take the logarithm of a pure number, and we take the logarithm of Q to get ΔG. This discussion also provides a rigorous explanation of why pure liquids and solids do not appear in equilibrium constant expressions; any pure liquid or solid is in its standard state, by definition, and its concentration enters into the K_{eq} expression as unity.

Chapter 18   Entropy, Free Energy, and Equilibrium

Answers

18.1 Energy and spontaneity: the need for a second law

1. cooler
2. neither
3. may either increase or decrease
4. (a) w = 0
   (b) q = 0
   (c) ΔE = 0
5. (a) w < 0
   (b) q > 0
   (c) ΔE = 0

18.2 Reversible processes

1. reversible
2. maximum
3. True

18.3 Derivation of the reversible work of expansion of an ideal gas and
18.4 The maximum work that can be done by an isothermal expansion of an ideal gas

1. (a) $P_{int} = P_{ext}$
   (b) $P_{int} > P_{ext}$
   (c) $P_{int} < P_{ext}$
   (d) –
2. (a) $-nRT\ln 2$
   (b) $-nRT\ln x$
   (c) $nRT \ln x$
   (d) $-nRT$
   (e) 0
   (f) $-w$

18.5 The entropy change for an isothermal process

1. added to
2. (a) False
   (b) False
   (c) False
   (d) False
   (e) False (for all of these, ΔS is a state function, and depends only on the initial and final states.)
3. boiling point
4. melting point
5. about the same for both
6. bigger
7. increase.

18.6 The second law of thermodynamics

1. (a) and (f)

2. increased

3. increased (For 2 and 3, *any* spontaneous process in an isolated system is associated with positive ΔS.)

4. can't say

5. of the universe

18.7 The Gibbs Free Energy Function

1. (a) can't say
   (b) is negative
   (c) True
   (d) True
   (e) is positive
   (f) $\Delta H - T\Delta S$
   (g) is negative

2. True

18.8 Calculation of $\Delta G°$

1. They must be in their standard states

2. (a) False
   (b) True
   (c) True
   (d) False

3. $S°=0$ at 0K for a pure crystalline substance

4. X and its constituent elements must be in their standard states

5. (a) increases
   (b) increases
   (c) increases

6. $\Delta G°$

18.9 $\Delta G°$ and the equilibrium constant

1. False (can calculate $\Delta G°$; for ΔG need concentrations as well.)

2. (a) $\Delta G° = -R (300) \ln (1.8 \times 10^{-5})$
   (b)
      (i) 1
      (ii) $\Delta G = \Delta G°$
   (c)
      (i) $Q = K_{eq}$
      (ii) $\Delta G = 0$

3. <, <, forward

4. (b) or (d); for (e), need concentrations as well.

5. No

6. smaller

18.10 $\Delta G°$ and the Nernst Equation

1. on, reversible, P and T

2. When the reactants and products are in their standard states.

3. True

Chapter 18

4. (a), (c), (e)

18.11 The temperature dependence of ΔH, ΔS, ΔG, and K_{eq}

1.

	$\Delta G°$	K_{eq}
(a)	gets more negative	gets bigger
(b)	gets more negative	gets bigger
(c)	gets more positive	gets smaller
(d)	gets more positive	gets smaller

2. (a) ΔH > 0, ΔS > 0
   (b) ΔH > 0, ΔS > 0
   (c) ΔH = 0, ΔS > 0
   (d) ΔH = 0, ΔS > 0
   (e) ΔH < 0, ΔS < 0
   (f) ΔH < 0, ΔS > 0

   Note on (c) and (d). We show at the beginning of the Perspective that ΔH=0 for the isothermal expansion of an ideal gas. For (c), which corresponds to Process 2 in the text, q>0; but here q≠ΔH$_{sys}$ because even though P$_{ext}$ is constant, P$_{sys}$ is not constant.

CHAPTER 19: CHEMICAL KINETICS

Scope

Thermodynamics tells what will happen *eventually*, but it doesn't tell us *when* anything will happen. In fact, if thermodynamics were the only truth, the essential human activities — such as eating, sleeping and learning chemistry — would be impossible; the chemical and physical processes responsible for life would all be in equilibrium, and nothing would be happening on a macroscopic scale. Instead, however, the inexorable increase of S_{total} takes time.

Although, as you saw in Chapter 18, it is fairly easy to calculate in which direction a reaction will go spontaneously, there is no simple and general scheme to calculate in advance how fast it will go. To devise such a scheme is the ultimate goal of *chemical kinetics*; we are still far from it. For now, classical chemical kinetics, the study of reaction rates, consists of a set of tools for the classification of rate behaviors, and for the elucidation of *reaction mechanisms*, the step-by-step sequence of bond making and breaking that results in an overall process. In a particular system, the products may be the ones that thermodynamics predicts; this is called a *thermodynamically controlled* process. Some processes, however, obviously including many of the ones responsible for life, are *kinetically controlled*, meaning that the products that result are the ones which are produced most quickly.

The same sorts of things that affect the positions of chemical equilibria affect the positions of chemical reactions.

— chemical nature of the reacting system: reactants and solvent.

— chemical concentrations

— the temperature

In addition, there is a factor that can alter a reaction rate without affecting the equilibrium properties of a system:

— presence of a catalyst

Chapter 19 — Chemical Kinetics

We will consider all of these in this chapter.

Rates, Rate constants, Rate Equations and Reaction Order

It is very important to understand what a *chemical rate* is and what a *rate constant* is, and how they differ. A chemical rate is the rate of change of a concentration with time, and has units of [concentration/time], such as [moles/L·sec]. Suppose I have the overall reaction:

$$\alpha A \rightarrow \beta B$$

The rate of this reaction is defined as either

$$-\frac{1}{\alpha}\frac{d[A]}{dt}$$

or as

$$\frac{1}{\beta}\frac{d[B]}{dt}$$

depending on preference. These are not necessarily equal throughout the progress of the reaction, but they are often close to equal. The definition using a reactant is the more common. Also, sometimes the factor $(1/\alpha)$ or $(1/\beta)$ is omitted. The rate is always a positive quantity; since for a reactant the concentration decreases with time, $d[A]/dt$ is a negative quantity. Figures 19.1 and 19.2 in the text should clarify this.

A *rate equation* is an equation which tells either how a chemical rate changes with concentration (a *differential rate equation*), or else how a concentration changes with time (an *integrated rate equation*). An example of a differential rate equation is

$$\frac{-d[NO_2]}{dt} = k[NO_2]^2$$

An example of an integrated rate equation is

$$[A] = [A]_o e^{-kt}$$

A *rate constant* is a constant that appears in a rate equation.

Most of our work will involve differential rate equations. The power that a concentration of a component is raised to in such an equation is called the *reaction order* in that component. The sum of these is called the *overall reaction order*. For the above differential rate equation, the reaction order is two in $[NO_2]$ and also two overall. Reaction orders can be positive, negative, zero or fractional, and it is not guaranteed that a reaction will even have an order in every component.

To give a complex example, for the reaction

$$H_2 + Br_2 \rightarrow 2HBr$$

it is found experimentally that

$$\frac{d[HBr]}{dt} = \frac{k_a[H_2][Br_2]^{3/2}}{k_b[Br_2] + [HBr]}$$

This reaction is first order in $[H_2]$, but is not any order in $[Br_2]$ or $[HBr]$, since these occur in more complicated algebraic expressions than simply $[X]^n$. However, at the very early stages of the reaction, where $[HBr] \approx 0$, this equation goes to

$$\frac{d[HBr]}{dt} = \frac{k_a}{k_b}[H_2][Br_2]^{1/2}$$

which is one-half-order in $[Br_2]$ and three-halves-order overall; similarly, late in the reaction, when $[Br_2] \approx 0$, the full expression goes to

$$\frac{d[HBr]}{dt} = k\frac{[H_2][Br_2]^{3/2}}{[HBr]}$$

which is of order one in $[H_2]$, three-halves in $[Br_2]$, minus one in $[HBr]$, and three halves overall. It is important to know that the form of the rate equation, the various reaction orders and the values of the rate constants are strictly empirical; they must all be determined experimentally. Note, too, from this example, that products as well as reactants can appear in a rate equation.

One way to experimentally determine reaction order is to use the method of *initial rates*. For the reaction

$$A \to B$$

in some solvent, at the very beginning we can start with no [B] present. Therefore, if we run the experiment with differing initial concentrations $[A]_o$, we can see how $d[A]/dt$ varies with $[A]_o$. Similarly, by starting out with different values of $[B]_o$ and a single value of $[A]_o$, we can see how $d[A]/dt$ varies with $[B]_o$.

Let us consider some possible experimental findings for this reaction, and see how they give rise to corresponding reaction orders. Each line in each table corresponds to a single experiment.

Possible Data I

$[A]_o$	$[B]_o$	Initial Rate
0.1	0.05	3
0.2	0.05	6
0.2	0.10	6

Here, looking at the first two lines, note that $[A]_o$ changes whereas $[B]_o$ does not; in fact, when $[A]_o$ doubles, the rate doubles, so that

$$\text{rate} = C[A]_1 \qquad \text{(constant [B])}$$

where C is some constant; that is, the rate is proportional to [A] to the first power; thus the reaction is first order in [A]. Comparing the second two lines, these two experiments were done holding $[A]_o$ constant and varying $[B]_o$. Since there is no change in the rate, the reaction must be zeroth order in [B]; that is,

$$\text{rate} = C'[B]^0 = C' \qquad \text{(constant [A])}$$

since $[B]^0 = (\text{anything})^0 = 1$. The overall equation is then

$$\text{rate} = k[A]$$

and is first-order overall (since 1+0=1). Note that k can now be calculated from any line in the table; for example, using the first line,

$$k = \frac{\text{rate}}{[A]} = \frac{3}{0.1} = 30$$

we will discuss the units of the rate constant later.

Now let us consider another possible set of experimental findings for the same reaction

Possible Data II

$[A]_o$	$[B]_o$	Initial Rate
0.1	0.05	3
0.2	0.05	12
0.2	0.10	24

Now, comparing the first two lines, note that doubling [A] makes the rate go up by four; that is

$$\text{rate} = C[A]^2 \qquad \text{(constant [B])}$$

If instead of doubling [A] we had tripled it, then the rate would have gone up by a factor of nine. The reaction is thus second-order in [A]. Looking at the second and third lines in the table, doubling [B] at constant [A] doubles the rate, so that

$$\text{rate} = C'[B] \qquad \text{(constant [A])}$$

and the reaction is first-order in [B], and third-order overall; the rate law is then

$$\text{rate} = k[A]^2[B]$$

and again k can be determined from any line in the table; again we pick the first.

$$k = \frac{\text{rate}}{[A]^2[B]} = \frac{3}{(0.1)^2(.05)} = 6000$$

Note several things so far. First, we determine k from the overall rate equation; C and C' are not rate constants in our sense of the term, because the equations in which they appear are not overall rate equations. Second, note that for the same overall reaction stoichiometry (A→B), different rate laws are possible; it is not possible to determine a rate law from inspection of the overall reaction stoichiometry; rate laws are purely empirical. Finally, note that in the second example the product influences the rate. Some students are under the misapprehension that this cannot occur; in fact, it is very common.

We consider still another possibility for the same reaction.

Possible Data III

$[A]_o$	$[B]_o$	$[H^+]_o$	Initial Rate
0.10	0.05	0.10	3
0.20	0.05	0.10	6
0.20	0.10	0.10	6
0.40	0.20	0.20	6

Here the first three lines are similar to Case I. Thus we have

$$\text{rate} = C[A] \qquad \text{(constant [B] and [H}^+\text{])}$$

$$\text{rate} = C' \qquad \text{(constant [A] and [H}^+\text{])}$$

so that, as before, the reaction is first-order in [A] and zeroth-order in [B]. Things get tricky when trying to determine the order in [H$^+$], since in line 4 everything changes at once; however, we know the orders in [A] and [B], so that the effects of their changes are known. Between the third and fourth line in the table, [B] doubles, but we know that this should not affect the rate, since the reaction is zeroth-order in [B]. [A] also doubles, and the reaction is first-order in [A], so that if [H$^+$] had no effect we would expect the rate to double as well, to 12. Instead, it remains at 6; thus, [H$^+$], which doubled, must have *halved* the rate. In other words,

$$\text{rate} = C'' \frac{1}{[H^+]} \qquad \text{(constant [A] and [B])}$$

If [H$^+$] gets multiplied by two, then the rate gets multiplied by one-half; but $(1/2) = 2^{-1}$, so that the reaction is of order minus-one in [H$^+$]. The overall rate equation is then

$$\text{rate} = k \frac{[A]}{[H^+]}$$

$$k = \frac{(\text{rate})[H^+]}{[A]} = \frac{(3)(0.1)}{(0.1)} = 3$$

where once again we picked the first line of the table to solve for k. Incidentally, if your rate law and value for k are correct, then if you calculate the rate from k and the concentrations the result should always be equal to the value exhibited in the table, plus or minus experimental error. Thus picking line four for Case III,

$$\text{rate} = k\frac{[A]}{[H^+]} = 3\left(\frac{0.40}{0.20}\right) = 6$$

as we expect.

This example shows how negative reaction orders come about, and also demonstrates that concentrations of substances that do not participate in the overall reaction can affect the rate; here $[H^+]$ exerted an inhibitory influence. Some texts define a *catalyst* as anything which appears in the rate law but not in the equilibrium-constant expression, so by this definition $[H^+]$ is a catalyst; however, it is also common to reserve the term "catalyst" for a substance which in addition speeds up the reaction; this is the definition used in the text, and by this definition $[H^+]$ is not a catalyst for our reaction.

First-Order Reactions

For a first-order reaction,

$$\text{rate} = \frac{-d[A]}{dt} = k[A]$$

This gives (though it takes calculus to show it)

$$\ln\frac{[A]}{[A_o]} = -kt$$

so that a plot of $\ln([A]/[A]_o)$ against t gives a straight line of slope -k. In fact, if one measures [A] as a function of t over the course of a reaction, and then makes this plot, it will be a straight line if and only if the reaction is first-order in [A]. This is another way (besides the initial method) of determining if a reaction is first-order.

This equation is an example of an integrated rate law. By raising e (the base of natural logarithms) to the power of both sides of the equation, we get

$$\frac{[A]}{[A]_o} = e^{-kt}$$

$$[A] = [A]_o e^{-kt}$$

Plots of these behaviors are given in Figures 19.3 through 19.5.

Consider the logarithmic form of the rate equation. If we wait until [A] has diminished to just one-half of its initial value, $[A]_o$, then we have

$$\ln(1/2) = -kt_{1/2}$$

$$t_{1/2} = -\frac{\ln(1/2)}{k} = \frac{0.693}{k}$$

$t_{1/2}$ is called the *half-life* of the reaction, and is defined as the time it takes for half the initial reactant to disappear. Note that neither [A] nor $[A]_o$ appears in the equation for $t_{1/2}$. This means that for a first-order reaction, the half-life is a constant. Thus, if I start with [A] = 1M, say, and after, say, 10min [A] = 1/2M, then after another 10min [A] will be 1/4M, and so on. For any ten-minute interval I choose, the value of [A] at the end of the interval will be half that at the beginning. This is true not only for the half-life, but also for the "one-third-life", the "90% life" and any other fractional life you may care to define, but only for first-order reactions. Radioactive decay is always first-order. Reactions taking place in dilute solution, where the concentration of solvent changes little, may appear to be first-order overall, when the solvent really does enter into the rate equation. This is called *pseudo-first-order* behavior, and is sometimes difficult to confirm or disprove.

Second-Order Reactions

Sometimes it is found experimentally that a process is second order

$$\frac{-d[A]}{dt} = k[A]^2$$

Using calculus it can be shown that if this is true, then

$$\frac{1}{[A]} = kt + \frac{1}{[A]_o}$$

Thus, a plot of 1/[A] against t gives a straight line with slope k. This is a test for second-order-behavior. Here the half-life changes as the reaction proceeds. Again, we start with the definition that $t_{1/2}$ is the time at which $[A] = (1/2)[A]_o$. This gives

Chapter 19 — Chemical Kinetics

$$\frac{1}{(1/2)[A]_o} = kt_{1/2} + \frac{1}{[A]_o}$$

which is easily solved, giving

$$t_{1/2} = \frac{1}{k[A]}$$

As the reaction proceeds, [A] gets smaller, and it takes longer for the concentration to diminish to $(1/2)[A]$.

The units of k are easily determined by inspection of the differential or integral rate equation. Recall that a chemical rate always has units of concentration divided by time. For a first-order reaction

$$\text{rate} = \left[\frac{\text{mol}}{\text{L} \cdot \text{s}}\right] = k[A] = k\left[\frac{\text{mol}}{\text{L}}\right] \quad \text{(first-order)}$$

$$k = [s^{-1}]$$

For second-order

$$\text{rate} = \left[\frac{\text{mol}}{\text{L} \cdot \text{s}}\right] = k[A]^2 = k\left[\frac{\text{mol}}{\text{L}}\right]^2 \quad \text{(second-order)}$$

$$k = \left[\frac{\text{L}}{\text{mol} \cdot \text{s}}\right]$$

and so on.

So far all we have done is to define some basic terms, to tell how to determine reaction order, and how to determine rate constants and half-lives for first- and second-order processes. Next we will discuss how the rate is affected by temperature, and then discuss how kinetics gives insight into *reaction mechanism*.

Arrhenius Equation

It is found empirically (the *Arrhenius equation*) that, usually, for a given reaction

$$k = Ae^{-E_{act}/RT}$$

E_{act} is called the *activation energy* and A is called the *frequency factor*. Taking the logarithm of both sides

Chapter 19 — Chemical Kinetics

$$\ln(k) = \ln(A) - \frac{E_{act}}{RT}$$

Thus, if $\ln(k)$ is plotted against $1/T$, the result is a straight line with slope $-E_{act}/R$ and y-intercept $\ln(A)$. E_{act} is almost always positive, so that higher temperatures almost always give faster rates.

Let us see how this equation can be used. For most processes observed in the laboratory, the rate approximately doubles when the temperature is raised by 10°C. "In the laboratory" implies room temperature, which, for simplicity, we can take as 300K (27°C). Let us calculate E_{act} for a process observing this general rule. Suppose at 300K a process exhibits rate constant k_1, and at 310K it exhibits k_2, which by the rule, must be $2k_1$. Then the Arrhenius equation gives

$$\frac{k_2}{k_1} = 2 = \frac{Ae^{-E_{act}/R(310)}}{Ae^{-E_{act}/R(300)}}$$

$$= e^{-\frac{E_{act}}{R}\left(\frac{1}{310} - \frac{1}{300}\right)}$$

$$\ln 2 = +\frac{E_{act}}{R}\left(\frac{1}{300} - \frac{1}{310}\right) = +\frac{E_{act}}{R}(1.08 \times 10^{-4})$$

$$E_{act} = \frac{R \ln 2}{(1.04 \times 10^{-4} \, K^{-1})} = 53.6 \, kJ/mol$$

We will rationalize this behavior below.

Although the Arrhenius equation is an empirical relationship rather than a rigorously justifiable law, it can be interpreted in terms of a simple mental picture that is essentially correct. In this picture, the reactant(s) must pass through a state of high energy, called the *activated complex* or *transition state*, in order to form the products. This state can be formed by chance collisions with other molecules, or by chance redistribution of the energies of the bonds in a single molecule so as to give extra energy to a bond that needs to be broken. E_{act} is the difference between the energy of the activated complex and that of the reactants. Recall from Chapter 18 that $\Delta G° = -RT \ln K_{eq}$. Solving this for K_{eq}

$$K_{eq} = e^{-\Delta G°/RT}$$

This looks similar to the term $e^{-E_{act}/RT}$, and it is plausible to identify $e^{-E_{act}/RT}$ with something like K_{eq} for

Chapter 19　　　　　　　　　　　　　　　　　　　　　　　　　　　　　　　　Chemical Kinetics

the reaction

$$\text{reactants} \to \text{activated complex}$$

Now, at least when only a single species is being activated, as in

$$X \to X^*$$

if K_{eq} is very small then $[X] \approx [X]_o$, and

$$e^{-E_{act}/RT} \approx K_{eq} = \frac{[X^*]}{[X]} \approx \frac{[X^*]}{[X]_o}$$

so that $e^{-E_{act}/RT}$ is the fraction of reactant molecules with enough energy to form the transition state; that is, to react. A, the frequency factor, is the number of potentially reactive events occurring per unit time. For example, if forming the activated complex involves collisions, A includes both the number of collisions per second and the fraction of these that are in the right geometric orientation to react (the *orientation factor*). In this picture a "potentially reactive event" will result in a successful reaction if it is sufficiently energetic. THe exponential term gives the fraction of total events that are.

The reason the Arrhenius equation is so useful is that A and E_{act} do not change much with temperature. Furthermore, A remains fairly constant from reaction to reaction, except that it tends to be smaller for gas-phase reactions than for liquid-phase reactions, since gas-phase reactions are more dilute, so that there are fewer collisions per unit-time. The relative constancy of A explains why reactions observed in the laboratory exhibit a fairly consistent temperature dependence. Usually, the reactions measured in a laboratory have half-times somewhere between minutes and hours. For E_{act} more than a few kilojoules smaller than the value which we calculated, half-times go down to seconds or less, too fast to observe without specialized equipment; for E_{act} more than a few kilojoules greater, half-times go up to days or weeks, exhausting the patience of all but the most diligent student. Thus, it is the laboratory time scale, not chemical laws, that determines E_{act} for common laboratory processes.

Figure 19.14 exhibits a potential energy profile for an exothermic reaction, and Figure 19.15 exhibits one for an endothermic reaction. Note that whether a reaction is exothermic or endothermic depends on $\Delta H_{overall}$, that is, the enthalpy difference between reactants and products, whereas E_{act}

depends on the potential energy difference between reactants and the activated complex. E_{act} is essentially always positive, regardless of the sign of $\Delta H_{overall}$. Note that this "picture" of a chemical reaction is in accord with what we have said about the Arrhenius equation. Forming the activated complex is always an endothermic process, ($E_{act} > 0$), and endothermic processes proceed further to the right when the temperature is raised; therefore the reason a higher temperature virtually always raises a reaction rate is simply that this causes more activated complex to form.

You may have noticed that we have blurred the distinction between ΔH and ΔE in this discussion: in our potential energy profiles we measured both ΔH and E_{act} along the same y-axis. In fact, it is possible to derive a more complicated equation than the Arrhenius equation that allows ΔH, ΔG and even ΔS of activation to be defined; this is called the Eyring equation, but it is beyond the scope of this course. The E_{act} in the Arrhenius equation is not a true ΔE in the thermodynamic sense. The equation was not derived starting with the definition of the energy difference between the activated complex and the starting compounds; it is a purely empirical relationship, and E_{act} is a purely empirical parameter that happens to have the units of energy. E_{act} has no rigorous thermodynamic basis, and we can compare it loosely with any kind of energy we choose; note that it is somewhat analogous $\Delta H°$, because in the same way that $\Delta H°$ determines the temperature variation of K_{eq}, E_{act} determines the temperature variation of k.

Mechanisms

A *reaction mechanism* is a series of *elementary steps* which purport to tell exactly which molecules must interact with which, and in what sequence, for the overall reaction to occur. The steps must add up to the overall reaction, but other substances than just the reactants and products may — and in general do — appear in mechanisms. These fall into two categories: *reactive intermediates*, molecules that are produced and rapidly consumed as the reaction proceeds, and *catalysts*, which will be discussed later.

Chapter 19 — Chemical Kinetics

As an example of a mechanism, consider the reaction cited in the text.

$$NO_2(g) + CO(g) \rightarrow CO_2(g) + NO(g) \qquad \text{(overall)}$$

The observed rate law is

$$\text{rate} = -\frac{d[NO_2]}{dt} = k[NO_2]^2$$

A proposed mechanism is

$$2NO_2 \rightarrow NO + NO_3 \qquad \text{(step 1: slow)}$$

$$NO_3 + CO \rightarrow NO_2 + CO_2 \qquad \text{(step 2: fast)}$$

We are going to show that this mechanism is constent with the observed rate law. Note first of all that the two equations add up to the overall reaction. NO_3 is produced in the first step and used up in the fast second step, so it is a reactive intermediate.

When "slow" and "fast" are used to describe elementary steps, it means that the fast step occurs so quickly that the time it takes can be neglected. For example, let us suppose that step 1 takes place about once an hour, and step 2 takes place in, say, a tenth of a second. Suppose we let two NO_2 molecules collide in the presence of CO. One hour later a molecule of NO and one of NO_3 appear, but a tenth of a second later one NO_2 is regenerated and a CO_2 appears as a CO disappears. Thus every hour (to one significant figure) one NO_2 and one CO have disappeared, and a CO_2 and an NO have appeared. If the first reaction took en hours,, then the overall process would happen only every ten hours; on the other hand, if step 2 were ten times faster the rate would not be materially affected; we would release products one hour plus 0.01s after the start of the experiment, rather than one hour plus 0.1s. Thus the rate of the overall process is determined by the rate of step 1. Of course, if step 1 speeded up to close to 0.1s, then the rates of both steps would affect the overall rate, but then we would not call step 1 "slow." If one step in a mechanism is much slower than the others, it is called the *rate-determining step*, and of course it determines the rate, as we have shown.

Chapter 19   Chemical Kinetics

Molecularity

Now that we know that step 1 is rate determining, how can we predict the rate law, given the mechanism? Recall that elementary processes such as step 1 or step 2 indicate the actual molecular interactions that take place; step 1 tells us that the reaction starts off when two NO_2 molecules collide. This can be used to predict the rate law for step 1. First, a simpler case. Suppose we have, as an elementary step,

$$A \rightarrow 2B$$

This means that every so often — say, once a minute, on the average — an A molecule will turn into a B molecule, without needing to collide with anything else. Clearly, if we have 60 A molecules present, we can expect to lose an A molecule and form two B molecules in the first second, on average, and if we have 600 A molecules present, we can expect to lose ten A molecules and form twenty B molecules in the first second. Thus, for this elementary process, the rate is proportional to the amount of A present, so that

$$\text{rate} = -\frac{d[A]}{dt} = \frac{1}{2}\frac{d[B]}{dt} = k[A]$$

Two things are notable. First of all, this is a first-order rate law, which we were able to derive purely from the form of the elementary step; second, in an elementary step it is rigorously true that the rate defined by the reactant(s) is equal to the rate defined by the product(s); an A disappears and two B's appear; nothing happens in between. The first of these observations is the important one.

—   For an elementary step, the rate law is given by the stoichiometry of the process.

The reaction order of an elementary process is called its *molecularity*, and can be determined by inspection. For $A \rightarrow 2B$ the process is said to be *unimolecular*, and gives a first-order rate law.

Now let us consider a *bimolecular* process.

$$A + B \rightarrow C$$

This says that A and B have to collide to form C. If I double [A] I double the likelihood of a collision, and this is true if I double [B] as well, so that the rate law for this process is

$$\text{rate} = -\frac{d[A]}{dt} = \frac{-d[B]}{dt} = \frac{d[C]}{dt} = k[A][B]$$

Chapter 19 — Chemical Kinetics

By the same reasoning, if I have

$$2A \to B$$

I have

$$\text{rate} = -\frac{1}{2}\frac{d[A]}{dt} = \frac{d[B]}{dt} = k[A]^2$$

In fact, if I have the elementary step

$$\alpha A + \beta B \to \gamma C + \delta D$$

then the rate law is

$$\text{rate} = -\frac{1}{\alpha}\frac{d[A]}{dt} = -\frac{1}{\beta}\frac{d[B]}{dt} = \text{etc.} = k[A]^\alpha [B]^\beta$$

You should know, however, that elementary steps with a molecularity greater than two are rare, since the probability of three or more molecules colliding in the same split-second with precisely the right geometry to react is very small.

Now let us return to our NO_2/CO reaction. The rate (not the rate law!) *defined* by the stoichiometry of the overall process as

$$\text{rate}_{\text{overall}} = -\frac{d[NO_2]}{dt} = -\frac{d[CO]}{dt} = \text{etc.}$$

The rate of step 1 is *defined* as

$$\text{rate}_1 = -\frac{1}{2}\frac{d[NO_2]}{dt} = \frac{d[NO]}{dt} = \frac{d[NO_3]}{dt}$$

and, using our observations about molecularity, we can write

$$\text{rate}_1 = -\frac{1}{2}\frac{d[NO_2]}{dt} = 2\text{rate}_1 = 2k_1[NO_2]^2$$

Thus, since step 1 is rate determining, it would appear that

$$\text{rate}_{\text{overall}} = \frac{d[NO_2]}{dt} = 2\text{rate}_1 = 2k_1[NO_2]^2$$

but this is not quite correct. We said that every hour (using our entirely fictitious numbers) step 1 takes place, and that then step 2 takes place immediately. Step 2 regenerates one of the two NO_2's that got used up in step 1, so that the disappearance of NO_2 is only half as rapid as step 1, taken alone, makes it out to be. This gives

$$\text{rate}_{\text{overall}} = \frac{d[NO_2]}{dt} = \frac{1}{2}(2\text{rate}_1) = k_1[NO_2]^2$$

Thus the overall rate law is the same as that of the rate-determining step.

There are two very common reaction mechanisms that you should be familiar with. We have just examined one of them: a slow step followed by a fast step. The second, exemplified by the HI reaction, involves a fast equilibrium followed by a slow step.

$$H_2 + I_2 \rightarrow 2HI \qquad \text{(overall)}$$

The observed rate law is

$$\text{rate} = k[H_2][I_2]$$

but there is abundant evidence that the overall reaction is not an elementary step. One theoretical reason to doubt that it is is that the H_2 and I_2 molecules would have to collide in a parallel orientation, and given the number of collisions per second and the probability of such an orientation, the observed reaction rate seems too fast for this mechanism. An alternative mechanism is

$$I_2 = 2I \qquad \text{(step 1: fast equilibrium)}$$
$$H_2 + 2I \rightarrow 2HI \qquad \text{(step 2: slow)}$$

Note that step 2 is a *termolecular* step. If step 2 is slow, then we can assume that [I] is given by the equations

$$K_1 = \frac{[I]^2}{[I_2]}$$

$$[I] = \sqrt{K[I_2]}$$

Then, from step 2, the rate-determining step,

$$\text{rate} = k_2[H_2][I]^2$$
$$= K_1 k_2 [H_2][I]$$

Which reveals that k, the experimental rate constant, is given by

$$k = K_1 k_2$$

Note that two hypothetical mechanisms — the one we discussed, plus the mechanism that postulates the overall reaction to be an elementary step — give the same rate law. This demonstrates an important point: kinetic data can prove a mechanism wrong, but they cannot prove it right. Other kinds of data are necessary to support a proposed mechanism; theoretical arguments, such as the one mentioned above, can help, as can other experimental data, such as Sullivan's work, cited in the text. Ultraviolet light is known to cause the dissociation of I_2; thus it should speed up the two-step mechanism, but have no effect on the one-step mechanism. This speed-up is, in fact, observed, lending credence to the two-step mechanism.

There is one more point to make before we go on. You should understand the difference between a reactive intermediate and a transition state. A reactive intermediate is not usually an inherently unstable species. For example, the I generated as an intermediate in the second reaction we have discussed would be quite content at very low pressures, where there were not many other I's to bump into; as LeChatelier's principle predicts, the $I_2 = 2I$ equilibrium shifts to the right at low pressures. On the other hand, the transition state separating I_2 from 2I *is* inherently unstable; it is an I_2 molecule with its bond stretched so tightly that it is about to break. If something should act to bring the atoms closer together, the transition state it would reform I_2; if something should pull them apart, it would form 2I. Thus the formation of 2I from I_2 looks very much like Figure 19.15. If we were to proceed further to HI, the potential energy curve in the figure would rise again to the right, as the reaction proceeded toward the transition state for step 2 ($2I + H_2 \rightarrow 2HI$), and would drop again as products were formed.

Catalysts and Chain Reactions*

Sometimes substances appear in a rate law that do not appear in the overall equation. These are called *catalysts*. A catalyst affects the rate without being used up in the reaction. A catalyst can be consumed in one step of a mechanism, just as long as it is regenerated in another. This, in fact, is what happens. You should know what *homogeneous* and *heterogeneous* catalysts are; homogeneous catalysis takes place in the same phase as the reactants and products — such as the liquid state — and heterogenous catalysis takes place over phase boundaries, as in the use of powdered platinum to catalyse the reaction of dissolved hydrogen gas with an organic molecule. An *enzyme* is a protein whose function

---

* There is a Perspective on these subjects.

is to catalyse a specific reaction in the body.

One common and complex type of reaction mechanism is the *chain reaction*. This is a reaction which may be difficult to start, but once a reactive intermediate is generated subsequent steps keep the intermediate replenished even while product is being formed, as in the reaction of H_2 plus Br_2 to form HBr (Equations 19-39 through 19-40 in the text).

Reaction Rates and Equilibrium

We have often said that when a system is in equilibrium nothing happens macroscopically. Microscopically, however, lots is happening! If the reaction

$$A = B$$

is in equilibrium, we know that [A] and [B] do not change with time. This does not require that no A molecules be reacting to form B and vice-versa; all it requires is that for every A that forms B, there is a B that forms A; in other words, for a system at equilibrium,

$$\text{forward rate} = \text{reverse rate}$$

For $A = B$, if both the forward and reverse rate are elementary steps

$$\text{forward rate} = k_f[A]$$

$$\text{reverse rate} = k_r[B]$$

$$k_f[A] = k_r[B]$$

$$\frac{[B]}{[A]} = K_{eq} = \frac{k_f}{k_r}$$

Even where the forward and reverse reactions are not elementary steps, K_{eq} may be obtained in terms of the rate constants for the forward and reverse reactions by setting

$$\text{forward rate} = \text{reverse rate}$$

This has been experimentally confirmed many times.

We have said that a catalyst affects the rate without affecting the equilibrium. Since we must have $K_{eq} = k_f/k_r$, if a catalyst increases k_f, it must also increase k_r, to keep K_{eq} constant. Figure 19.16 shows how this comes about. The catalyst operates by lowering the activation energy. But if E_{act} is lowered for the forward reaction, it must also be lowered for the reverse reaction, as the figure makes clear.

In summary, there is no scheme to predict reaction rates, in the same sense that tables of $\Delta H°_f$ and $S°$ let us predict reaction equilibria; however, reaction rates can be classified using the concept of reaction order, and kinetics is a great help in unravelling the detailed mechanisms of chemical processes; furthermore, elementary processes must dictate reaction rates. Although not discussed in the text, modern chemical kinetics, in contrast to the "classical" kinetics we have been discussing, seeks to start out with a molecular description of a molecule — such as a description in terms of its quantum states — and use that description to predict reaction rates. This field is highly mathematical, and, as with the rigorous description of chemical bonding, has been most successful in describing very small systems, such as a hydrogen molecule and a tritum atom, T, reacting to form HT+H in the gas phase. Only now is progress beginning to be made in understanding solution processes at a fundamental molecular level. As we write, this is one of the most exciting and active frontiers of chemical research.

Chapter 19 — Chemical Kinetics

Questions

19.1 Factors affecting the rates of chemical reactions and
19.2 The rate expression

1. Suppose for some reaction it is found that

$$\text{rate} = \frac{-d[A]}{dt} = \frac{k[A]^2}{[B]}$$

   (a) What are the units of the rate?

   (b) What are the units of the rate constant, k?

   (c) the sign of the rate is (+, -, 0).

   (d) the sign of k is (+, -, 0)

   (e) the reaction order is _____ in A, _____ in B and _____ overall.

2. For the reaction of ethylene bromide and potassium iodide in methanol

$$C_2H_4Br_2 + 3KI \rightarrow C_2H_4 + 2KBr + KI_3$$

   The rate can be defined in terms of ethylene bromide as

$$\text{rate} = \frac{-d[C_2H_4Br_2]}{dt}$$

   Write similar expressions that define the rate in terms of each of the other components.

Chapter 19 — Chemical Kinetics

3. For an ideal gas, the partial pressure is often used instead of the molarity as a unit of concentration. Starting with $P\Delta V = nRT$, write an expression relating molarity to P.

19.3 Determination of the order of reaction by the method of initial rate

1. If the concentration of a product is plotted against time for a rate process,

    (a) what is the slope of the curve equal to?

    (b) what is the slope of the curve at t=0 equal to?

2. For the reaction

$$A + B \rightarrow C$$

   the following data were collected

$[A]_o$	$[B]_o$	$[C]_o$	initial rate
.001	.001	.001	1
.002	.001	.001	4
.002	.002	.001	8
.002	.002	.002	4

   (a) The reaction order is _____ in A, _____ in B, _____ in C and _____ overall.

   (b) Give the rate law.

   (c) k= _____ .

Chapter 19 — Chemical Kinetics

19.4 First-order reactions

1. A first-order rate constant has units of _____.

2. For a first order reaction, if [A] is tripled, the rate gets multiplied by _____.

3. In the expression $[A] = [A]_o e^{-kt}$

    (a) kt has units of _____

    (b) e^{-kt} has units of _____

    (c) If [A] is plotted against t, what will be the value of [A] where the curve hits the y-axis?

    (d) what value of t does this correspond to?

    (e) what value does [A] have at $t = \infty$?

4. Suppose you have values for [A] vs. t for a reaction you suspect is first-order. What should you plot against what to confirm your suspicion, and how will you know if you are right?

5. Starting with the expression $\ln([A]/[A]_o) = -kt$, write an expression for the $\tau_{.90}$, the time for 90% of the starting material to react in a second-order process.

19.5 Second-order reactions

1. A second-order rate constant has units of _____.

2. For a second-order reaction, if [A] gets doubled, the rate gets multiplied by _____.

3. Answer Question 4 in the last section, but assuming you suspect the reaction is second-order.

Chapter 19 — Chemical Kinetics

4. Starting with the expression $1/[A] - 1/[A]_o = kt$, write an expression for $t_{.90}$, the time for 90% of the starting material to react in a second-order process.

5. Suppose, for a process second-order in reactant [A], $k = 0.05 s^{-1}$. If I start my experiment with $[A]_o = 0.1 M$,

    (a) What will be the concentration of [A] after 200s?

    (b) after 400s?

19.6 The temperature dependence of rate constant

1. Generally, when the temperature goes up, reaction rates _____ .

2. In general, if the number of collisions per second goes up, the reaction rate _____ .

3. In general, if the energy of collision goes up, the probability of reaction _____ .

4. In general, if E_{act} goes up, the reaction rate _____ .

5. In general, when A (the frequency factor) goes up, the reaction rate _____ .

6. In the equation $k = Ae^{-E_{act}/RT}$

    (a) E_{act} has units of _____ .

    (b) RT has units of _____ .

    (c) E_{act}/RT has units of _____ .

    (d) $e^{-E_{act}/RT}$ has units of _____ .

(e) If k is first-order, A has units of _____ .

7. In the same equation,

   (a) If T increases, 1/T _____ .

   (b) If T increases, E_{act}/RT _____ .

   (c) If T increases, $e^{-E_{act}/RT}$ _____ .

   (d) If 1/T increases, $e^{-E_{act}/RT}$ _____ .

8. For two reactions, ln(k) is plotted against 1/T. The one that has the steeper slope has the _____ value of E_{act}.

9. (a) $e^{-E_{act}/RT}$ can be (positive, negative, either) and is generally a (large, small, either) number.

   (b) A can be (positive, negative, either) and is generally a (large, small, either) number.

10. If E_{act} were equal to zero for some reaction, what would k be equal to? _____

19.7 Transition state theory and the activated complex

1. E_{act} is the same for the forward and the reverse reactions. (True, False).

2. $\Delta H_{forward} = -\Delta H_{reverse}$ (True, False).

3. The forward and reverse reactions proceed through the same activated complex (True, False).

4. The process of forming the activated complex is *exothermic or endothermic, depending on* whether one is dealing with the forward or the reverse reaction.

5. A catalyst *lowers* E_{act} for a reaction, and speeds up *both the forward and the reverse reaction* .

19.8 The relation between the mechanism of a reaction and kinetics data

1. The slowest step in a mechanism is called the

Chapter 19   Chemical Kinetics

2. Each step in a mechanism is called

3. *Just as for an overall reaction*, the rate law for an elementary step can be easily derived from the stoichiometry.

4. If a mechanism predicts the correct rate law, it *must be* correct.

5. The overall reaction order of an elementary step is called its _____ .

6. Write down the rate expressions for the following elementary steps:

   (a) $A \rightarrow B$ rate =

   (b) $B \rightarrow A$ rate =

   (c) $2A \rightarrow B$ rate =

   (d) $2A + B \rightarrow C$ rate =

7. A catalyst can appear in an elementary step (True, False).

8. A catalyst can appear in an overall reaction (True, False).

9. For the following mechanism, tell whether each ingredient is a reactant, a product, a catalyst, or a reactive intermediate.

$$A + B \rightarrow C$$
$$C + D \rightarrow E$$
$$E \rightarrow F + B$$

(a) A is a _____ .

(b) B is a _____ .

(c) C is a _____ .

(d) D is a _____ .

(e) E is a _____ .

(f) F is a _____ .

19-25

Chapter 19   Chemical Kinetics

19.9 Chain Reactions

1. What is a free radical?

2. Give two ways they can be produced.

3. A reaction is a chain-reaction if in at least one step

19.10 Reaction Rates and Equilibrium

1. For an equilibrium the forward rate is equal to *the negative of* the reaction rate.

2. A catalyst concentration *may appear* in an equilibrium constant expression.

3. If a system A=B is in equilibrium, it means that no A atoms are reacting to form B and vice versa. (True, False).

Chapter 19 — Chemical Kinetics

Perspective: The Steady State Approximation and Michaelis-Menten Kinetics

The Steady-State Approximation

Consider the mechanism

$$A \rightarrow B \quad \text{(step 1)}$$

$$B \rightarrow C \quad \text{(step 2)}$$

where B is a reactive intermediate, so that the overall reaction is

$$A \rightarrow C \quad \text{(overall)}$$

For B, we can write

$$\frac{d[B]}{dt} = \text{rate of formation} - \text{rate of destruction} = k_1[A] - k_2[B]$$

This is just a straightforward application of the rules for writing rate laws from elementary steps.

Now, however, note that if B is a reactive intermediate, it must react quickly to form C; thus there is never much of it around. Because [B] is small, $d[B]/dt$ must also be small. The *steady state approximation* says that if B is a reactive intermediate, then $d[B]/dt \approx 0$. The approximation is called "steady state" because $d[B]/dt = 0$ means that [B] is not changing with time — it is approximately constant. (It does *not* mean that [B]=0!). This assumption is reasonable, because if B is indeed a reactive intermediate it will be used up as fast as it is formed; its rate of destruction will be equal to its rate of formation.

This approximation gives

$$\frac{d[B]}{dt} = 0 = k_1[A] - k_2[B]$$

$$= \text{rate of formation of B} - \text{rate of destruction}$$

$$[B] = \frac{k_1[A]}{k_2}$$

Now, to find the rate of formation of [C], we can write (from the second step of the mechanism)

$$\frac{d[C]}{dt} = k_2[B] = \frac{k_2 k_1[A]}{k_2} = k_1[A]$$

so that k_1 will be equal to the experimentally determined k in the rate law

$$\text{rate} = k[A]$$

What have we accomplished? First, note that the fact that B is a reactive intermediate is equivalent here to stating that the first step is rate-limiting: B gets formed slowly and used up quickly. From this fact the rate law just derived could have been written down immediately, so that this particular application of the steady state approximation does not enable us to derive a rate law that we couldn't have derived otherwise; there are, however, instances where it does, as we shall see. Suppose, though, that we had an instrument capable of measuring the very low concentration of [B] present. If we knew [B], we could use this value, together with our known values of [A] and k_1, to calculate k_2. Of course, by doing the same experiment at several temperatures we could get E_{act} for steps 1 and 2, and perhaps relate them to bond energies or make other interesting and useful chemical correlations.

We will now see how the steady-state approximation is used to calculate the rate law that results from the complicated chain mechanism for the reaction

$$H_2 + Br_2 \rightarrow 2HBr \qquad \text{(overall)}$$

According to the text, the mechanism is

$$Br_2 \rightarrow 2Br \qquad \text{(step 1: initiation)}$$

$$Br + H_2 \rightarrow HBr + H \qquad \text{(step 2: propagation)}$$

$$H + Br_2 \rightarrow HBr + Br \qquad \text{(step 3: propagation)}$$

$$H + HBr \rightarrow H_2 + Br \qquad \text{(step 4: inhibition)}$$

$$2Br \rightarrow Br_2 \qquad \text{(step 5: termination)}$$

The empirically determined rate law is

$$\text{rate} = \frac{1}{2}\frac{d[HBr]}{dt} = \frac{k_a[H_2][Br_2]^{3/2}}{k_b[Br_2] + [HBr]} \qquad \text{(empirical rate law)}$$

Where k_a and k_b are *empirical* rate constants. What we wish to do is to show how the above mechanism gives rise to this rate law. You will see that this will also show how k_a and k_b are to be interpreted in terms of the k's for the elementary steps.

From the overall reaction and the definition of the rate,

$$\text{rate} = \frac{1}{2}\frac{d[HBr]}{dt}$$

From the mechanism

$$\text{rate} = \frac{1}{2}\frac{d[HBr]}{dt} = \frac{1}{2}\Big(k_2[Br][H_2] + k_3[H][Br_2] - k_4[H][HBr]\Big)$$

$$= \frac{1}{2}\Big(k_2[Br][H_2] + [H](k_3[Br_2] - k_4[HBr])\Big)$$

For each step in which [HBr] is created, $d[HBr]/dt$ gets a positive term; for each step in which [HBr] is destroyed, $d[HBr]/dt$ gets a negative term. The terms themselves are given directly from the stoichiometry of the respective steps. This equation shows that in order to derive the rate law we need to know [H] and [Br]. Since there are two reactive intermediates we will have to apply the steady-state approximation to both of them. (Incidentally, given the above mechanism, it should be clear that H and Br are in fact reactive intermediates, since they are formed from reactants, but do not appear in the overall reaction. Of course the scientist, given only the rate law and the overall equation, has to be clever enough to figure out what the intermediates, as well as the mechanism, are likely to be.) Applying the steady-state approximation is lengthy but straightforward. We start exactly the same way as we did when we wrote an expression for $d[HBr]/dt$, but here of course we set $d[H]/dt$ and $d[Br]/dt$ to zero.

$$\frac{d[Br]}{dt} = 0 = 2k_1[Br_2] - k_2[Br][H_2] + k_3[H][Br_2] + k_4[H][HBr] - k_5[Br]^2$$

$$\frac{d[H]}{dt} = 0 = k_2[Br][H_2] - k_3[H][Br_2] - k_4[H][HBr]$$

These are two equations in two unknowns, [H] and [Br], which can be solved simultaneously. Simply adding the two together gives

$$0 = k_1[Br_2] - k_5[Br]^2$$

$$[Br] = \left(\frac{k_1[Br_2]}{k_5}\right)^{1/2} = (k_1/k_5)^{1/2}[Br]^{1/2}$$

The equation for $d[H]/dt$ above can be solved for [H]

$$[H] = \frac{k_2[Br][H_2]}{k_3[Br_2] + k_4[HBr]}$$

Substituting our value for [Br] gives

$$[H] = \frac{k_2(k_1[Br]/k_5)^{1/2}[H_2]}{k_3[Br_2] + k_4[HBr]}$$

$$= \frac{k_2(k_1/k_5)^{1/2}[Br_2]^{1/2}[H_2]}{k_3[Br_2] + k_4[HBr]}$$

What we have done — and what one strives in general to do by using the steady-state approximation — is to solve for the concentration of intermediates in terms of the concentrations of stable substances. That way, when we use these values to determine the rate law, the intermediates will not appear. We now substitute our expression for [H] and [Br] into the equation for the rate.

$$\text{rate} = \frac{1}{2}\frac{d[HBr]}{dt}$$

$$= \frac{1}{2}\Big(k_2[Br][H_2] + [H](k_3[Br_2] - k_4[HBr])\Big)$$

$$= \frac{1}{2}\left\{k_2(k_1/k_5)^{1/2}[Br_2]^{1/2}[H_2] + \Big(k_2(k_1/k_5)^{1/2}[Br_2]^{1/2}[H_2]\Big)\left[\frac{k_3[Br_2] - k_4[HBr]}{k_3[Br_2] + k_4[HBr]}\right]\right\}$$

$$= \frac{1}{2}\left\{\Big(k_2(k_1/k_5)^{1/2}[Br_2]^{1/2}[H_2]\Big)\left[1 + \frac{k_3[Br_2] - k_4[HBr]}{k_3[Br_2] + k_4[HBr]}\right]\right\}$$

$$= \frac{1}{2}\left\{\Big(k_2(k_1/k_5)^{1/2}[Br_2]^{1/2}[H_2]\Big)\left[\frac{2k_3[Br_2]}{k_3[Br_2] + k_4[HBr]}\right]\right\}$$

$$= \frac{k_2 k_3 (k_1/k_5)^{1/2}[H_2][Br_2]^{3/2}}{k_3[Br_2] + k_4[HBr]}$$

$$= \frac{(k_2 k_3/k_4)(k_1/k_5)^{1/2}[H_2][Br_2]^{3/2}}{(k_3/k_4)[Br_2] + [HBr]}$$

This is the same as the empirical rate law, with

$$k_a = (k_2 k_3/k_4)(k_1/k_5)^{1/2}$$

$$k_b = (k_3/k_4)$$

It is worth noting a few things. First of all, step 5 is the reverse of step 1, so that $(k_1/k_5)=K_{eq}$ for the dissociation of $[Br_2]$. The statement that

$$[Br] = (k_1/k_5)^{1/2}[Br_2]^{1/2}$$

This means that $[Br]$ is at its equilibrium concentration. This is similar to what occurred in simpler mechanisms when we assumed a fast pre-equilibrium step, but here no assumption was made about relative rates; it's just that other terms involving $[Br]$ cancelled out when we solved for this intermediate; thus $[Br]$ has its equilibrium concentration (referred to $[Br_2]$ even if the equilibrium is slow. The second point is just to clarify how we got rid of the "1" in the fourth line of the derivation. We simply replaced it using

$$1 = \left(\frac{k_3[Br_2] + k_4[HBr]}{k_3[Br_2] + k_4[HBr]}\right)$$

then combined terms within the square brackets over the common denominator. Algebra can get complicated, and it takes persistance and practice to solve a problem as complicated as this one.

This example is quite complex, but involves only algebra, and shows how very complicated rate expressions can arise from innocent-looking mechanisms. It would certainly be much harder to come up with the mechanism given only the rate law and the overall reaction; in fact, it took some of the best chemists of the early 20th century over a decade to do so!

Note again, as we did in the Scope, that the rate expression is first-order in $[H_2]$, but that the order is not defined for $[Br_2]$ or $[HBr]$. When $[HBr]<<k_b[Br_2]$, however, as would occur at the very beginning of the reaction, the expression becomes

$$\text{rate} = \frac{k_a}{k_b}[H_2][Br_2]^{1/2}$$

Toward the end of the reaction we will have $k_b[Br_2]<<[HBr]$, and the rate becomes

$$\text{rate} = k_a\frac{[H_2][Br_2]^{3/2}}{[HBr]}$$

Both of these expressions have well-defined orders in all components. As reaction proceeds, the order in $[Br_2]$ goes from (1/2) to (3/2), and the order in $[HBr]$ goes from 0 to -1. The full rate law exhibits a smooth transition between these respective values.

Michaelis-Menten Kinetics

Enzyme reactions in solution, as well as other homogeneous catalytic processes, often fit a simple reaction scheme known as the Michaelis-Menten mechanism. This example is instructive because

(a) it can be easily solved and understood using the tools developed in this chapter

(b) it is in daily use in laboratories all over the world as a tool for understanding enzymatic reactions

(c) it is the epitome of the useful chemical model, in that it correlates so much diverse data under a common rubric.

An enzyme, E, is a protein which catalyses the reaction of a *substrate*, S, to form a product, P. The Michaelis-Menten mechanism is

$$E + S = ES \quad \text{(step 1)}$$

$$ES \rightarrow E + P \quad \text{(step 2)}$$

Note that E is used up in the first step, but regenerated either in the reverse of the first step or in the second step. Thus E fits the definition of a catalyst: it does not appear in the overall reaction

$$S \rightarrow P$$

ES is an intermediate called the *enzyme-substrate complex*.

The physical picture of what is going on is that E and S first bind reversibly to form ES. There "something happens" to S that makes it easier for it to form P than would be possible without E present. Just what that "something" is depends on the particular reaction, and finding out is often the object of scientific investigation.

From step 2, we can write

$$\text{rate} = V = \frac{d[P]}{dt} = k_2[ES]$$

(It is common in enzyme kinetics to use the symbol V for the rate.) We use the steady-state approximation to find [ES]. (Note that k_{-1} stands for the rate constant for the reverse of step 1.)

$$\frac{d[ES]}{dt} = 0 = k_1[E][S] - k_{-1}[ES] - k_2[ES]$$

$$[ES] = \frac{k_1[E][S]}{k_{-1} + k_2}$$

This is well and good; unfortunately, although we can measure [S], all we know about [E] is $[E]_o$, the concentration of enzyme we started with. Some of $[E]_o$ exists as [E], the free enzyme, some as [ES], so that

$$[E] = [E]_o - [ES]$$

$$[ES] = \frac{k_1([E]_o - [ES])[S]}{k_{-1} + k_2}$$

which can be solved for [ES]

$$[ES] = \frac{k_1[E]_o[S]}{k_{-1} + k_2} - \frac{k_1[ES][S]}{k_{-1} + k_2}$$

$$[ES]\left(1 + \frac{k_1[S]}{k_{-1} + k_2}\right) = \frac{k_1[E]_o[S]}{k_{-1} + k_2}$$

$$[ES]\left(\frac{k_{-1} + k_2 + k_1[S]}{k_{-1} + k_2}\right) = \frac{k_1[E]_o[S]}{k_{-1} + k_2}$$

$$[ES] = \frac{k_1[E]_o[S]}{k_{-1} + k_2 + k_1[S]}$$

$$= \frac{[E]_o[S]}{\left(\dfrac{k_{-1} + k_2}{k_1}\right) + [S]}$$

$$= \frac{[E]_o[S]}{K_M + [S]}$$

From step 2,

$$V = k_2[ES] = \frac{k_2[E]_o[S]}{K_M + [S]}$$

This is the Michaelis-Menten rate equation, in which we have defined a new constant $K_M = (k_{-1}+k_2)/k_1$.

Let us consider a series of experiments in which V, the initial rate, is measured as a function of [S], substrate concentration, at constant $[E]_o$. At very low [S], the [S] in the denominator can be ignored, giving

$$V = \frac{k_2}{K_M}[E]_o[S] \qquad \text{(low [S])}$$

so that the reaction is first-order in [S]. At high [S], k_M can be ignored, so that

$$V = k_2[E_o] = V_{max} \qquad \text{(high [S])}$$

In other words, when [S] gets high, the reaction is zeroth-order in [S], and a maximum velocity is achieved.

To understand how this behavior comes about, consider how the expression for [ES] changes with [S].

$$[ES] = \frac{[E]_o[S]}{K_M} \qquad \text{(low [S])}$$

$$[ES] = [E]_o \qquad \text{(high [S])}$$

At very high [S] all the enzyme initially added is present in the form of ES; therefore increasing [S] cannot increase the rate: there is no more enzyme present to react with the added S. At very low [S], on the other hand, added molecules of S immediately form new molecules of ES, which can then react to form P; therefore V increases linearly with [S].

There are a number of ways to plot Michaelis-Menten data so as to obtain values of K_M and k_2 from the plot. Like many methods of plotting, they are all designed so as to linearize the data. One of the most useful methods is the "double reciprocal" or "Eadie-Hofstee" method. Taking the inverse of the Michaelis-Menten rate law

$$\frac{1}{V} = \frac{K_M + [S]}{k_2[E]_o[S]}$$

$$= \left(\frac{K_M}{k_2[E]_o}\right)\frac{1}{[S]} + \left(\frac{1}{k_2[E]_o}\right)$$

Thus, if 1/V is plotted against 1/[S], a straight line will result which has slope $K_m/k_2[E]_o$ and y-intercept $1/k_2[E]_o$. Dividing the slope by the intercept gives K_M, and k_2 can be calculated directly from the intercept, if $[E]_o$ is known.

Chapter 19 — Chemical Kinetics

Answers

19.1 Factors affecting the rates of chemical reactions and 19.2 The rate expression

1. (a) concentration/time
    (b) 1/time
    (c) +
    (d) +
    (e) 2, -1, 1

2. $\text{rate} = -\dfrac{1}{3}\dfrac{d[KI]}{dt} = \dfrac{d[C_2H_4]}{dt} = \dfrac{1}{2}\dfrac{d[KBr]}{dt} = \dfrac{d[KI_3]}{dt}$

3. n/V is molarity if V is in liters. So n/V = P/RT if R is in L·atm.

19.3 Determination of the order of reaction by the method of initial rate

1. (a) the rate, times the stoichiometric coefficient of the product
    (b) the initial rate, times the stoichiometric coefficient of the product

2. (a) 2, 1, -1, 2
    (b) rate = $k[A]^2[B]/[C]$
    (c) $k = 10^{-6}$

19.4 First-order reactions

1. 1/time
2. 3
3. (a) (dimensionless)
    (b) (dimensionless)
    (c) $[A]_o$
    (d) 0
    (e) 0
4. Plot ln[A] or log[A] against t. This will be a straight line if the reaction is first-order.
5. After $\tau_{.90}$ passes, $[A] = 0.10\,[A]_o$ so $\ln(0.10) = -k\tau_{.90}$
    $$\tau_{.90} = \dfrac{-\ln(0.10)}{k} = \dfrac{2.30}{k}$$

19.5 Second-order reactions

1. 1/(time)(concentration)
2. 4
3. Plot 1/[A] against t. This will be a straight line if the reaction is second-order.
4. As before, after $\tau_{.90}$, $[A] = 0.10[A]_o$. Then
    $(1/0.10[A]_o) - (1/[A]_o) = k\tau_{.90}$
    $9/[A]_o = k\tau_{.90}$
    $\tau_{.90} = 9/k[A]_o$
5. (a) [A]=0.05 at t=200s
    (b) [A]≈0.03 at t=400s

19.6 The temperature dependence of rate constant

1. increase
2. increases

3. increases
4. decreases
5. increases
6. (a) energy
   (b) energy
   (c) (dimensionless)
   (d) dimensionless)
   (e) 1/time
7. (a) decreases
   (b) decreases
   (c) increases
   (d) decreases
8. greater
9. (a) positive, tiny
   (b) positive, huge
10. A

19.7 Transition state theory and the activated complex

1. False
2. True
3. True
4. Endothermic, regardless of
5. (True), (True)

19.8 The relation between the mechanism of a reaction and kinetics data

1. rate-limiting step
2. an elementary step
3. Unlike for an overall reaction
4. may be
5. molecularity
6. (a) $k[A]$
   (b) $k[B]$
   (c) $k[A]^2$
   (d) $k[A]^2[B]$
7. true
8. false
9. (a) reactant
   (b) catalyst (regenerated at end)
   (c) intermediate
   (d) reactant
   (e) intermediate
   (f) product

Chapter 19 — Chemical Kinetics

19.9 Chain Reactions

1. ... a species with one or more unpaired electrons
2. ... with short-wavelength light and with heat
3. ... one intermediate is used up and another is generated

19.10 Reaction rates and equilibrium

1. the same value as
2. may not appear
3. False

CHAPTER 20: COORDINATION COMPOUNDS

Scope

Inorganic Chemistry

Inorganic chemistry has changed over the past few decades perhaps more than any other subfield of chemistry. The main focus of this change has been an increased interest in coordination compounds and the role they play in both industrial processes (as homogeneous catalysts) and in biological systems, where a metal is often found bound in the active site of some enzyme or other protein (iron in hemoglobin, magnesium in chlorophyl, zinc in carboxypeptidase). What has made this revolution in inorganic chemistry possible has been the descent of *crystal field theory* and *ligand field theory* from the esoteric domain of the theoretician to the laboratory of the working chemist.

Recall (Chapters 12 through 14) that quantum chemistry was developed by physicists in the 1930's, but that the theory they constructed provides powerful conceptual models that every chemistry student today can and must learn to use, whether or not he understands the theory in depth. The ideas behind our understanding of periodic properties, hybrid orbitals, VSEPR and so on are all quantum mechanical in origin. In the same way that quantum chemical ideas rationalize and predict the properties of small molecules, crystal and ligand field theory rationalize and predict the behavior of coordination compounds. Historically, crystal field theory considered how the energies of the orbitals of a central atom (such as a metal ion in a complex) would change if it were surrounded by a symmetrical array of point charges or dipoles; only electrostatic forces were considered. This symmetrical array is a simple model for, say, the two chloride ions ("point charges") and four water molecules ("dipoles") surrounding chromium in the complex ion $[Cr(H_2O)_4Cl_2]^+$. Such a model turns out to be *too* simple for many purposes, and ligand field theory is an elaboration which takes into account the covalent nature of the metal-to-ligand bond. In this Scope, when we wish to refer to the two theories as a class, we will use the term "ligand field theory".

Chapter 20 leads up to a discussion of these theories (Section 20.11), but first a number of basic ideas and observations about coordination compounds are presented. On the way, several topics (Lewis acids and bases, geometric and optical isomerism) are introduced which are important not only in coordination chemistry, but in other areas as well.

A *coordination compound* or *complex ion* consists of a central metal atom or ion surrounded by *ligands*, which may be anions, neutral species or (rarely) cations. The difference between a coordination compound and a complex ion is that a complex ion has an overall charge. The nature of the bond between the metal and the ligands in these species depends on the specific metal and ligands. Usually the terms "coordination compound" and "complex ion" are reserved for situations in which the metal has unoccupied orbitals of low energy, such as the d-orbitals of the transition metals (Section 13.9). In this situation, the bonds generally have appreciable covalent character, which results from an overlap of the low-lying unoccupied orbitals of the metal with occupied orbitals on the ligand; the special name, *coordinate bond*, is given to the typical case in which both electrons in the bond come from the ligand.* Even so, as we shall see, many of the properties of these species can be explained just on the basis of crystal field theory, which assumes no covalence.

The *coordination number* is the number of ligands surrounding a central ion. This term is also used when discussing crystal structure (Chapter 21), where it refers to the number of nearest neighbors an atom or ion in the crystal has. You should know what *monodentate*, *bidentate* and *tridentate* ligands are (from the Latin *dent*, meaning tooth). When more than one site of a multidentate ligand is coordinated to the same central ion, the species is called a *chelete*. When a multidentate ligand is shared between two or more central ions, the ligand is termed a *bridging ligand*.

Table 20.1 in the text lists a number of common ligands. Note that the same species has different names depending on whether it is a ligand or an independent entity: water=aquo, ammonia=ammino, chloride=chloro, carbon monoxide=carbonyl, etc. On a similar subject, though it is jumping ahead a bit, Section 20.6 gives a fairly complete set of rules for naming complex ions and coordination compounds. Some of the chief things to remember are that a complex anion name always ends with the metal name in Latin, followed by " -ate," and that coordination compounds and complex cations end with a metal

---

* Coordinate bonds are not restricted to coordination compounds. In the structure :C:::O: one of the bonds between C and O is coordinate, since O supplies both electrons. (C starts out with only four.)

name in English, with no suffix. In both cases, the oxidation state of the metal is given in Roman numerals, in parentheses, at the end of the name.

PtF_4^{2-} is tetrafluoroplatinate(II)

$Pt(NH_3)_4^{2+}$ is tetraamminoplatinum(II)

$Fe(H_2O)_6^{3+}$ is hexaaquoiron(III)

$Fe(CN)_6^{3-}$ is hexacyanoferrate (III)

The formation of a coordinate bond as we have defined it (one species donating an electron pair into an overlapping vacant orbital on another species) is an example of a *Lewis acid-base reaction*, where the electron-pair donor is the base, and the electron pair acceptor is the acid. For review, here is a summary of the Arrhenius, Bronsted and Lewis' concepts of acids and bases. Each successive definition extended the concept of acids and bases. For Bronsted, NH_3 is a base even when it reacts with HCl in the gas phase. For Lewis, NH_3 is a base even when it reacts with BF_3, which has no protons, but which accepts electrons.

The Lewis notion appears very different from the way we usually think of acids and bases, but in fact it is a logical extension of the Bronsted and Arrhenius ideas. When the salt of a small, highly charged cation — $AlCl_3$ or $FeCl_3$ for example — is dissolved in water, the solution becomes strongly acidic, even though these compounds are not acids in the Bronsted or Arrhenius sense. The acidity of these compounds can be explained by the fact that the free electron pairs of the oxygen in H_2O are

Concept	Acid	Base	Acid-Base Reaction
Arrhenius	gives excess H^+ in aqueous sol'n	gives excess OH^- in aqueous sol'n	$HCl(aq) + NH_4OH(aq)$ $\rightarrow NH_4Cl(aq) + H_2O$
Bronsted	donates proton	accepts proton	$HCl(g) + NH_3(g) \rightarrow NH_4Cl(s)$
Lewis	accepts electron pair	donates electron pair	$BF_3(g) + NH_3(g) \rightarrow BF_3NH_3(s)$

highly attracted to such cations, and especially to ions like Fe^{3+}, which have vacant d-orbitals, enabling the Fe-O bond to be strengthened by covalent contributions. This is a Lewis acid-base interaction, in which Fe^{3+} is the acid and H_2O the base.

According to the text, and historically, a Lewis acid must have available low-lying unoccupied orbitals which can accept an electron pair from (that is, form a coordinate covalent bond with) a base. By this definition, Fe^{3+} is a Lewis base, but Al^{3+} is not, since the latter has no low-lying d-orbitals; yet, as we have noted, $AlCl_3$ forms acidic solutions. In practice, the term "Lewis acid" is often applied to Al^{3+} as well as to Fe3+. Although not used in the text, the terms *hard acid* and *hard base* are used for species such as Al^{3+} and F^-, respectively, which interact mainly by ionic (Coulombic) interactions. *Soft acids* and *soft bases* are species such as Hg^{2+} and CO, respectively, whose interaction is largely covalent. In this language, Fe^{3+} is a softer acid than Al^{3+}. There is a rule of thumb that hard acids tend to form the strongest interactions with hard bases, and likewise for soft acids and bases. The concept of hard and soft acids and bases, which can be thought of as an elaboration of the Lewis theory, is due to the American inorganic chemist, R. G. Pearson.

Your text explains how the formation of the metal-oxygen bond causes protons to be released in solution. The Lewis theory explains this acidity easily, whereas the Bronsted and Arrhenius theories do not. Incidentally, if you read the **Perspective** in Chapter 15 of this study guide, you may recall that the acidity of the oxides of the elements was explained on the basis of oxidation state, formal charge and electronegativity. Now we can say that the strength of cations as Lewis acids goes up as the oxidation state, the formal change and the electronegativity all increase.

Let us now summarize some of the features of the chemistry of coordination compound and complex ions.

Labile and Inert Complexes

This refers to how quickly the species can exchange ligands. In a solution containing $Cu(NH_3)_4^{2+}$ and excess NH_3, ammonia molecules are being exchanged between the complex ions and the solution extremely quickly, whereas this is not true for $Cr(NH_3)_6^{3+}$. We say that $Cu(NH_3)_4^{2+}$ is labile, whereas $Cr(NH_3)_6^{3+}$ is inert. Labile complexes are generally in thermodynamic equilibrium with their surroundings, whereas inert complexes may not be. For example, consider the reaction

$$4H_2O + Cu(NH_3)_4^{2+} = Cu(H_2O)_4^{2+} + 4NH_3$$

If the ammonia concentration is high, $Cu(NH_3)_4^{2+}$ is stable, and since the equilibrium is fast (this is what lability means) the ion will be formed. If the NH_3 concentration is lowered, Le Chatelier's principle shifts the equilibrium to the right, and the tetraamminocopper(II) ion dissociates. In contrast, $Cr(NH_3)_6^{3+}$ once formed will persist even in solutions where free NH_3 is completely absent, because it is inert. Of course, given enough time, it too will dissociate, because under these conditions the complex is unstable.

As your text mentions, the two pure substances $[Cr(H_2O)_4Cl_2]Cl\cdot 2H_2O$ (green) and $[Cr(H_2O)_6]Cl_3$ (violet) have identical empirical formulas, yet the first will give two ions per molecule when dissolved in water, whereas the second will give four. Clearly, a technique which tells how many moles of ions a mole of substance gives in water solution will be a useful tool in coordination chemistry. *Colligative properties* (Chapter 6), such as freezing point depression, are obvious candidates for such techniques, since they depend only on the total concentration of particles in dilute solution. *Molar conductivity*, which goes up as the concentration of ions in solution rises, is another useful method.

Still another guide to structure, important in organic chemistry as well as here, is *isomer number*; this method is how the geometry of many structures was determined, before *x-ray diffraction* (which, though laborious, gives precise depictions of the locations of atoms in molecules) became routine. To illustrate isomer number, consider two conceivable geometries of a planar complex with a metal ion and four ligands, a square one and a rectangular one. For the compound MX_2Y_2 the square arrangement gives two possible isomers (the X's either along on edge (*cis*) or on a diagonal (*trans*)); the rectangular arrangement gives three (the X's could be along a short edge or a long edge or on a diagonal). If only two isomers are found, the rectangular arrangement can be ruled out (it does not, in fact, occur in nature). If only one is found, the square arrangement can also be ruled out, and the compound must have the non-planar tetrahedral configuration.* By extended reasoning of this sort, it was deduced that coordination compounds with formula ML_6 are octahedral, rather than, say, hexagonal planar or triangular prismoidal.

Whether a 4-coordinate complex has *square-planar* or *tetrahedral* geometry, and whether a 5-coordinate complex has *trigonal-bipyramidal* or *square-based-pyramidal* geometry, depends on VSEPR considerations (Section 14.3); square-based pyramidal and square-planar geometries are really octahedra with one and two positions, respectively, occupied by lone pairs.

Linkage isomerism is a form of geometric isomerism which depends on the orientation of a given ligand; for example, the nitrite ion (NO_2^-) will become a nitro ligand if the nitrogen points toward the metal, and a nitrito ligand if the oxygen points toward the metal.

Optical-isomerism is harder to visualize than the *geometric isomerism* we have just been discussing. It arises when a compound's mirror image is not superimposable on itself, in the same way that a left and right glove are not superimposable. This condition is equivalent to the absence of a *plane of symmetry*. One member of such a mirror-image pair is called an *enantiomer*. The two enantiomers rotate the plane of any *plane-polarized light* that they encounter equal amounts, but in opposite directions. The two enantiomers will also interact differently with a pure enantiomer of the same or another compound, in the same way that a left glove interacts differently with a left and a right hand, and in the same way that a right-handed screw interacts differently with a left- and a right-handed nut. The presence of optical isomerism can be helpful in determining geometries; for example, a complex with the formula MWXYZ, where M is the central metal, will not exhibit optical isomerism if it is square-planar, but will if it is tetrahedral.

---

* As you can imagine, this technique is laborious. If only one MX_2Y_2 compound is found, it does not rule out the possibility that another might be found under different reaction conditions. Only if a given M *always* gives a single MX_2Y_2 compound, under many reaction conditions and choices of X and Y, can one infer from this technique that 4-coordinate M complexes are tetrahedral.

Stoichiometry

The EAN (*effective atomic number*) rule sometimes, but not always, predicts the coordination number of a central ion. The idea is that enough ligands, each donating two electrons, get coordinated to a central ion to bring the total number of electrons up to that of the next bigger inert gas configuration. For example, Zn^{++} has the electronic configuration $[Ar]\,d^{10}$. In order to bring this up to the configuration of $[Kr]$, eight electrons must be added, to fill the s and p orbitals. This can be accomplished by adding four ligands. Only when the ligand is CO, however, can you be fairly certain that the rule will hold.

Magnetic Moments

Atoms and ions are *diamagnetic* if they do not have unpaired electrons, and *paramagnetic* if they do. Diamagnetic substances are repelled by magnetic fields, whereas paramagnetic substances are attracted. For the latter, the greater the number of unpaired electrons, the stronger the attraction; thus, the strength of the paramagnetism is a measure of the number of unpaired electrons. As we shall see, this number can be predicted from ligand field theory.

Color

Atomic spectra were discussed in Chapter 12. Recall that a substance will absorb light of frequency ν if there is available to it a quantum state with energy $\Delta E = h\nu$ above that of its initial state.* Most electronic transitions take place in the near or far ultraviolet; ν is greater here than for light in the visible region. For many coordination compounds, however, ν is in the visible, meaning that ΔE is smaller than it is for most other materials. Generally, the lowest-energy electronic transition excites an electron in the highest occupied orbital to the lowest previously unoccupied orbital. As we shall see, ligand field theory shows why transition metal complexes have such low-lying orbitals available. It also provides an

---

* Several units are used to express this value ΔE. The frequency, ν, in [1/seconds], can be used, as can an *energy* in joules. In addition, the *wavelength* of the light, $\lambda = c/\nu$ is sometimes used, as is the *wave-number*, $\bar{\nu} = 1/\lambda$, commonly in units of cm^{-1}. The wave-number is the number of wavelengths that will fit into a length of one centimeter. We have
$$\Delta E = h\nu = \frac{hc}{\lambda} = hc\bar{\nu}$$

understanding of how ΔE between the highest occupied and lowest unoccupied orbitals varies with the nature of the ligand. Since this ΔE is directly related to the frequency of the light absorbed, ligand field theory explains the colors of transition metal complexes. It also reveals a deep connection between this ΔE and the number of unpaired electrons; that is, between the two seemingly unrelated measurable properties, color and paramagnetism.

Crystal Field Theory

We will start by using crystal field theory to explain certain observations about the color and paramagnetism of transition metal complexes, and then we will show where this model is inadequate, and the more complicated ligand field theory must be used.

The basic point of view of crystal field theory is that the orbitals on the metal start out as described in Chapter 13 for an isolated cation, but are perturbed (that is, somewhat altered) by the presence of negative charges (or negative ends of dipoles) surrounding it. These, of course, represent the ligands. Crystal field theory does not involve covalent bonding, and is not concerned with exactly why the ligands adopt their observed geometries, but rather with what happens to the orbitals on the metal once the ligands are there.*

Consider a transition metal with its d-orbitals, surrounded by an octahedral array of negative charges. In crystal field theory the d-orbitals are strictly non-bonding (Chapter 14), since the theory does not allow for covalent interactions. As Figure 20-16 in the text shows, the $d_{x^2-y^2}$ and the d_{z^2} orbitals point toward the ligands, and the d_{xy}, d_{xz} and d_{yz} orbitals point between the negatively-charged ligands. Electrons in $d_{x^2-y^2}$ or d_{z^2} orbitals will therefore have higher energy than electrons in d_{xy}, d_{xz} or d_{yz} orbitals, simply for Coulombic reasons. Thus the orbital *degeneracy*, (that is, the energetic equality) of the d-orbitals in the isolated metal ion is *lifted* (that is, removed) in the *octahedral field* of the negative ligands. The $d_{x^2-y^2}$ and d_{z^2} orbitals in an octahedral complex are thus less stable than the d_{xy}, d_{xz} and d_{yz}

---

* One can stay within crystal field theory and say that the metal attracts the ligands by Coulombic forces, and that the most stable arrangement of six negative charges around a positive charge is an octahedron and that the most stable arrangement of four negative charges around a positive charge is a tetrahedron, but not why some four coordinate complexes are square planar and some tetrahedral.

orbitals (text, Figure 20.17). The energy separation between these two sets of orbitals is called Δ_o ("o" for "octahedral"), the low-lying three orbitals are termed t_{2g}, and the two that are raised in energy are called the e_g orbitals. Since in crystal field theory the energy holding the complex together is due entirely to the coulombic attraction between the negative or dipolar ligands and a spherically symmetric positive charge centered at the cation, this d-orbital splitting cannot affect the total energy: the total amount of energy the e_g's are raised* must equal the total amount that the t_{2g}'s are lowered. This causes the two e_g's to be raised by $0.6\Delta_o$ and the three t_{2g}'s to be lowered by $0.4\ \Delta_o$, since $2(0.6\Delta_o)-3(0.4\Delta_o) = 0$. Geometries other than octahedral result in different splitting patterns than this one; however, we will consider only octahedral complexes.

What are the implications of this splitting? First, recall (Chapter 13) that *Hund's rule* states that when electrons are added to a set of orbitals of the same energy, first a single electron is placed in each orbital, then a second, with opposite spin, is added to each one; since electrons repel each other, it is energetically advantageous to avoid placing more than one in the same orbital, if possible. As we built up the periodic chart, we almost always filled one degenerate (equivalent) set of orbitals before starting to fill the next higher set; there were exceptions, however, as in chromium and copper. You can imagine that if two sets of orbitals are very close in energy, Hund's rule would apply to both sets together; that is, both sets might be filled with a single electron per orbital, before returning to the lower-lying set to add the second electrons. On the other hand, if one set were much more stable than the other, then it would be advantageous to fill it completely before any electrons were added to the set of higher energy.

This is exactly what happens (text, Figure 20.19). For complexes with low Δ_o, d-electrons go, one at a time, into both the t_{2g} and the e_g orbitals before a second electron is added to any orbital. If Δ_o is high, first the three t_{2g} orbitals are completely filled, then the two e_g's. If there are between four and seven d-electrons, then complexes with low values of Δ_o will have more unpaired electrons than complexes with high Δ_o, and will therefore exhibit stronger paramagnetism.

---

* Above that of a hypothetical spherical ligand field (text, Figure 20.17).

Δ_o is, in general, small compared with the splitting between successive subshells in a free atom; that is, it takes less energy to excite an electron from the t_{2g} to the e_g orbital in an octahedral complex than to excite one from, say, the 1s to the 2s orbital in something like a hydrogen atom. This is why coordination complexes tend to absorb light in the visible, rather than the ultraviolet region. Naturally, the greater the value of Δ_o, the shorter the wavelength absorbed. We see the complementary colors to the ones absorbed, so that as Δ_o changes the colors we see also change. We saw in the previous paragraph that high values of Δ_o are also associated with low-spin complexes, and this is the connection, alluded to earlier, that crystal field theory makes between paramagnetism and color.

The Spectrochemical Series and Ligand Field Theory

In crystal field theory, Δ_o depends only on coulombic repulsion between the ligands and the d-electrons. Anything that brings the ligands closer to the d-orbitals should increase Δ_o. Two examples are a greater charge on the cation and a smaller diameter for an anionic ligand. These trends are sometimes found; for example, given the same metal in the same oxidation state, Δ_o tends to increase as the ligand goes from I^- to Br^- to Cl^- to F^-. On the other hand, consider the *spectrochemical series*

$$I^- < Br^- < Cl^- < F^- < OH^- < C_2O_4^{2-} \approx H_2O < NCS^- < py\,NH_3 < en < bipy < NO_2^- < CN^- < CO$$

ligands toward the right of this list cause greater values of Δ_o than those toward the left. Certainly, nothing we have said so far would lead us to believe that H_2O should produce a greater ligand field strength than F^-, or that the neutral molecule, CO, should produce the greatest one of all.

The fact is that crystal field theory does not predict these observations, because it does not take into account covalent interactions. The theory that does, ligand field theory, is really a molecular orbital treatment similar to that described for diatomics in Chapter 14, but more complicated. The species toward the right of the spectrochemical series all have "loose" lone pairs of electrons; for example, the carbon of CO has a formal charge of -1, making its lone pair very reactive. The halides also have formal charges of -1, but since they are highly electronegative, their electron clouds are tightly bound. The lone pair on carbon monoxide can form a coordinate covalent σ-bond with orbitals directed toward it from the metal. In addition, CO can *accept* electrons into its antibonding π-orbitals from the metal's d_{xy}, d_{xz} and d_{yz} orbitals. This process, called *π-back-bonding*, increases Δ_o by lowering the energy of these d-orbitals. Recall (Chapter 14) that putting electrons into the π-antibonding orbital should lower the bond order and weaken and lengthen the CO bond. Crystallographic and spectroscopic studies show that this does occur, lending support to this explanation of CO's remarkable effects as a ligand.

Chapter 20   Coordination Compounds

Questions

20.1 Ligands and the coordinate bond and
20.2 The coordination number

1. For each of the complex ions given, write the equations for its formation from the metal and the ligands. Where an abbreviation is used, write out the structure of the ligand

    Example: $Zn(en)_2^{2+}$

    $$Zn^{2+} + 2NH_2CH_2CH_2NH_2 \rightarrow Zn(en)_2^{2+}$$

    (a) $Zn(pn)_2^{2+}$

    (b) $Cr(NH_3)_6^{3+}$

    (c) $CoCl_6^{3-}$

    (d) $Cu(H_2O)_2^{2+}$

    (e) $Fe(CN)_6^{3-}$

    (f) $Fe(CN)_6^{4-}$

    (g) $Ca(EDTA)^{2-}$

    (h) $Ag(py)_2^{+}$

Chapter 20   Coordination Compounds

(j) Pt(acac)$_2$

(k) AgF$_4^-$

(l) Ru(CO)$_5$

(m) Rh(bipy)$_3^{3+}$

2. For each of the species above, fill in the table, giving the name of the ligand, the name of the metal, the coordination number and the number of the d-electrons

		Metal	Ligand	Coordination Number	d-Electrons
(Ex)	Zn(OH)$_4^{2-}$	zinc(II)	hydroxo	4	10
(a)	Zn(pn)$_2^{2+}$				
(b)	CrNH$_{36}^{3+}$				
(c)	CoCl$_6^{3-}$				
(d)	CuH$_2$O$_2^{2+}$				
(e)	Fe(CN)$_6^{3-}$				
(f)	Fe(CN)$_6^{4-}$				
(g)	Ca(EDTA)$^{2-}$				
(h)	Ag(py)$_2^+$				
(j)	Pt(acac)$_2$				
(k)	AgF$_4^-$				

20-13

(l) $Ru(CO)_5$

(m) $Rh(bipy)_3^{3+}$

20.3 Lewis acids and bases

1. Give the definition of an acid according to

    (a) Arrhenius

    (b) Bronsted

    (c) Lewis

2. Give the definition of a base according to

    (a) Arrhenius

    (b) Bronsted

    (c) Lewis

Chapter 20 Coordination Compounds

3. Any reaction in which a coordinate bond is formed is a Lewis acid-base reaction. (true, false)

4. "Other things being equal" (which they rarely are), how would you expect each of these to affect the strength of a cation as a Lewis acid:

   (a) greater charge

   (b) larger size

   (c) availability of law lying orbitals

5. The stronger the metal-oxygen bond in a hydrated cation, the *more* acidic solution

20.4 Inert vs. labile complexes

1. "Stability" and "inertness" are terms sometimes used to describe complexes. Explain the difference between them, using a single sentence with the word "whereas" in the middle.

2. Which of these can the predominent species in solution never be?

   _____ stable and inert

   _____ stable and labile

   _____ unstable and inert

   _____ unstable and labile

Chapter 20 — Coordination Compounds

3. Why does the addtion of acid break up a labile metal-ammi= .....

4. You have seen in the text that acid breaks up a labile complex where the ligand is NH_3. What would you add to break up a labile complex involving these following ligands?

    (a) OH^-

    (b) Cl^-

5. Adding acid to a solution of the inert complex $Cr(NH_3)_6^{3+}$ makes the complex unstable. (true, false)

20.5 Evidence for the existence of complex ions

1. Consider aqueous solutions of the salts $[Cr(H_2O)_4Cl_2]Cl$ (A) and $[Cr(H_2O)_6]Cl_3$ (B). The complex cations are inert.

    (a) On adding excess Ag^+, how many moles of AgCl will be precipitated per mole of each salt?

    (b) Which solution will exhibit the greater freezing-point depression?

    (c) Which solution will exhibit the greater molar conductivity?

2. Answer the same questions for aqueous solutions of these salts, but this time assume the salts (A and B) are labile.

    (a) On adding excess Ag^+, how many moles of AgCl will be precipitated per mole of each salt?

    (b) Which solution will exhibit the greater freezing-point depression?

Chapter 20 — Coordination Compounds

(c) Which solution will exhibit the greater molar conductivity?

3. Still assuming the salts A and B are labile, suppose a single solution contains both salts in equilibrium. What might you add to the solution to increase the concentration of A at the expense of B?

20.6 Nomenclature

1. Name each of the complex ions given in Question 20.1.1.

   (a) $Zn(pn)_2^{2+}$

   (b) $Cr(NH_3)_6^{3+}$

   (c) $CoCl_6^{3-}$

   (d) $CuH_2O_2^{2+}$

   (e) $Fe(CN)_6^{3-}$

   (f) $Fe(CN)_6^{4-}$

   (g) $Ca(EDTA)^{2-}$

   (h) $Ag(py)_2^+$

   (j) $Pt(acac)_2$

   (k) AgF_4^-

   (l) $Ru(CO)_5$

   (m) $Rh(bipy)_3^{3+}$

2. Write the formula for each of the following

   (a) hexaamminocobalt(III)

   (b) tetraaquoplatinum(IV)

   (c) dicyanoargentate(I)

   (d) lithium ethylenediaminetetraacetatoplumbate(II)

Chapter 20 Coordination Compounds

(e) tris (2,2'-bipyridine)iron(III)

(f) sodium dicholorodifluoroplatinate (II)

(g) biscyclopentadienyliron (III) nitrate

(h) dichlorotetraaquochromium (III) chloride

(j) hexaaquochromium (III) chloride

20.7 The geometry of coordination compounds; isomers

1. Fill in this table with the number of geometric isomers possible with the given geometry and stoichiometry

Stoichiometry	Square-Planer	Tetrahedral
(a) MX_4		
(b) MX_3Y		
(c) MX_2Y_2		
(d) MX_2YZ		
(e) $MXYZW$		

2. Do the same for this table

Stoichiometry	Octahedral	Prismatic
(a) MX_6		
(b) MX_5Y		
(c) MX_4Y_2		
(d) MX_3Y_3		

3. Do the same for this table

Stoichiometry	Square-based Pyramid	Bipyramid

(a) MX_5

(b) MX_4Y

(c) MX_3Y_2

(d) MX_2Y_2Z

4. For square-planar MX_2Y_2, draw the *cis* and *trans* isomers.

5. For octahedral MX_4Y_2, draw the *cis* and *trans* isomers.

6. Give an example of linkage isomerism.

Chapter 20 Coordination Compounds

20.8 Optical isomerism

1. Which of the following every day objects is an enantiomer? (that is, not superimposable on its mirror image).

    — (a) a coffee cup

    — (b) a shoe

    — (c) a tee-shirt

    — (d) a sports jacket

    — (e) a screw

    — (f) a nail

    — (g) a pair of shoes standing next to each other

2. Which of the following chemical structures can exist in enantiomeric forms?

    (a) square-planar MXYZW

    (b) tetrahedral MXYZW

    (c) *cis* $Co(NH_3)_4Br_2^+$

    (d) *trans* $Co(NH_3)_4Br_2^+$

    (e) *cis* $Co(bipy)_2Br_2^+$

    (f) *trans* $Co(bipy)_2Br_2^+$

20.9 The effective atomic number (EAN) rule

1. For each of the complexes in Question 20.1.1, tell whether the EAN rule is satisfied (a) $Zn(pn)_2^{2+}$

    (b) $Cr(NH_3)_6^{3+}$

    (c) $CoCl_6^{3-}$

    (d) $Cu(H_2O)_2^{2+}$

    (e) $Fe(CN)_6^{3-}$

(f) $Fe(CN)_6^{4-}$

(g) $Ca(EDTA)^{2-}$

(h) $Ag(py)_2^+$

(j) $Pt(acac)_2$

(k) AgF_4^-

(l) $Ru(CO)_5$

(m) $Rh(bipy)_3^{3+}$

2. If each ligand donates two electrons to a complex, the EAN rule can never be satisfied for a metal ion with an odd number of electrons. (true, false)

20.10 Magnetic moments of transition metal complexes

1. Tell whether each of the following atoms or ions is diamagnetic or paramagnetic.

   (a) Zn^{2+}

   (b) Na^0

   (c) Na^+

   (d) Fe^{3+}

   (e) Fe^{2+}

   (f) Co^{3+}

   (g) Pt^{2+}

   (h) Mn^{3+}

   (j) Ti^{3+}

2. The greater the number of unpaired electrons, the *lower* the magnetic moment.

Chapter 20   **Coordination Compounds**

3. The number of unpaired electrons in a complex is *at least* as great as that in the free metal ion.

4. Looking through a colored solution, the color we see is the color absorbed by the solution.

5. If a substance absorbs light in the visible region, it will appear colored (true, false)

20.11 Theories of bonding in coordination complexes

1. According to crystal field theory, the d-orbitals that point toward the ligands become *destabilized*.

2. In an octahedral field, there are *three* e_g orbitals and they are *less* stable than the t_{2g} orbitals.

3. What is the degeneracy of

   _____(a) the d-orbitals in a free ion

   _____(b) the t_2g orbitals in an octahedral field

   _____(c) the e_g orbitals in an octahedral field

4. Δ_o is called the _____

5. According to crystal field theory, if more electrons are added to the d-orbitals in a complex, Δ_o (increases, decreases, remains the same).

6. If the ligand field gets stronger, Δ_o (increases, decreases, remains the same).

7. Hund's rule is sometimes violated in complex species. (true, false)

8. The Pauli principle is sometimmes violated in complex species. (true, false)

9. If Δ_o is large, the number of unpaired electrons may be *greater* than for the free metal ion.

10. Give the number of unpaired electrons for the following situations:

Number of d-electrons	Large Δ_o	Small Δ_o
2		
4		
5		
7		
10		

12. Predict the number of unpaired electrons in each of the following species

    (a) CoF_6^{3-}

    (b) $Co(CN)_6^{3-}$

    (c) $Fe(Cl)_6^{4-}$ (d) $Fe(bipy)_3^{2+}$

13. Which would you expect to be more labile, CoF_6^{3-} or $Co(CN)_6^{3-}$?

14. $Co(H_2O)_3^{3+}$ absorbs greenish-yellow light. It therefore is colored reddish-violet. When the ligand is changed to NH_3, the color becomes (redder, more violet)

Chapter 20 — Coordination Compounds

Answers

20.1 Ligands and the coordinate bonds and
20.2 The coordination number

1. (a) $Zn^{2+} + 2NH_2CH_2CH(CH_3)NH_2 \rightarrow Zn(pn)_2^{2+}$
   (b) $Cr^{3+} + 6NH_3 \rightarrow Cr(NH_3)_6^{3+}$
   (c) $Co^{3+} + 6Cl^- \rightarrow CoCl_6^{3-}$
   (d) $Cu^{2+} + 2H_2O \rightarrow Cu(H_2O)_2^{2+}$
   (e) $Fe^{3+} + 6CN^- \rightarrow Fe(CN)_6^{3-}$
   (f) $Fe^{2+} + 6CN^- \rightarrow Fe(CN)_6^{4-}$
   (g) $Ca^{2+} + [(OOCCH_3)_2NCH_2CH_2N(CH_3COO)_2]^{4-} \rightarrow Ca(EDTA)^{2-}$

   (h) $Ag^+ + 2 \quad \rightarrow Ag(py)_2^+$

   (j) $Pt^{2+} + 2CH_3C(O)CH = C(O)CH_3^- \rightarrow Pt(acac)_2$
   (k) $Ag^{3+} + 4F^- \rightarrow AgF_4^-$
   (l) $Ru + 5CO \rightarrow Ru(CO)_5$

   (m) $Rh^{3+} + 3 \quad\quad \rightarrow Rh(bipy)_3^{3+}$

2.

	Metal	Ligand	Coordination Number	d-Electrons
(a)	zinc(II)	propylenediamine	4	10
(b)	chromium(III)	ammine	6	3
(c)	cobalt(III)	chloro	6	6
(d)	copper(II)	aquo	2	9
(e)	iron(III)	cyano	6	5
(f)	iron(II)	cyano	6	6
(g)	calcium(II)	ethylenediaminetetraacetato	6	0
(h)	silver(I)	pyridine	2	10
(j)	platinum(II)	acetoacetonato	4	8
(k)	silver(III)	fluoro	4	8
(l)	ruthenium	carbonyl	5	8
(m)	rhodium(III)	bypyridine	6	6

20.3 Lewis acids and bases

1. (a) creates extra H^+ in water solution
   (b) donates proton
   (c) accepts electron pair

2. (a) creates extra OH^- in water solution
   (b) accepts proton
   (c) donates electron pair

3. True

4. (a) increases strength
   (b) decreases strength
   (c) increases strength

20-24

Chapter 20 — Coordination Compounds

5. True

20.4 Inert vs. labile complexes

1. If a complex is stable then it will predominate at equilibrium, whereas if it is inert it will persist for relatively long times even under conditions where it is unstable.

2. unstable and labile

3. $H^+ + NH_3 \rightarrow NH_4^+$, so that $[NH_3]$ decreases. The complex then dissociates by Le Chatelier's principle.

4. (a) H^+
   (b) Ag^+

5. True

20.5 Evidence for the existence of complex ions

1. (a) A=1, B=3
   (b) B
   (c) B

2. (a) A and B: 3
   (b) same
   (c) same
   (The reason for (b) and (c) is that $[H_2O]$ is so high that equilibrium will greatly favor B, by Le Chatelier's principle; therefore both will have the value characteristic of B.)

3. Cl^-

20.6 Nomenclature

1. (a) bis(propylenediamine)zinc(II)
   (b) hexaamminechromium(III)
   (c) hexachlorocobaltate(III)
   (d) diaquocopper(II)
   (e) hexacyanoferrate(III)
   (f) hexacyanoferrate(II)
   (g) ethylenediaminetetraacetatocalcate(II)
   (h) dipyridinesilver(I)
   (j) diacetoacetonateplatinum
   (k) tetrafluoroargentate(III)
   (l) pentacarbonylruthenium
   (m) tris(bipyridine)rhodium(III)

2. (a) $Co(NH_3)_6^{3+}$
   (b) $Pt(H_2O)_4^{4+}$
   (c) $Ag(CN)_2^-$
   (d) $Li_2[Pb(EDTA)]$
   (e) $Fe(bipy)_3^{3+}$
   (f) $Na_2[PtCl_2F_2]$
   (g) $[Fe(C_5H_5)_2]NO_3$
   (h) $[CrCl_2(H_2O)_4]Cl$
   (j) $[Cr(H_2O)_6]Cl_3$

20.7 The geometry of coordination compounds; isomers

1.
   Square-Planar   Tetrahedral

(a)	1	1
(b)	1	1
(c)	2	1
(d)	2	1
(e)	1	1

2.

	Octahedral	Prismatic
(a)	1	1
(b)	1	1
(c)	2	3
(d)	2	3

3.

	Square-based Pyramid	Trigonal Bipyramid
(a)	1	1
(b)	2	2
(c)	2	3
(d)	6	5

4.

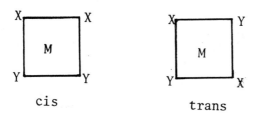

5.

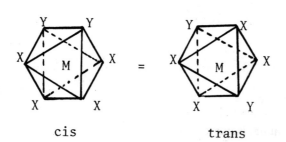

Chapter 20 — Coordination Compounds

6. SCN⁻ is thiocyanato if it links through S, isothiocyanato if it links through N.

20.8 Optical isomerism

1. (b)
   (d) because of the way it buttons (woman's is mirror image of man's)
   (e)
   Not (g), because contains a plane of symmetry between the shoes.

2. (b)
   (f)

20.9 The effective atomic number (EAN) rule

1. Rule is satisfied for (a), (c), (f), (l), (m)
2. True

20.10 Magnetic moments of transition metal complexes

1. Following are paramagnetic: Na^0, Fe^{3+}, Ti^{3+}

   Remainder are diamagnetic

2. True
3. at most
4. transmitted (or not absorbed)
5. True

20.11 Theories of bonding in coordination complexes

1. True
2. two, True
3. (a) 5-fold
   (b) 3-fold
   (c) 2-fold
4. octahedral crystal-field splitting
5. remains the same
6. increases
7. True
8. False
9. fewer
11.

Number of d-electrons	Large Δ_o	Small Δ_o
2	2	2
4	2	4
5	1	5
7	1	3
7	1	3
10	0	0

12. (a) 4
    (b) 0

     (c) 4
     (d) 0
13. CoF_6^{3-} (because of occupied e_g orbitals)
15. redder (because Δ_o goes up, absorbed light shifts into green, and we see the complement, red)

CHAPTER 21: PROPERTIES AND STRUCTURES OF METALLIC AND IONIC CRYSTALLINE SOLIDS

Scope

Types of Crystalline Solids

What *metallic* and *ionic crystals* have in common is the fact that the entities which occupy the sites on the *crystal lattice* are atoms or ions, rather than, for example, molecules, and furthermore these entities interact strongly with their near, and to some extent, distant, neighbors in a manner generally dependent on distance, but not direction. They can be viewed as spheres attracting or repelling other spheres. For ionic crystals the force has a particularly simple form: it is purely Coulombic. For metallic crystals a force law can not be written down in a simple manner, since the attraction depends on properties of all the atoms taken together, not on a sum of pairwise Coulombic interactions. These features — strong forces independent of bond angle — are in contrast to what is exhibited by *molecular crystals* and by *covalent network crystals*.

In molecular crystals the entities geometrically arranged on the lattice are molecules, which have shape. Within each molecule the bonding is covalent, and depends strongly on direction; for example, the H-O-H angle of water is 105°. The molecules may interact by means of hydrogen bonding and dipole-dipole forces, which are directional but of rather low strength, and by London dispersive forces, which are very weak in comparison. Therefore molecular crystals tend to be low-melting and soft, especially when the molecules are non-polar.

The covalently bonded network crystals are intermediate between the ionic and the molecular crystals. Atoms occupy the lattice sites, but they are bonded to their neighbors covalently, and exhibit bond angles typical of the atoms in question. For example, diamond exhibits the 109° angle typical of sp^3 carbon, and graphite the 120° angle typical of sp^2 carbon. The arrangement of neighbors around a given atom results from these bond angles. In metallic and ionic crystals, in contrast, the spacial arrangement of the atoms is often due to purely geometric factors, such as the way spheres pack in space.

Chapter 21 — Properties and Structures of Metallic and Ionic Crystalline

Crystalline and Amorphous Solids

Crystalline solids in general must be distinguished from *amorphous solids* or *glasses*. In a liquid, the molecules are constantly in motion, as they are in a gas, but the attractive forces between the molecules holds the mass of liquid together. There may be a certain amount of "short-range order" in a liquid — for example, an average water molecule in the liquid is surrounded by about four other water molecules, due to hydrogen bonding — but there is no long-range lattice structure. The positions of the atoms are largely random. A glass is a substance in which these random positions have been "frozen in", and the random motion of the molecules has been largely stopped. This is in contrast to the crystalline solid, in which the order is long-range: if I "sit" on an atom in a crystal and see another atom at a lattice site a distance D away, I know that there will be another atom lying on that line a distance 2D away, and still another 3D away, and so on. This is not true for a liquid or a glass.

How can one tell if a given apparently solid substance is a crystal or a glass?* On the macroscopic level, crystal properties are *anisotropic* (different in different directions). This is clearly manifested by the shapes of crystals that grow by nucleation (e.g., snowflakes), by the tendency of crystals to crack or cleave along planes, which, moreover, make constant interfacial angles with each other, and by the fact that mechanical properties, such as the compressibility, and electrical properties, such as the conductivity, generally vary depending on which faces of the crystal they are measured between. This is not true for amorphous solids; these tend to be *isotropic* (exhibiting the same properties in all directions).

Furthermore, the melting of a crystal is a true phase transition, characterized by a temperature, the melting point, where the solid and liquid are in equilibrium. To make a liquid out of a crystalline solid at the melting point, one adds a well-defined quantity, the heat of fusion, and the process is isothermal. In contrast, a glass really is, thermodynamically, a *supercooled liquid*. Most liquids will form glasses if they are cooled sufficiently quickly. What happens is that the viscosity increases so rapidly that the molecules slow down almost completely before they can rearrange into the energetically more favorable positions

---

* Incidently, the fine glassware sold as "crystal" or "lead crystal" is actually not crystalline at all, but rather a particular composition of silicate glass.

characteristic of a crystal lattice. On heating, most glasses will simply become more fluid, eventually becoming liquids, rather than melt at a constant temperature, and there is no well-defined heat of fusion.

On a microscopic level, the regularity of the crystalline arrangement manifests itself in very well-defined x-ray diffraction patterns, which vary with the relative orientation of the crystal and the beam.

Crystal Defects

It is said that in this world nothing is perfect. Be that as it may, it is certainly true of crystals; in fact, *crystal defects* are responsible for some of the most useful and interesting aspects of these materials. (It is possible that this is also true of the world.) The properties of semiconductors, which are used to make transistors and the "chips" and "wafers" in the integrated circuits in our radios and calculators, are carefully tailored by introducing substitutional impurities, small amounts of a foreign atom or ion, into the lattice sites of an otherwise pure crystal. Some non-stoichiometric compounds can be viewed as pure crystals with substitutional impurities; for example, $Fe_{0.9}S$ can be viewed as FeS with some sites occupied by Fe(III) rather than Fe(II). In order to balance charge, *vacancies* must occur at other iron sites. Other non-stoichiometric compounds exhibit *interstitial defects*; for example, in the graphite intercalation compounds, metal ions take up positions between the planes of the graphite crystal.

We will now summarize, in sequence, crystal structure (that is, geometry), x-ray diffraction, which is used to deduce crystal structure, and crystal energetics (that is, bonding). The text discusses these three topics in an interleaved fashion; the following table should help you correlate this summary with the treatment in the text.

Topic	*Sections in the Text*
Structure (geometry)	21.4, 21.5, 21.6
x-ray diffraction	21.6
energetics (bonding)	21.4, 21.8, 21.9

The summary of structure will be brief here, since there is a **Perspective** which discusses the most common structures in some detail.

Structure

You should know what a *unit cell* is, and what a *lattice* is. You may view unit cells as bricks which can fill space when packed together; in two dimensions they would be tiles which would have to fill the plane without spaces between them. Furthermore, they have to be able to fill the plane in the same angular orientation: to get one unit cell to superimpose on another it will have to be translated, but it may not be rotated. This requires that the sides of the cells come in parallel pairs. In practice, parallelepipeds (the three-dimensional analogs of parallelograms) are used. The unit cell is the smallest such unit that can generate the entire lattice. The choice of unit cell is not always unambiguous.

Although your text lists the seven crystal systems (Table 21.1) and illustrates the fourteen Bravais lattices (Figure 21.14), most instructors in freshman courses only expect students to know a few of these by heart. (Make sure you find out if this is true of yours!) The most important structures are described in the Perspective.

X-ray Diffraction

To understand *Bragg's law* you need first of all to understand *constructive* and *destructive interference* of light. The basic idea is simple: light "looks like" a sine wave. At any instant in time, along a light beam emitted from a small or "point-" source, a snapshot of the beam, if we could take one (we can't), would reveal that the electric field along the beam points up, goes through zero, points down, goes through zero, points up, etc., in a sinusoidal manner. The wavelength, λ, of the light is the repeat distance of the sine wave. *Monochromatic light* is light all with the same wavelength, and we will assume this of the beam in this discussion.

Now, suppose that by some system of mirrors (for example) the beam is split into two beams, one of which is bounced around by more mirrors, and then recombined with the first. Let Δ be the difference in the path length of the two beams. If $\Delta=\lambda$, or one wavelength, it should be clear that the peaks and valleys of the two beams will line up. The positive peaks of the two beams line up and the electric fields add, and the same is true of the negative peaks, or throughs. The resultant beam will have the same intensity as the original, unsplit beam. The same will be true if $\Delta=2\lambda$, 3λ or $n\lambda$, where n is any integer. This is constructive interference. On the other hand, if $\Delta=\lambda/2$, once the beams are recombined the peaks of one will line up with the *valleys* of the other. The electric fields still add, but they add vectorially: one is positive and one is negative, so they cancel out. The resultant beam has zero intensity. This is destructive interference, and will also occur if $\Delta=3\lambda/2$, $5\lambda/2$, or $(n + 1/2)\lambda$, where again n is any integer. If Δ is between $\lambda/2$ and λ, then the beam has an intermediate intensity.

In addition to understanding interference, to derive Bragg's law it is also necessary to understand that planes of atoms act like partly transparent mirrors to x-rays. This is not obvious; in fact, an x-ray impinging on a free atom gets scattered equally in all directions; however, it turns out to be true, for reasons too complicated to explain here. You will have to accept it for now to understand the derivation. Figure 21.17, and the accompanying discussion in Section 21.6, simply show that the difference in path traveled by two beams reflecting from two succesive parallel layers is given by

$$\Delta = 2d \sin\Theta$$

where d is the spacing between the layers, and Θ is the angle of the impinging x-ray beam, on one side, and of the detector, on the other. Since constructive interference occurs when $\Delta = n\lambda$, it occurs when

$$n\lambda = 2d \sin\Theta$$

$$\sin\Theta = \frac{n\lambda}{2d}$$

The minimum diffraction angle is given by n=1, and this is virtually always assumed in freshman chemistry problems.

By looking at the unit cells in Figure 21.14 in the text, it should be clear that x-ray diffraction can be used to distinguish between them. For example, simple cubic, tetragonal and orthorhombic cells all have 90° unit-cell axes, but in the cubic structure the spacing of the planes is the same in all three directions, in tetragonal the spacing is equal in two directions, and in orthorombic it is different in all three directions. Using monochromatic x-radiation of known λ, the Bragg equation can be used to determine the three values of d, and thus distinguish between the structures. For more complicated structures, one must also consider spacings of planes of atoms not parallel to the faces of the unit cells; for example, along the varous diagonals.

Chapter 21 — Properties and Structures of Metallic and Ionic Crystalline

Energetics

First we will discuss ionic bonding, for which the forces and energies are Coulombic

$$F \propto \frac{q_1 q_2}{r^2}$$

$$E \propto \frac{q_1 q_2}{r}$$

As you can see from these equations, Coulombic forces are *pairwise*: they are defined for a pair of charges (ions, in our case) interacting. The difference between the energy of charges q_1 and q_2 when they are a distance r_1 apart and when they are a distance r_2 apart is

$$\Delta E = \frac{q_1 q_2}{r_2} - \frac{q_1 q_2}{r_1}$$

(Depending on the units of q, r and E a proportionality constant may be necessary in these equations as well, but we will ignore it for now.) If r_1 is "infinite separation" — a hypothetical state in which the particles are "too far apart to interact" — the second term goes to zero. If we arbitrarily define infinite separation as the zero of Coulombic potential energy (as we arbitrarily defined the standard state of an element to coincide with the zero of the standard enthalpy), then we can write for any two charges a distance r apart

$$E = \frac{q_1 q_2}{r}$$

Note that if q_1 and q_2 are opposite, E is negative, corresponding to a more stable state than that of infinite separation, whereas if q_1 and q_2 are the same E is positive.

Now let r_o be the spacing observed in the crystal, and consider the reaction

$$MX \text{ (ion pair distance } r_o \text{ apart)} \rightarrow M^{z+}(g) + X^{z-}(g)$$

Z+ and Z- are the *magnitudes* of the charges on the ions (positive numbers!). In the gas the separation is huge compared to the crystal spacing, so the energy for this reaction is

$$E = \frac{Z+ \; Z-}{r_o}$$

For a mole of ion pairs the energy is

$$E = \frac{N_A Z+ \; Z-}{r_o}$$

where N is Avogadro's number. Now, how does a crystal differ from a mole of ion pairs? In a sodium chloride crystal, each sodium is attracted by six chloride nearest neighbors, but also repelled by twelve sodium second nearest neighbours... and also attracted by eight chloride third nearest neighbours, and so on. When all these pairwise interactions are added up, the *lattice energy*, U, comes out to our last equation multiplied by a constant, A, called the Madelung constant

$$U = \frac{N_A A \, Z+ \, Z-}{r_o}$$

A is different for each particular crystal structure, but is always greater than 1, so that crystals are more stable than ion pairs. U is the energy for the reaction

$$MX(\text{crystal}) \rightarrow M^{z+}(g) + X^{z-}(g)$$

What is responsible for the spacing (r_o) in ionic crystals? The simplest answer is that ions tend to have fairly constant radii, so two ions will approach each other so that the centers are the sum of the ionic radii apart. What, then, is responsible for ionic radius? Recall from Chapter 5 (Figure 5.2) that even neutral atoms repel at close distances. If this repulsion were due to the Coulombic interaction between the electron clouds or nucleii of the two atoms, it would have the form $E = q_1 q_2 / r$, just like the interactions we have been discussing; in fact the repulsion is much steeper, more like $E \propto 1/r^{12}$. This is due to the Pauli principle, one statement of which is that electrons cannot occupy the same space at the same time. This fundamental physical principle — which would hold even if electrons were uncharged! — is responsible for the relative "hardness" of ions; for ionic radii, in short.

Lattice energies can be determined through the application of the *Born-Haber cycle*, which is analysed in detail in Section 21.9. It is helpful to check your understanding by noting which steps in the process are endothermic, and which are exothermic. For example, to sublimate sodium is endothermic by ΔH_{vap}, to ionize the gaseous atom is endothermic by IE. To dissociate enough chlorine gas to make a mole of chlorine *atoms* is endothermic by $(1/2)D_{Cl_2}$. The most confusing thing to realize is that to ionize a Cl atom

$$Cl(g) + e^- \rightarrow Cl^-(g)$$

The energy is -EA, the *negative* of the electron affinity. For Cl, EA is positive, so the ionization is exothermic, but many EA's are negative, so this step can be endothermic or exothermic, depending on the anion. (The ionization of a neutral species to form a cation, however, is always exothermic.) As the text shows, the above processes, together with the heat of formation, which can be exothermic or endothermic, can be used to calculate the lattice energy, which is always endothermic.

Your book discusses factors that cause changes in the lattice energy. The purely Coulombic model, which we have been using, says that U will change only if the crystal structure changes (which changes A), or if $Z+$, $Z-$ or r_o change. More highly charged ions increase the lattice energy, as do smaller ions for the same structure. Beyond this, however, if the cation has unfilled low-lying orbitals — in other words, if it is a soft acid, a term we used in the Scope of the previous chapter — then its interaction with the anions will be stronger than the Coulombic treatment implies, leading to larger lattice energies. This is especially true if the anion is a "soft base", one whose electron pairs are easily polarizable. In Example 21.6 in the text, the soft acid Ag^+ forms a more covalent bond with I^- than with Br^-, because I^- is the softer base; thus AgI has a lattice energy which exceeds that of AgBr by more than Coulombic considerations predict.

You should understand that a *heat of hydration* is ΔH for a reaction like

$$M^+(g) \rightarrow M^+(aq)$$

Hess' law cannot be used in a straightforward way to determine ΔH_{hyd}, because there is no experimental way to examine a process involving a single ion; however, consider the reaction

$$Na^+(g) + Cl^-(g) \rightarrow Na^+(aq) + Cl^-(aq)$$

δH for this reaction will be equal to $(\delta H_{hyd}^{Na^+} + \delta H_{hyd}^{Cl^-})$. This reaction can be expressed as the sum of the following two reactions.

$$Na^+(g) + Cl^-(g) \rightarrow NaCl(s)$$

$$NaCl(s) \rightarrow Na^+(aq) + Cl^-(aq)$$

For the first of these $\Delta H = -U_{NaCl}$, and for the second $\Delta H = \Delta H_{sol}^{NaCl}$, so that for the overall process

$$\Delta H_{hyd}^{Na^+} + \Delta H_{hyd}^{Cl^-} = \Delta H_{sol}^{NaCl} - U_{NaCl}$$

Note that ΔH_{sol} and U can be measured or calculated. Similarly, if we were to consider the same processes for KCl, we would have

$$\Delta H_{hyd}^{K^+} + \Delta H_{hyd}^{Cl^-} = \Delta H_{sol}^{KCl} - U_{KCl}$$

Subtracting this equation from the previous one for NaCl gives

$$\Delta H_{hyd}^{Na^+} - \Delta H_{hyd}^{K^+} = (\Delta H_{sol}^{NaCl} - \Delta H_{sol}^{KCl}) - (U_{NaCl} - U_{KCl})$$

Thus, heats of solution and lattice energies can be used to calculate *differences* in ΔH_{hyd} for pairs of ions. By comparing many such reactions a consistent set of hydration energies can be derived for the commonly occurring ions.

Metals

We now turn to a discussion of the bonding in metals. First let us note that we were able to discuss the energetics of ionic crystals in purely classical (non-quantum) terms: the forces holding these crystal together are purely Coulombic, pairwise forces. The Madelung constant is a purely geometric factor. The ionic separation in the crystal, r_o, cannot be calculated classically, but it can be determined by x-ray diffraction. It turns out that for an ionic crystal, when U is determined by means of the Born-Haber cycle, and then compared with calculated value

$$U = \frac{N_A A Z^+ Z^-}{r_o}$$

the results are in fair to good agreement. The anomalously high values of U observed for crystals like AgI are due to covalence; in other words, AgI is not a purely ionic crystal, so that we should not expect our simple treatment to hold completely.

We mentioned that a covalently bonded network crystal may be viewed as one big molecule. In such a molecule, the directed-bond notion of hybrid atomic orbitals (such as the overlap of sp^3's or sp^2's of adjacent atoms) is a good approximation. Metallic crystals can also be viewed as single big molecules, but for them the directed-bond picture is insufficient; a molecular orbital treatment is required. Recall that in Chapter 14 molecular orbitals were described for the second-row diatomics. These orbitals encompassed the entire molecule; for example the single π^*-orbital had lobes near both atoms. These orbitals ultimately came from the solution of the Schroedinger equation for a system with two fixed nuclei (the H_2^+ molecular ion). On the other hand, the sp^2 and sp^3 hybrid orbitals which gave, respectively, the trigonal and tetrahedral geometries of carbon compounds, came ultimately from the solution to the Schroedinger equation for a system with a single nucleus (the H atom). For most chemical substances — for example, molecules and covalently-bonded networks — this latter approach is a good approximation. It is an approximation because it starts with quantum-mechanical solutions to single-centered species (atoms), and constructs a multi-centered species (a molecule) out of them. The fact that VSEPR works so well shows that this approach is a good one.

On the other hand, this approach does not work for metals. The bonding in metals is described by the solution of the Schroedinger equation for the entire lattice — an infinite geometric array of nuclei. The resulting molecular orbitals span the entire lattice. There are no directed bonds from one atom to another, just orbitals which are delocalized over the entire crystal. These orbitals are then filled with as many electrons as were contained in the contributing orbitals of all the constituent atoms. In these calculations the inner orbitals, such as the 1s for lithium, are viewed as non-bonding and localized on each atom. The 2s orbitals from all the atoms then interact to form the *valence band*. It is a band because it turns out that when many, many orbitals of the same energy interact there is a virtually continuous set of molecular orbitals formed — continuous in terms of their energy. As the text explains, the valence band is only half full for Li.

The same is true of sodium, where the valence bands are made from the 3s orbitals. For magnesium, however, the valence band is full, since each 3s orbital contributes two electrons. The 3p orbitals, however, also combine to form a band, called the *conduction band*. Its energetic continuum overlaps that of the valence band, so that even though the 3p orbital is empty in a ground-state isolated magnesium atom, in the crystal the conduction band is partly filled with electrons that came from the valence band. Since the conduction band is only partly filled, electrons move about freely; thus, the electrical conductivity of magnesium — and of most metals — is due to the presence of a conduction band which overlaps the valence band. For lithium and sodium, the conduction band also overlaps the valence band, but since the valence band is only half full, both bands contribute to the conductivity.

Although not discussed in the text, in some crystals, such as those of Be and Si, the conduction band does not overlap the valence band, but is separated from it by a small *band gap*. These crystals are the semiconductors. In them, a small excitation energy is required for an electron to move to the conduction bond; this is supplied by thermal energy for materials that are semiconductors near room temperature. Heating such a material will increase the number of electrons in the conduction band, and raise the conductivity. For true metals, on the other hand, the conductivity goes down when the metal is heated. This is because thermal vibrations disturb the regularity of the lattice. When the lattice becomes irregular, the band structure gets disturbed, and can become discontinuous momentarily in local regions. This lowers the mobility of the electrons. Of course, this eventually affects semiconductors, too, so that for them the conductivity really goes through a maximum and then decreases as the temperature is increased.

Chapter 21   Properties and Structures of Metallic and Ionic Crystalline

Questions

21.1 Crystalline and amorphous solids

1. Tell whether each of the following is associated with a crystal or with an amorphous solid.

    _____(a) constant interfacial angles

    _____(b) isotropic properties

    _____(c) supercooled liquids

    _____(d) regular lattice

21.2 Crystal defects

1. What kind of defect is responsible for the non-stoichiometry of Fe_9S?

2. If there is an empty space in a crystal where an ion or atom should be, this is termed a _____.

3. If there is an atom or ion in a crystal where an empty space should be, this is termed a _____.

21.3 The structures of metals

1. In a single close-packed layer, a sphere is surrounded by _____ others.

2. In a close-packed solid, a sphere is surrounded (in three dimensions) by _____ others.

3. The number of nearest neighbors of an atom or ion has in a crystal is called its _____.

4. Which of these is not a close-packed structure?

    _____(a) hexagonal close-packed

    _____(b) face-centered cubic

    _____(c) body centered cubic

5. Give another name for the face centered-cubic structure _____.

6. An atom or ion in the corner of a cubic unit cell contributes _____ atoms to the cell.

7. The body-centered cubic unit cell contains _____ atoms.

Chapter 21   Properties and Structures of Metallic and Ionic Crystalline

8. Of the two common close packed structures, which has the pattern of layers

    _____(a) ABABAB...

    _____(b) ABCABC...

9. An atom in the body-centered cubic structure is surrounded by _____ neighbors.

21.4 Metallic bonding

1. For each of the following isolated atoms fill in the table

	Is it a metal?	Number of Valence Orbitals	Number of Volume Electrons
(a) Na			
(b) Mg			
(c) Al			
(d) Si			
(e) P			
(f) S			
(g) Cl			
(h) Ar			
(i) Fe			

2. The ionization energies of metals tend to be *higher* than for non-metals.

21-13

3. For the following, what fraction full is each band?

	Valence Band	Conduction Band
(a) Na		
(b) Mg		
(c) Al		
(d) Fe		

21.5 The seven crystal systems

1. An atom or ion on the face of a cubic unit cell contributes _____ atoms to the cell.

2. An atom or ion on the edge of a cubic unit cell contributes _____ atoms to the cell.

3. How many atoms are contained in the unit cell of the following structures?

    _____(a) simple cubic

    _____(b) face-centered cubic

    _____(c) body-centered cubic

4. In the NaCl structure, the Cl^- ions form an fcc lattice, and the Na^+ ions form a _____ lattice.

21.6 Atomic dimensions and unit-cell geometry

1. What kind of interference (constructive or destructive) is exhibited by two waves which are

    _____(a) in phase

    _____(a) out of phase

2. Two in-phase waves of the same frequency and amplitude combine to give a wave whose amplitude is _____ times that of either one alone.

3. Two out of phase waves of the same frequency and amplitude combine to give a wave whose amplitude is _____ times that of either one alone.

4. If two beams of wave-length λ traverse paths whose length *differs* by $x\lambda$, give a value of x consistent with the recombined beams being

Chapter 21 Properties and Structures of Metallic and Ionic Crystalline

_____(a) in phase

_____(b) out of phase

5. If the *minimum* value of Θ for which x-ray diffraction is observed from the face of a crystal is given, then (in the Bragg equation) must be equal to _____.

21.7 The lattice energy of ionic crystals

1. For which of these crystal structures is the edge of a unit cell equal to an atomic diameter?

    (a) _____ simple cubic

    (b) _____ body-centered cubic

    (c) _____ face-centered cubic

2. For each pair of ions, which has the larger diameter?

    _____(a) Fe^{2+}, Fe^{3+}

    _____(b) Cl^-, Br^-

    _____(c) Li^+, Be^{2+}

    _____(d) O^{2-}, Cl^-

21.8 The Born-Haber cycle

1. Write the chemical equation whose enthalpy charge corresponds to the lattice energy, U.

2. Suppose I compare two crystals with composition $M^{2+}X^{2-}$ and M^+X^- respectively. If the structure and unit-cell dimensions are the same the (first, second) crystal should exhibit a lattice energy that is greater than that of the other by a factor of _____.

3. For a purely ionic crystal of composition M^+X^-, if X^- is replaced by Y^-, which is larger, the lattice energy will go (up, down).

4. The Madelung constant makes the lattice energy (greater, less) than the energy which would hold together a mole of ion pairs.

5. In a mostly ionic crystal, some covalent character in the bonds tends to (increase, decrease) the lattice energy.

21.9 X-ray diffraction

1. The Born-Haber cycle is on application of _____ Law.

2. Consider the calculation of the lattice energy using the Born-Haber cycle. Tell whether each of these will increase or decrease the calculated value of U, or leave it the same.

    _____(a) increased ΔH_{sub} of the metal

    _____(b) increased IE of the metal

    _____(c) increased EA of the non-metal

    _____(d) increased $\Delta H°_f$

Chapter 21 — Properties and Structures of Metallic and Ionic Crystalline

Perspective: Crystal Structure and the Close Packing of Spheres

Note: Visualization of the three dimensional structures described in this **Perspective** is made much easier if you are able to purchase a few dozen styrofoam balls (from a craft or variety store — about a one-inch diameter is best). If you can get the balls, also get toothpicks (the round kind are good) to hold the balls together.

Geometry has been a source of curiosity and of inspiration since the dawn of history — witness the early work of Pythagoras and Euclid, and early knowledge of the "five Platonic solids", not to mention the pyramids of Egypt. Crystals are naturally occurring geometric forms, and their geometry of course arises from the arrangement of the atoms in space. The close packing of spheres — for example, the pyramid of cannonballs sometimes seen on village greens — is an important part of crystal structure. It turns out that even when the bonding in crystals is covalent rather than ionic, the lattice is often related to a close-packed structure.

Metals commonly crystallize in one of four structures:

    (a) hexagonal close-packed

    (b) cubic close-packed (also called face-centered cubic)

    (c) body-centered cubic

    (d) simple cubic.

Of these, (a) and (b) are instances of the close-packing of spheres. We will consider these four structures, with special concentration on (a) and (b). Of course, in all these structures all the lattice sites are occupied by the same species; for example, a metal atom. The close-packed structures, however, are closely related to structures of ionic crystals, such as LiCl, where different ions occupy different lattice sites. As the text discusses, in the LiCl structure the Cl^- ions form a face-centered cubic lattice, and the Li^+ ions "snuggle into" the spaces left between the chlorides. You may recall (or else consult Figure 21.21 in the text) that these spaces or "holes" occupied by the lithium ions form *another* face-centered cubic lattice, so that the unit cell can be chosen either with chlorides or with lithium ions on the corners. As we shall see, it is typical for anions to form a close-packed structure, and for the cations to fill certain holes in the structure, depending on stoichiometry and ionic size. Considering how simple this picture is, it is amazing how much it explains. (Of course, it doesn't explain everything.)

To understand structures in three dimensions, it is often helpful to start with two. Consider the two structures shown below.

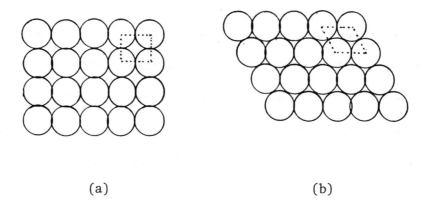

(a) (b)

These can be viewed as "two-dimensional crystals" of disks whose diameter we will be calling d. One possible choice of a unit cell is shown for each structure (there are others).

First consider structure (a), which we will call "square packing". Clearly the unit cell contains $4(1/4)=1$ disk. Since the area of the one disk is $A_{disk} = \pi(d/2)^2$ and the area of the unit cell is $A_{cell} = d^2$, the fraction of two-dimensional space filled by this structure, which we will call f_{square}, is

$$f_{square} = \frac{A_{disk}}{A_{cell}} = \frac{\pi}{4} \approx 0.785$$

In structure (b), which we will call "hexagonal packing", the unit cell also contains one disk (though this is a little harder to see), so its area is also $A_{disk} = \pi(d/2)^2$, but here A_{cell} is the area of two equilateral triangles each with side d, which turns out to be $A_{cell} = (\sqrt{3}/2)d^2$ *, giving

$$f_{hex} = \frac{\sqrt{3}\pi}{6} \approx 0.907$$

Clearly, hexagonal packing fills two-dimensional space better than square packing. In fact, hexagonal

---

* In this Perspective, whenever you see an asterisk it means that you should be able to derive the result given yourself.

packing fills two dimensional space better than any other infinite lattice.

Note that both the square and the hexagonal arrangements leave holes between the disks. For the square arrangement the largest disk that can fit into a hole (just touching its four nearest neighbors) has diameter $(\sqrt{2}-1)d \approx 0.414d$*. For the hexagonal structure, the diameter of a hole is $(2\sqrt{3}/3 -1)d \approx 0.155d$.** Also, note that for the hexagonal structure, there are two holes for each disk; for the square structure there is only one. Our results are summarized in this table, in which I have also given names to the holes. The names are based on the geometry of the tangential disks: when three disks make contact and form a hole, the centers of the disks form a triangle, so the hexagonal structure has triangular holes; the square structure has square holes. In this table I have also given names to the holes. The name is based on the geometry of the tangential disks: when three disks make contact and form a hole, the center of the disks form a triangle, so the hexagonal structure has triangular holes; the square structure has square holes.

You may recall that anions have larger diameters than cations. This is because anions have extra electronic charge. The electron clouds are directly responsible for ionic radius, and the greater the nuclear charge, the more tightly the electrons will be "pulled in" toward the nucleus. This is why it is the anions that often close-pack in a crystal, leaving the cations to fill the holes.

Let us apply this idea to our two-dimensional crystals, using two-dimensional ions. If we had an ionic substance MY (M^{z+} the cation, Y^{z-} the anion), then, according to our hypothesis, the Y's would form one of the arrays illustrated above, and the M's would fit into the holes. If d_M (the diameter of M)

Some Two-Dimensional Crystals

Lattice	Type of Hole	f	d_{hole}/d_{disk}	Holes/Disk
square	square	0.785	0.414	1
hexagonal	triangular	0.907	0.155	2

** A double asterik in this Perspective means that although only simple geometry need be used to derive the result, the derivation is fairly challenging. If you remember your geometry, you can have fun with these.

were about $0.414d_Y$ then the square structure would probably be formed, because this would allow each M to approach four Y's as nearly as possible. On the other hand, if d_M were about $0.155d_Y$, the hexagonal structure would probably form, since even though each M could only be surrounded by three Y's, it could get closer to them in this geometry. Since the hexagonal structure has two holes per disk, only half the holes would be filled by the M's — presumably alternate holes. If the stoichiometry of the compound were M_2Y, then clearly the compound would "try" to crystallize in the hexagonal structure. On the other hand, if it were MY_2, ionic size might once again be the determinant: the M's could fill either half the holes in the square lattice or every fourth hole in the hexagonal lattice.

In real crystals, the metal ions tend to fit quite snugly into the anionic lattice, which may even expand somewhat from the close-packed structure (as in RbCl, Example 21.4 in the text). Therefore, pursuing the two-dimensional analogy, even if d_X somewhat exceeds $0.155d_Y$, the hexagonal structure might be expected. In addition, ions are not hard spheres, so that ionic radius is difficult to define precisely. Of course, the bonding in real crystals often has some covalent nature as well. These three factors mean that the purely geometric and Coulombic picture we have been developing is only an approximate one.

Now let us consider some three-dimensional structures. The text illustrates the *body-centered cubic* (bcc) structure (Fig. 21.8). In it the holes are octahedral, since each hole is surrounded by six spheres forming the vertices of an octahedron, but the octahedra are greatly squished: a hole is centered on a Bcc face, and its center is far closer to the body centered spheres than to the corners of the unit cell. There is one hole per sphere, so the natural stoichiometry which would arise from filling holes in a bcc lattice is MY; however, due to the lack of symmetry just alluded to, such structures never occur, to my knowledge.

Recall that the *simple cubic-packed* (scp) structure consists of spheres centered on an ordinary cartesian lattice — like cubic boxes stacked. The holes are cubic, since each is surrounded by eight spheres forming the vertices of a cube, and there is one hole per sphere, giving the natural stoichiometry MY. The holes are relatively large ($d_M = (\sqrt{3}-1)d_X \approx 0.732d_X$),* and so this structure tends to form if anions and cations are of comparable size. This structure is easy to envision, since the centers of the spheres and the holes are exactly where they are in a bcc unit cell; the only difference is that here the corner spheres have expanded till they make contact, and the central sphere has shrunk accordingly. This is called the *cesium chloride structure*, and is exhibited by the cesium halides. The *fluorite structure* has stoichiometry MY_2, and is formed by filling in alternate holes in the scp lattice. Examples are CaF_2 (fluorite), and the fluorides of Ba, Sr and Pb.

We now turn to ionic crystals based on close packing of the anion. First we consider the close-packed structures. Consider a close-packed plane of spheres — our hexagonal two-dimensional structure — and see what happens when we put a second layer of spheres on top of it.

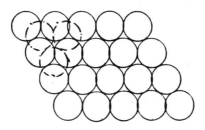

Note that *two* kinds of holes are formed. Directly beneath each sphere in the second layer there is a tetrahedral hole, surrounded by four spheres, and directly under each triangular hole in the second layer there is an octahedral hole, surrounded by six spheres. The octahedral holes are larger than the tetrahedral holes.

Call the bottom layer A, and the second layer B. If you look carefully (models are very helpful here) there are two ways to put on a third layer: either directly above the first — giving the pattern ABA — or directly above the octahedral holes formed by the first two layers — a completely new horizontal orientation, C, which gives the pattern ABC. Repeating the scheme ABABAB... gives *hexagonal close packing* (hcp), whereas repeating the scheme ABCABC... gives *cubic close packing* (ccp). Both fill space equally efficiently (see table below); in fact, although the mathematicians have been unable to prove that these sorts of packing fill all space better than any other, no one has found a better way. Incidentally, space is filled equally well by other patterns — such as ABCBABCB... (called *double hexagonal packing*) — but we will only be concerned with hcp and ccp.

For each lattice sphere in hcp or ccp, there is one octahedral hole and two tetrahedral holes. If the stoichiometry of an ionic crystal is MY, the M's tend to occupy either all the octahedral holes or half the tetrahedral holes in such a lattice, depending on the ratio of ionic sizes. A ccp structure with all the octahedral holes filled in is called the *rock-salt structure* and is exhibited by lithium chloride (as illustrated in the text), by the sodium halides and by the alkaline earth oxides. Rock salt is cystalline NaCl. Ccp with half the tetrahedral holes filled is the *zinc blende* structure (α-ZnS, one allotrope of ZnS, is called zinc blende). Note that Zn^{2+} is far smaller relative to S^{2-} than Na^+ is relative to Cl^-; hence the zinc ions fill tetrahedral, rather than octahedral holes.

Naturally, there are a lot of crystal structures which can be formed by filling in combinations of octahedral and tetrahedral holes in ccp and hcp lattices, but the details, though fascinating, are not our aim here. The aim has simply been to show how simple ideas can take one fairly far into an understanding of which structures will arise. One more point of interest, however: for purely ionic crystals, structures based on close packing tend to be those based on the ccp lattice, due to its great symmetry. Structures built on the hcp lattice tend to involve transition metals whose d-orbitals can give rise to covalent interactions; thus, for example, the hcp lattice with the octahedral holes filled in is the *chromium sulfide* structure (also called the *nickel arsenide* structure) and is exhibited by the sulfides of the +2 oxidation states of Fe, Co and Ni, as well as of Cr.

This table summarizes the geometries of the three-dimensional lattices we have discussed. Again, it takes only simple geometry to derive these results, but some of the derivations are tricky — particularly those involving the close-packed structures.

Some Three-Dimensional Crystals

Lattice	f	Type of Hole	d_{hole}/d_{sphere}	Holes/Sphere
scp	.51	cubic	.732	1
bcc	.68	octahedral*	.077, .633	1
ccp	.74	tetrahedral	.223	2
		octahedral	.414	1
hcp	.74	tetrahedral	.225	2
		octahedral	.414	1

* Greatly squished; both diameters given.

Chapter 21   Properties and Structures of Metallic and Ionic Crystalline

Answers

21.1 Crystalline and amorphous solids

1. (a) crystal
   (b) amorphous solid
   (c) amorphous solid
   (d) crystal

21.2 Crystal defects

1. substitutional impurity
2. vacancy
3. interstitial site

21.3 The structures of metals

1. six
2. twelve
3. coordination number
4. (c)
5. cubic close-packed
6. one eighth
7. two
8. (a) hexagonal close-packed
   (b) cubic close-packed
9. eight

21.4 Metallic bonding

1.

	Is it a metal?	Number of Valence Orbitals	Number of Valence Electrons
(a) Na	yes	4	1
(b) Mg	yes	4	2
(c) Al	yes	4	3
(d) Si	no	4	4
(e) P	no	4	5
(f) S	no	4	6
(g) Cl	no	4	7
(h) Ar	no	4	8
(j) Fe	yes	4	2

2. lower

3.

	Valence Bond	Conduction Bond
(a) Na	1/2	0
(b) Mg	1	0
(c) Al	1	1/6
(d) Fe	1	0

Chapter 21 — Properties and Structures of Metallic and Ionic Crystalline

21.5 The seven crystal systems

1. one half
2. one quarter
3. (a) 1
   (b) 4
   (c) 2
4. fcc

21.6 Atomic dimensions and unit-cell geometry

1. (a) constructive
   (b) destructive
2. two
3. zero
4. (a) 0 (or 1, or 2, or 3, etc.)
   (b) 1/2 (or 3/2, or 5/2, or 7/2, etc.)
5. one

21.7 The lattice energy of ionic crystals

1. (a) only
2. (a) Fe^{2+}
   (b) Br^-
   (c) Li^+
   (d) O^{2-}

21.8 The Born-Haber cycle

1. $M^+X^-(s) \rightarrow M^+(g) + X^-(g)$
2. first, four
3. down
4. greater
5. increase

21.9 X-ray diffraction

1. Hess'
2. (a) increase
   (b) increase
   (c) decrease
   (d) decrease

Chapter 22   Radioactivity and Nuclear Chemistry

CHAPTER 22: RADIOACTIVITY AND NUCLEAR CHEMISTRY

Scope

There are relatively few subatomic particles with which we will be concerned; they are listed in Table 22.1 in the text. A *nucleon* is a *proton* or a *neutron* (weight ≈ 1amu); a *nuclide* is some bound combination of these, such as an alpha particle, or a carbon nucleus. The number of nucleons in a nuclide is its *mass number*, A, and the number of protons is its *atomic number*, Z. N, the number of neutrons, is of course A-Z. To signify a specific nuclide we write $^A_Z El$, where El is the chemical symbol for the element whose nucleus this nuclide corresponds to. Thus $^{14}_6 C$ is carbon-14. The notation is redundant, because Z defines El; thus ^{14}C is sometimes written for carbon-14, since carbon *must* have six protons; if not, it's not carbon.

Isotopes are nuclides which differ in A, but not Z; they behave almost identically chemically, and thus correspond to the same element. *Isobars* are nuclides with the same A but differing Z; they correspond to different elements.

Binding energy

For the reaction

$$\text{isolated nucleons} \rightarrow \text{nuclide}$$

the binding energy (BE) is defined as the *negative* of the energy change (ΔE) for the process. For stable nuclides this reaction is, of course, exothermic, so that ΔE is negative and BE is positive. BE's are determined using the Einstein equation

$$\Delta E = (\Delta m)c^2$$

ΔE is the energy change of the reaction, and Δm is the *mass defect*, or difference in mass between the right and left sides of the above nuclear equation. For example, $^{14}_6 C$ is stable, so that

$$6n + 6p \rightarrow {^{12}_6 C}$$

is exothermic, the energy change is negative, and Δm is negative: $^{12}_6 C$ weighs less than the unbound protons and neutrons. BE is positive. Although Δm is tiny when expressed in terms of grams per mole or in terms of amu per atom, it is huge in terms of energy, because c is so large. The text shows that BE for $_6^{14}C$ is equal to 8.9×10^9 kJ/mole. This is about 10^6 times greater than ΔE for typical chemical processes. Physicists commonly use *Mev*, or *millions of electron volts*, to report the energies of nuclear processes. The value is per molecule, rather than per mole, so that in converting kJ/mole to Mev, one must first convert kJ to Mev, then divide by Avogadro's number. The text shows how to do this; the binding energy of $_6^{14}C$ is shown to be 92.1Mev.

Nuclear forces

Many students absorb the misimpression that there is something unique about nuclear processes that makes the equation $\Delta E = \Delta mc^2$ apply to them. This is incorrect. *Any* exothermic process results in an annihilation of mass, and *any* endothermic process results in the creation of mass. When coal is burned, the total weight of the products is less than the total weight of the reactants, by $\Delta m = \Delta E/c^2$, where ΔE is the energy change of combustion. When water evaporates, the mass of the vapor is greater than the mass of the liquid by $\Delta m = \Delta E/c^2$, where ΔE is the energy of vaporization. What is unique about nuclear processes is that the ΔE's are so large. For chemical processes, the ΔE's are so small that Δm can usually not be measured.

The reason that ΔE's are so much larger for nuclear than for chemical reactions is that the forces holding the nucleus together are much stronger than the forces binding the electrons to the nucleus. Virtually all of chemistry as we know it is due to changes in electronic configurations. The forces involved are strictly Coulombic, though it takes quantum mechanics to work out the detailed implications of the Coulombic interactions on a microscopic level. The forces holding the nucleus together are not Coulombic; they are completely new kinds of forces, whose precise description is the subject of active research in elementary particle physics. But the magnitude of the binding energy of a nuclide such as $_6^{14}C$ indicates how strong these forces must be, compared to the Coulombic forces involved in ordinary chemical interactions. Another way of seeing this is to remember that in ordinary chemical processes only a negligible fraction of the mass changes in the course of reaction; in nuclear processes the fraction is already noticible, and in elementary particle interactions, such as the ones observed in particle accelerators like those at Fermilab and CERN, a large fraction of the initial rest

Chapter 22 Radioactivity and Nuclear Chemistry

mass is often annihilated or created.

The Curve of Binding Energy

The *binding energy per nucleon* (BE/A) is used as a measure of nuclear stability; it has the same role in nuclear chemistry that the standard molar heat of formation has in ordinary thermochemistry. Figure 22.1 in the text shows that, with some glitches, BE/A increases for A's up through about 60 (near iron), then decreases. Attempts are made to explain the overall trend and the glitches by using the concept of *nucleon pairing*. This is a bit like trying to explain chemistry by using only Lewis dot diagrams; not everything falls into place. The idea is that protons and neutrons like to pair with themselves, but not with each other; thus species with even Z and N tend to be most stable, and species with odd Z and N tend to be least stable. This is roughly borne out by the data in Table 22.3.

The thing to remember about the BE/A versus A curve (Figure 22.1) is that the most stable A is about 60. Exothermic nuclear reactions tend to be ones in which the product nuclides have A's closer to this value than the parent nuclides. Thus, heavy elements undergo *fission*, which decreases A, and light elements undergo *fusion*, which increases A.

Fission and Fusion

As the text indicates, fusion requires temperatures of the order of 4×10^7 K in order to overcome the Coulombic repulsion of the positively charged nuclei that are coming together. It is still a further indication of the strength of the nuclear forces that they *can* overcome the huge Coulombic repulsion of the protons at short distances and hold the nucleus together. The fact that there is such a high barrier to overcome as two low-Z nuclei come together to fuse also shows that the nuclear forces, though strong, are short in range. Not until the protons are pushed almost together do the nuclear forces take effect to form a stable system.

Fission is the mechanism whereby heavy nuclei can break apart to form lighter, more stable ones. If a heavy nuclide is hit by a projectile such as a *thermal neutron*, it can break apart into fragments. A variety of fragmentation reactions are, in general, possible. Certain of the reactions release more neutrons, so that chain reactions can occur, ones very similar in concept to those discussed in Chapter 19. If the mass of the sample is either small or else in an elongated shape, such as thin cylinder, most of these product neutrons escape; however, if the mass is large and compact (for example, spherical) there is a greater chance that they will be captured and that the chain will propagate. The *critical mass* is the minimum mass necessary to sustain a chain reaction. Clearly, it depends on shape; two subcritical bricks of fissile material could be put together to make a *supercritical* mass.

In a conventional nuclear reactor, the mass is kept subcritical by means of control rods made of neutron-absorbing, non-fissile materials. It should be noted that the fission reaction in such a reactor is merely used to produce heat. This heat is used to boil water, which produces steam and turns turbines which turn generators, just as in a coal-fired power plant.

The breeder reactor (whose further development is apparently dead in the U.S., but very much alive in France and elsewhere) is supposed to work on a fuel cycle that produces more fuel for further use. This does not violate the first law of thermodynamics; it is just that fuels are used for which the usable ΔE through a cascade of nuclear transformations is greater than for the conventional cycle. Unfortunately, the breeder fuels are also more dangerous than the conventional ones, both in terms of direct exposure and in terms of the possibility of their being used for weapons.

Nuclear Stability

We now return to the question of nuclear stability. It is found that as Z increases, more neutrons per proton are needed for a nuclide to be stable (N/Z increases for stable nuclides). This is illustrated by the *belt of stability* (Figure 22.3), and can be explained by observing that putting more protons into the nucleus increases the Coulombic repulsion they experience. Adding more neutrons "dilutes" the protons, and also must contribute to the cohesive nuclear interactions. Overly neutron-rich nuclei, on the other hand, are supposed to be unstable because nucleons seem to observe a *shell structure* analogous to the electronic shells. This is described in the text. The shell model of the nucleus is not nearly as successful in explaining nuclear phenomena as the periodic chart of the elements is in explaining chemical phenomena.

Neutron-rich nuclei can decay by *beta emission*, whereas neutron-poor nuclei may decay by *electron capture, position emission* or *alpha decay*. These are well described in the text, and the **Questions** below should provide a good test of your understanding. You should understand that the reason that neutron-rich and neutron-poor nuclides decay by the paths they do is that this brings them into the "band of stability;" that is, it readjusts the N/Z ratio to values that are stable for the Z of the products.

Gamma emission differs from the other decay modes in that neither N nor Z changes. Gamma rays are a form of electromagnetic radiation — light (photons) of very high energy. In the same way that an atom emits light when an electron falls from an excited to a lower energy state, a nucleus emits gamma rays when it goes from an excited to a lower state. Naturally, observation of which wavelengths are emitted can yield information about the energy levels within the nucleus.

Radioactive Decay Series

Fission by neutron bombardment, as described above for man-made processes, is a rare event in nature; there is a very low rate of spontaneous neutron emission among long-lived nuclides, and neutrons from outer space usually get absorbed by the sea or by the surface layers of the earth before they can penetrate very far. The way materials like ^{235}U decay in natural ore deposits is by a chain of alpha- and beta-emission processes. Alpha-emission lowers A by four, and beta-emission does not change A, so that all elements along the chain have A's differing from the parent's value by some multiple of four. Therefore the remainder after dividing A by four is the same for the parent and for all the successive daughters in each series. Thus the *uranium series*, whose parent is ^{238}U, is the 4n+2 series, since 238/4 is 59 with a remainder of 2. For similar reasons, the *thorium series*, whose parent is ^{232}Th, is called the 4n series, and the *actinium series*, whose parent is ^{235}U (just to make it hard!) is the 4n+3 series.

Decay Kinetics

If you have been through Chapter 19, then this subject should be easy for you. Radioactive decay is always first order, because obviously if the concentration of the unstable nuclide is doubled, the number of disintegrations per unit time will also be doubled. The thing to remember is that disintegrations per unit time is a measure of the amount of unstable nuclide present. This is often expressed as *counts per minute* (cpm) where a count is a single click on a geiger counter or a single flash on a scintillation counter (a device which measures radioactivity by converting the energy of an emitted particle to a flash of light by passing it through a solution of a phosphor).

Following first order kinetics, we write

$$\ln\left(\frac{N}{N_o}\right) = -\lambda t$$

$$N = N_o e^{-\lambda t}$$

(We use λ, rather than k, to denote the rate constant of a radioactive disintegration.) In Chapter 19 you saw that

$$\tau_{1/2} = (\ln 2)/k = 0.693/k,$$

and here the same expression obtains, with the substitution of λ for k. In the description of radioactive decay processes, it is customary to refer to the half-life, rather than the rate constant.

One way in which radioactive decay processes differ from first-order chemical processes is that the radioactive processes are almost unaffected by changes of temperature, pressure or chemical environment of the nucleus; thus, as the text points out, ^{14}C has the same decay rate whether it is in graphite or in CO_2, and whether it is in a meteorite or on the surface of the earth.

Tunneling (not in text)

The fact that the temperature does not affect the radioactive decay rate is interesting. Recall (Chapter 19) that most reaction rates vary with temperature according to

$$k = A e^{-E_{act}/RT}$$

Generally, the bigger effect temperature has on the rate, the greater the value of E_{act}. If temperature has no effect on the rate, then $E_{act}=0$. Recall that E_{act} represents an energetic barrier that the reactants have to overcome to go to products. The energy to overcome this barrier is supplied by thermal energy; every so often there is a collision energetic enough to allow the activated complex to be formed. We have already seen that for radioactive processes barriers are due, at least in part, to the Coulombic interactions of the subatomic particles, and that they are very high. For example, to surmount the barrier for fusion takes huge temperatures, of the order of 4×10^7K. Therefore the radioactive decay taking place near room temperature cannot be an activated process! The barriers are just too high to be overcome by thermal means.

It turns out that there is a quantum mechanical effect called *tunneling* that comes into play here. If there are two states (reactants and products) separated by a barrier, and if a quantum system starts on one side of the barrier, there is a finite possibility of its winding up on the other side, even if it does not have enough energy to "get over the hill"! For most chemical processes tunneling makes only a small contribution to the total rate, but for radioactive decay it is the predominent mechanism. The rate of tunnelling increases if the barrier gets thinner, so if a very high barrier allows a significant degree of tunnelling it must be very thin. This is obviously true for radioactive decay, since here the barrier is no thicker than the diameter of a nucleus; once the emitted particle has escaped the nucleus, the product nuclide has already been formed.

Uses of Isotopes

Section 22.6 discusses three practical applications of isotopes. These all make use of the following two facts

1. The radioactivity of an isotope does not depend on its chemical environment, or on the temperature or pressure.

2. The chemical behavior of two isotopes of the same element is esentially identical*.

Isotope dilution is a method of determining the concentration of a substance which is difficult to isolate quantitatively. One adds a known amount of the chemically identical substance which has been isotopically labeled, and whose *specific activity* (cpm/g) is known. Then, after waiting long enough for

---

* This is *almost* true. The biggest exception by far is hydrogen. Going to deuterium doubles the mass, and going to tritium triples the mass. This leads to noticable effects, especially in kinetic phenomena. For example, the viscosity of D_2O is significantly greater than that of H_2O. Equilibrium properties are also affected; for example D_2O has a self-ionization constant of about 10^{-15}, instead of water's 10^{-14}.

complete mixing to occur, some of the substance is isolated from the system. It doesn't have to be all of the substance that's there, but it must be of known purity — hopefully 100% or very close. Its specific activity is then measured. Because the labeled and the unlabeled substance exhibit identical chemistry, they exist in the extract in the some ratio in which they existed in the system. Therefore, if the extract has one-tenth the specific activity of the added labeled substance, then the labeled substance added must constitute one-tenth of the total substance in the system (original plus labeled). Thus the amount of unlabeled substance originally present must have been nine times greater than the amount of labeled substance added. (This discussion assumes 100% purity of the extract.) A general formula is given in Equation 22-30 in the text.

Radiocarbon dating relies on the fact that the ^{14}C-to-^{12}C ratio of living matter stays at a constant level, equal to that of atmospheric CO_2, because living matter is always eating, breathing and/or photosynthesizing, all of which relatively rapidly exchange carbon with the environment. The ^{14}C-concentration of atmospheric CO_2 is kept elevated by a constant flux of cosmic rays. When living matter dies, carbon exchange with the atmosphere ceases, and the ^{14}C-content starts decreasing in first-order fashion ($\tau_{1/2}$ = 5730 years). By measuring the ^{14}C-activity per gram of carbon it can be determined how long the object has been dead. This can even be used for inner tree rings, which cease to exchange CO_2 with the atmosphere as the tree grows out. Also, the CO_2 dissolved in the ocean in the form of carbonates exchanges with the atmosphere across the ocean surface. By applying radiocarbon dating methods to ocean water samples from varying depths, it has been determined, as a function of depth, what the time-scale of mixing of the ocean is. This is very important in efforts to understand how increased CO_2 emissions from the burning of fossil fuels will eventually affect world climate, and how oceanic ecological systems evolve and persist.

Tracer methods are very useful in chemical kinetic studies. Recall (Chapter 19) that often two mechanisms may give the same rate law. Tracer studies are often useful in determining which one (if either!) is correct. As the book illustrates, one can sometimes insert an isotopically labeled atom into a known position on a reactant. By observing which product becomes radioactive, one knows where the labeled atom has gone. This sort of study has been extremely useful in defining the metabolic pathways of living plants and animals. The animal, say, is fed a labeled nutrient, and then various chemicals its body makes are isolated and checked for radioactivity. In this way the reaction paths involved in metabolizing the nutrient can be elucidated.

Chapter 22   Radioactivity and Nuclear Chemistry

Questions

22.1 Subatomic particles and nuclides

1. Fill in the missing data in the table:

Particle	Symbol	Approximate Atomic Weight (in amu)	Charge
(a)		+1	+1
(b) neutron			
(c)	β^-	1/1840	
(d)		1/1840	+1
(e)	α		

2. How much does Avogadro's number of nucleons weigh?

3. Fill in the following table

	Atomic Number	Atomic Weight (amu)	Number of Neutrons
(a) $^{19}_{10}$Ne			
(b) $^{238}_{92}$U			
(c) $^{56}_{26}$Fe			

4. Define

   (a) *isotopes*

   (b) *isobars*

Chapter 22

Radioactivity and Nuclear Chemistry

22.2 Nuclear binding energies

1. (a) In an exothermic nuclear reaction with energy release ΔE, the mass lost is given by the equation $\Delta E = (\Delta m)c^2$. (True, False).

    (b) In an exothermic chemical reaction with energy release ΔE, the mass lost is given by the equation $\Delta E = (\Delta m)c^2$. (True, False)

2. In nuclear reactions a mass change is generally detected. In chemical reactions, a mass change is generally not detected. Why is this?

3. Consider the reaction

    $$6n + 6p \rightarrow {}^{12}_{6}C$$

    (a) Is ${}^{12}_{6}C$ a stable nuclide?

    (b) Is the binding energy positive or negative?

    (c) Is the reaction exothermic or endothermic?

    (d) (Binding energy) = ΔE for this reaction. (True, False)

4. A neutron weighs more than a proton plus an electron. The reaction

    $$n \rightarrow p + \beta^-$$

    is therefore (exothermic, endothermic).

5. A typical ΔE for a nuclear reaction is about 10^6 times greater than that for a typical chemical reaction. Therefore the mass change of a nuclear reaction will be about 10^6 times greater than that for a chemical reaction. (True, False)

6. Name the parameter which measures the relative stability of a nuclide.

Chapter 22 Radioactivity and Nuclear Chemistry

7. $^{56}_{26}$Fe is one of the (most, least) stable nuclides.

8. Indicate which combination is generally the most stable, and which is generally the least stable.

	Z	N
_____	even	even
_____	odd	odd
_____	even	odd
_____	odd	odd

9. In general, fission occurs with release of energy for nuclear processes involving (light, heavy) nuclei, whereas fusion occurs with release of energy for (light, heavy) nuclei.

10. The stars are (fission, fusion) reactors.

11. For a fusion reaction, the Coulombic barrier to the approach of the nuclei may be seen as a sort of activation energy. (True, False)

12. It is stated in the text that no substance can withstand the temperatures of a fusion reaction, and that therefore magnetic fields are usually used to contain the particles in experimental thermonuclear reactors. What do you suppose holds the reacting particles together in a star?

13. A thermal neutron is one that has the same (momentum, speed, kinetic energy) as a gas molecule at the same temperature.

14. In order for neutrons to induce an explosive nuclear chain reaction, *one neutron* must be released for every neutron absorbed in a fission step.

15. When an atomic bomb is triggered

    (a) The fissile charge goes from subcritical to supercritical (True, False)

    (b) The mass of the fissile charge increases (True, False)

    (c) The shape of the fissile charge changes (True, false)

Chapter 22 Radioactivity and Nuclear Chemistry

16. Give one advantage of the breeder reactor over the conventional nuclear reactor.

17. Give one disadvantage of the breeder reactor compared with the conventional nuclear reactor.

22.3 Radioactivity

1. As Z increases, the number of neutrons per proton required for stability (increases, decreases).

2. All known nuclei are unstable (above, below) Z=_____.

3. According to the shell model of the nucleus, a nuclide will be especially stable if a "magic number" is fulfilled for

    (a) atomic number (True, False)

    (b) atomic mass (True, False)

    (c) atomic mass — atomic number (True, False)

4. Consider a nuclide $_Z^A$El. Fill in the following chart. (In the last column, tell whether the process predominates for neutron-rich or -poor nuclides.)

Process	Change in A	Change in Z	Does El Change?	Neutron Rich or Poor?
(a) beta decay				
(b) electron capture				
(c) position emission				
(d) alpha decay				
(e) gamma decay				

Chapter 22 — Radioactivity and Nuclear Chemistry

22.4 Naturally occurring radioactive substances

1. What models of decay operate in the three radioactive decay series?

2. In these series, in each step, A changes by either _____ or _____.

3. (a) The ^{238}U series is called the 4n + _____ series.

    (b) The ^{232}Th series is called the 4n + _____ series.

    (c) The ^{235}U series is called the 4n + _____ series.

22.5 The rate of radioactive decay

1. Radioactive decay is a _____-order kinetic process.

2. The rate of radioactive decay increases with temperature. (True, False).

3. The number of disintegrations per second is a measure of the concentration of _____.

4. The half-life is constant over the full decay process. (True, False).

22.6 The uses of radioactivity

1. Different isotopes of the same element usually have significantly different chemical behavior. (True, False)

2. In an isotopic dilution assay for substance X, radioactive X with a specific activity of 10^6 cps is added to the sample. Then 1g of pure X is extracted, and its specific activity is found to be 10^5 cps. How many grams of X were in the sample?

3. Which of the following is or are responsible for maintaining the ^{14}C concentration of living matter?

    (a) _____ cosmic ray processes in the atmosphere

    (b) _____ breathing

    (c) _____ photosynthesis

4. ^{14}C-dating can only be used on material that was once alive. (True, False).

5. What is the approximate age-range of materials that can be dated using ^{14}C methods?

Chapter 22 — Radioactivity and Nuclear Chemistry

Answers

22.1 Subatomic particles and nuclides

1.

Particle	Symbol	Approximate Atomic Weight (amu)	Charge
(a) proton	p	1	+1
(b) neutron	n	1	0
(c) electron	β^-	1/1840	-1
(d) positron	β^+	1/1840	+1
(e) alpha	α	4	+2

2. 1g

3.

	Atomic Number	Atomic Weight (amu)	Number of Neutrons
(a) $^{19}_{10}F$	10	19	9
(b) $^{238}_{92}U$	92	238	146
(c) $^{56}_{26}Fe$	26	56	30

4. (a) nuclides of the same element with the same Z but different A and N.
   (b) nuclides of different elements with the same A, but different values of N and Z.

22.2 Nuclear binding energies

1. (a) True
   (b) True

2. Chemical reactions generally release much less energy than nuclear reactions, so that Δm is too small to measure.

3. (a) yes (from general knowledge of chemistry)
   (b) positive
   (c) exothermic
   (d) False; (Binding energy)=$-\Delta E$

4. exothermic

5. True

6. binding energy per nucleon

7. most

8. (even, even) is most stable; (odd, odd) is least stable.

9. heavy, light

10. fusion

11. True (since it takes temperature to overcome it)

12. gravity

13. kinetic energy

Chapter 22 — Radioactivity and Nuclear Chemistry

14. more than one neutron
15. (a) True
    (b) False
    (c) True
16. Produces products useful as fuel, therefore makes more efficient use of fuel.
17. These products are more dangerous than the original fuel, in many ways.

22.3 Radioactivity

1. increases
2. above, 83
3. (a) True
   (b) False
   (c) True
4.

Process	Change in A	Change in Z	Does Element Change	Neutron Rich or Poor?
beta decay	0	+1	yes	rich
electron capture	0	-1	yes	poor
positron emission	0	-1	yes	poor
alpha decay	-4	-2	yes	poor
gamma decay	0	0	no	either

22.4 Naturally occurring radioactive substances

1. Alpha and beta decay
2. four, zero
3. (a) 2
   (b) 0
   (c) 3

22.5 The rate of radioactivity decay

1. first
2. False
3. the decaying nuclide
4. True

22.6 The uses of radioactivity

1. False
2. 9g
3. all three
4. True usually; however, the example given in the *Scope* about ocean mixing is an exception.
5. 1000-25000 years

CHAPTER 23: INTRODUCTION TO ORGANIC CHEMISTRY

Scope

History

Organic chemistry, as a discipline, has a different "flavor" from most of the other material in the text. To oversimplify, most of the other subfields of chemistry are largely quantitative: they come from efforts to explain chemical phenomena based on the laws of physics. Examples are the ideal gas laws and their justification through the kinetic theory of gases, and the equilibrium constant concept and its rationalization by means of the laws of thermodynamics (the free-energy concept).

Organic chemistry is the study of carbon compounds. The name comes from the fact that living matter is composed of such compounds, and up until the early nineteenth century it was believed that there was some mysterious "life-essence" which was responsible for their production. In 1828 Wohler synthesized the organic substance urea, $(NH_2)_2CO$, in the laboratory from the inorganic starting materials ammonia and carbon dioxide, and since then it has become clear that all the synthetic processes taking place in the body occur only because the appropriate starting materials, reaction conditions and catalysts (enzymes) are present; the same laws of nature operate in the body as operate in the laboratory.

The concept of *chemical structure* is central to organic chemistry. Organic chemists were inferring the connectivity of atoms within molecules long before physicists and physical chemists were even willing to admit that atoms and molecules were real. To quote chemist Linus Pauling's classic work, *The Nature of the Chemical Bond* (Cornell University Press, 1960)

> Most of the general principles of molecular structure and the nature of the chemical bond were formulated long ago by chemists by induction from the great body of chemical facts. During recent decades these principles have been made more precise and more useful through the application of the powerful experimental methods and theories of modern physics.

You saw in Chapter 20 how Werner inferred the geometry of the octahedral complexes from isomer number. In 1874, on the basis of similar evidence, van t'Hoff and le Bel inferred that when carbon is bonded to four atoms, the central geometry is tetrahedral; a compound like CX_2Y_2 gives only a single isomer. In 1858 August Kekule proposed that carbon atoms could join together to form chains, and in 1865 he proposed the ring structure for benzene. On the other hand, many physicists doubted the reality of atoms and molecules until Bragg's x-ray diffraction experiments in 1913.

The "great body of chemical facts" which Pauling refers to includes a huge quantity of data on *chemical synthesis*: what chemicals are formed when known starting materials are mixed under given conditions. Synthesis is central to organic chemistry, and the ability to figure out how to make a compound of a given structure from available starting materials (which usually requires many steps) is the central problem of synthesis. Solving such problems is primarily an inductive, rather than a deductive process, perhaps more like working out a chess problem than like solving an equation.

Of course, the importance to organic chemistry of the theory of bonding and of inherently physical techniques, such as the various forms of spectroscopy, should not be underestimated, and efforts to rationalize and predict complicated chemical behavior based on quantum chemistry are becoming more and more succesful. Lately, computers have been put to work to solve structural and synthetic problems, sometimes employing the techniques of "artificial intelligence". This effort is in its infancy, but shows great promise.

Bonding

Since carbon is of intermediate electronegativity, it forms covalent, rather than ionic bonds. Since it has four valence electrons, it forms four bonds. When these are to four separate atoms, the four bonds point out from the central carbon toward the vertices of a regular tetrahedron, as predicted by VSEPR (Chapter 14), and this corresponds to sp^3 hybridization (bond angle $\approx 109°$). Each sp^3 hybrid orbital can overlap with an orbital from an adjoining atom to form a σ-bond (Example 23.1 in the text), and this is what occurs in methane, ethane, and the other *alkanes*. Rotation about a σ-bond is somewhat free, so that ethane (CH_3–CH_3) can rotate internally about the C-C bond, forming a continuous set of *conformations*. These are sets of molecular structures which can be interconverted without breaking

Chapter 23   Introduction to Organic Chemistry

bonds; *isomers* are usually defined as molecules with the same molecular formulas, but with either differing connectivities or with the same connectivity, but no possibility of interconversion except through bond breaking. Usually one can isolate isomers but not conformers, but this is not a hard and fast rule. The *staggered* conformation of ethane (Figure 23.2) is the most stable of its continuous set; the *eclipsed* conformation (where the hydrogens tend to "bump") is the least stable, but the energy difference is only about 12 kJ/mole, so rotation is fairly free. The eclipsed conformation can be viewed as the transition-state between the more stable staggered conformations. At low enough temperatures, where $RT << 12$ kJ/mole, there would not be enough thermal energy to populate the transition state appreciably, so that virtually all the ethane would exist in the staggered form, and rotation would be much slower.

Carbon also exists in molecules where it is bonded to three neighbors, and here VSEPR successfully predicts trigonal geometry (planar, 120° angles). This involves the formation of three equivalent sp^2 orbitals, with the third p-orbital left alone. The sp^2 orbitals form σ-bonds just like the sp^3-orbitals of tetrahedral carbons, but the p-orbital forms a π-bond with an adjacent p-orbital of another carbon, or perhaps an oxygen or nitrogen. When it does so with an adjacent carbon the compound is called an *alkene*, and exhibits a carbon-carbon double bond (C=C), one bond of which is the sp^2–sp^2 σ-bond, and the other of which is the p-p π-bond. Ethylene ($CH_2=CH_2$) is the simplest example. Rotation about the double bond is not free.

Going one step further, carbon can exhibit linear geometry as well, involving only two bonded neighbors. Hybridization gives two linear sp-orbitals and two perpendicular p-orbitals. In acetylene, the simplest example of this (Figure 23.4), each carbon uses one of its sp-orbitals to form a σ-bond with the other carbon, and its two p-orbitals to form two π-bonds with the other carbon, so that a C≡C triple bond results. (The other sp-orbital from each carbon forms a σ-bond with a hydrogen 1s orbital.)

Note that as the bond-order increases from ethane through acetylene, the C-C bond shortens, and the bond dissociation energy increases. The first C-C bond takes 347kJ/mole to break; adding a π-bond increases this by 268kJ/mole, and adding the second π-bond gives a further increase of 197kJ/mole, so that subsequent bonds (beyond the first) between the same two atoms are successively weaker.

Geometric and Optical Isomerism

When the overlap between two orbitals is "endwise", forming a σ-bond, the only inhibition to rotation comes from neighboring substituents which can bump. Not so with π-bonds. When two p-orbitals overlap to form a π-bond, the overlap is "sideways" (Figure 14.19); rotation about the bond axis would eleminate the overlap and break the π-bond, since the p-orbitals would then be perpendicular instead of parallel. Since RT at room temperature is about 2.5 kJ/mole, far less than the 268 kJ/mole additional stability that a double bond confers over that of a single bond, this bond is not likely to break spontaneously at room temperature; the 268 kJ/mole it would take to break it can be viewed as the activation energy for rotation, and the probability of its being supplied by thermal energy at 300K is very small ($e^{-E_{act}/RT} = e^{-268/2.5} \approx 10^{-47}$). Because of this, alkanes can exist as geometric isomers, in *cis* and *trans* forms, as discussed and illustrated in the text (Example 23.2).

As discussed in Chapter 20, tetrahedral geometry results in *optical isomerism* when the four substituents are different. Rather than repeat the discussion of how this comes about and what it means, let me refer you to the Scope of Chapter 20 for a refresher on this subject. You should be able to understand the treatment given there even if you did not read chapter 20. Once you have done that, you should know what an enantiomer is. An equal mixture of two enantiomers is called a *racemic* mixture, and exhibits no optical activity. In living systems, almost every carbon compound capable of optical isomerism exists in only one of its mirror-image forms; proteins, for example, are made up only of the L-amino acids.

Two enantiomers of the same substance behave identically, chemically, except in their interactions with other optically isomeric substances. If you had a left glove in a room with a bunch of, say, racquet balls, there would be no way of describing the glove as being left- or right-handed in terms of the contents of the room. If you now were to add to the room a glove known to be right-handed, you could quickly determine that the two gloves were different. The gloves would interact with the racquetballs identically, but with another glove differently.

The question of how living systems came to discriminate between optical isomers has long been a subject of speculation, and provides for an amusing diversion (not in the text). Louis Pasteur was the first to recognize that the origin of optical activity (rotation of the plane of plane-polarized light) is in molecular handedness. He discovered in 1848 by chance microscopic examination that when a racemic solution of sodium ammonium tartrate is concentrated and allowed to crystallize, two sorts of crystals are formed, differing only in their handedness: one kind of crystal is the mirror image of the other. Pasteur laboriously separated the two types, and when he redissolved them separately, discovered that the optical activities of the two solutions were equal and opposite. (One rotated the plane of plane-polarized light to the left, the other to the right.) Actually, Pasteur was very lucky; most substances, even pure enantiomers, give symmetrical crystals.

Theories to account for the entantiomeric discrimination of living systems fall into two categories: extraterrestial theories and terrestrial theories. It turns out that at a very fundamental level of physics, the universe has a handedness, which is exhibited in phenomena like β-decay. (For this discovery, Yang and Lee won the Nobel prize for physics in 1957.) So far, attempts to show how this could possibly lead to chemical enantiomeric discrimination have been unconvincing, since processes taking place within the nucleus are essentially decoupled from the interactions of the nuclei and the electrons which give rise to chemistry. To make an analogy, one might just as easily expect the D- and L-enantiomers of a compound to behave differently when placed in a flask shaped like a left-handed glove. They do not. Furthermore, it appears that all the carbon compounds which have been extracted from carbon-containing meteorites (called carbonaceous chondrites) exist in racemic mixtures. If the origin of enantiomeric discrimination were due to the fundamental structure of the universe, then it should not be limited to the earth, or even to living things.

Most scientists believe that life started with a chance series of reactions, perhaps catalysed on a solid surface, someplace on earth. If the solid surface happened to be enantiomeric, just as Louis Pasteur's crystals were enantiomeric, then enantiomeric discrimination could have occurred in the very first such reaction. The products would then have been enantiomerically enriched, and the basis laid for all future life-processes. Of course, if the first life-reaction had occurred on the crystal of the other enantiomer, then today we would all be making proteins from D-amino acids, instead of L-. The plausibility of this hypothesis (or perhaps I should say, the lack of a better one), together with the fact that *all* living organisms — bacteria, viruses, plants, animals — use corresponding chemicals of the same handedness (for example, L-amino acids), argues for the common origin of all life on earth.

Aromaticity

The subjects of resonance and aromaticity were discussed in Chapter 14. Since sp^2-carbon bonds at $120°$ angles, six such carbons can form a regular hexagon. Each then has a half-filled p-orbital perpendicular to the ring. These combine to form *multicentered* π-molecular orbitals into which the six p-electrons are *delocalized* (Figure 14.41). This is the quantum-chemical explanation of what organic chemists had already been using the concept of *resonance* to discuss. Benzene is held together not by three single bonds and three double bonds, but by six equivalent bonds, each of order 3/2. Using the bond energies in Table 23.1 for three single and three double bonds, ΔH for the reaction

$$C_6H_6(g) \rightarrow 6CH$$

should be 2886 kJ/mole; in fact it is 152 kJ/mole greater than this. (Your text shows this using heats of hydrogenation.) Delocalizing the π-electrons confers additional stability, called *resonance stabilization*.

If you read Chapter 21, it may occur to you that this sort of delocalization is similar, though on a smaller scale, to what happens in the band theory of metals. In fact, graphite consists of planes of sp^2 carbons in a hexagonal lattice, with the p-electrons delocalized over the complete plane (Figure 14.42). Graphite does in fact exhibit metallic luster, and is a good conductor of heat and of electricity.

The *polynuclear aromatics* are smaller *condensed-ring* systems. As the text discusses, aromaticity can persist when a carbon in a benzene ring is replaced by certain other atoms, such as nitrogen (*pyridine*). Pyridine has a lone pair in an sp^2 orbital on nitrogen and is basic, in both the Lewis and

Bronsted senses.

Classes of Compounds and Functional Groups

We have already defined *alkanes*; these are sometimes called *paraffins*, and sometimes *saturated hydrocarbons*. *Unsaturation* refers to the existence of double bonds, as in *alkenes*, also defined earlier. As with most substances (for example, the inert gases), the boiling point of alkanes increases with increasing molecular weight. When *chain-branching* occurs, this tends to lower the boiling point, since a branch point (such as the central carbon in neopentane, $C(CH_3)_4$) is "shielded" by the surrounding atoms, and therefore cannot make strong van der Waals contacts with other molecules. This is also manifested in ΔH_{vap} (5650 cal/mole for neopentane, 6600 cal/mole for the *straight-chain* isomer, n-pentane) and in the densities of the pure liquids at room temperature (0.614 g/cm^3 for neopentane, 0.626 g/cm^3 for n-pentane); the straight-chain molecules can "interleave", making close contact, thus taking up less volume. For isopentane, $CH_3CH(CH_3)_2$, the boiling point, heat of vaporization and density are between the values for n-pentane and neopentane. Note that all non-cyclic alkanes have the formula C_nH_{2n+2}, so that the three pentanes just discussed all have the molecular formula C_5H_{12}.

Naming alkanes is straightforward (Example 23.3 in the text), and we will not go over it here. The same holds for the detailed naming schemes for the other classes of compounds.

Alkanes are fairly non-reactive. Organic chemists sometimes consider molecules to be made up of a relatively inert skeleton, with reactive functional groups attached to or embedded within it. A C=C double bond tends to be reactive, and is one kind of functional group. Alkanes undergo *addition reactions*, leading to the formation of a saturated carbon skeleton. If H_2 is added to an alkene, the product is an alkane, as in

$$CH_2=CH_2 + H_2 \rightarrow CH_3-CH_3$$

Other species can also be added, however, as in the synthesis of bromoethane

$$CH_2=CH_2 + HBr \rightarrow CH_3-CH_2Br$$

or of ethyl alcohol

$$CH_2=CH_2 + HOH \rightarrow CH_3-CH_2OH$$

These reactions are thermodynamically driven by the fact that a double bond is less stable than two single bonds. In the formation of ethane, the two C-H bonds formed exhibit a greater total bond energy than do the H-H bond and the C=C π-bond that were broken in the reactants. Perhaps more important, the π-electrons are loosely bound and are *nucleophilic*, so that the somewhat positively charged H of HBr, say, is attracted to it. This provides a facile kinetic path to addition reactions. As your text discusses, such reactions often go through a *carbocation* intermediate.

We have already given an example of an *alcohol*; the functional group is —OH, and it hydrogen-bonds with water or with another alcohol just as one would expect; therefore alcohols boil at much higher temperatures than alkanes of similar molecular weight.

If an alcohol can be viewed as a water molecule with one hydrogen replaced by an alkyl group, R, an *ether* can be viewed as a water molecule with both groups so replaced.

HOH	water
ROH	alcohol
ROR	ether

Of course, ethers can not act as hydrogen-bond donors, and are therefore less soluble in water than are alcohols. There is no hydrogen bonding at all in a pure ether, and therefore the boiling point of an ether is below that of an alcohol of similar molecular weight; on the other hand, the ether oxygen exhibits a bond angle of about 110°, and ethers therefore have permanent dipole moments. Dipole-dipole interactions make the boiling point of an ether higher than that of an alkane of similar molecular weight.

An *amine* is like an ammonia molecule with one or more hydrogens replaced by organic groups.

NH_3	ammonia
NRH_2	primary amine
NR_2H	secondary amine
NR_3	tertiary amine

Chapter 23 — Introduction to Organic Chemistry

If the R's are alkyl groups, then the amines are called aliphatic amines. There are also aromatic amines, such as *aniline*, which is like benzene with one of the H's replaced by $-NH_2$. Of course, amines are basic, for the same reason that ammonia is basic: they have a lone pair of electrons in an sp^3 orbital. Thus the amines can react with acids to form the corresponding ammonium salts. A *quaternary ammonium salt* is one of the form $NR_4^+X^-$.

If an sp^2 carbon forms a double bond to an oxygen, the resulting functional group, C=O, is called a carbonyl. Molecules with the structure CH_2=O or CHR=O are called *aldehydes*, and those with the structure CR_2=O are called *ketones*. The C=O bond is highly polarized, and the resulting permanent dipole moment is large, larger even than that of an ether; therefore ketones and aldehydes boil at higher temperatures than ethers or, of course, alkanes of similar molecular weights.

An *imine* is like a ketone, but with an NH instead of an O: CR_2=NH. They can be made by nucleophilic attack of ammonia upon the carbon of a carbonyl (Equation 23-11). The reaction is driven by the formation and removal of H_2O; however, when much H_2O is present, the equilibrium favors the ketone, by Lechatlier's principle.

A *carboxylic acid* has the formula CR(OH)=O. The reason they are acidic is that the *carboxylate anion* exhibits two resonance forms, as shown in the text, with the negative charge delocalized over both oxygens. This delocalization confers stability on the anion, making the compound acidic. As the text notes, the formation of hydrogen-bonded dimers in the liquid makes these compound high boilers.

A *condensation* is an organic reaction one of whose products is water; imine formation is an example we have already seen. Others are the reaction of a carboxylic acid and an alcohol to form an *ester*

$$CR(OH)C=O + R'OH \rightarrow CR(OR')=O + H_2O$$

and that of a carboxylic acid with an amine to form an *amide*

$$CR(OH)=O + R'NH_2 \rightarrow CR(NHR')=O + H_2O$$

Chapter 23 — Introduction to Organic Chemistry

These reactions are generally reversible in the presence of water and acid.

Oxidation States of Carbon

You saw in Chapter 15 that, using the standard rules for determining oxidation state, carbon in propane, $CH_3CH_2CH_3$, exhibits an oxidation state of -8/3. This is vaguely unsettling; in ethane, CH_3CH_3, each carbon exhibits an oxidation state of -3, using the same rules, and somehow it seems as if any carbon bonded to three hydrogens and another carbon should have the same oxidation state. There is a way to make this happen, and still make the sum of the oxidation states of all the atoms in the molecule come out to zero, while maintaining the familiar values of +1 for H and -2 for O. Starting with the structural formula, we consider a carbon-carbon bond not to contribute to the oxidation states of either carbon. Thus carbons in the end CH_3-'s in propane each have an oxidation state of -3, while the central carbon has an oxidation state of -2. Note that these add up to -8, which exactly balances the +8 contributed by the hydrogens. Table 23.8 shows how this works out for a series of functional groups.

Please note that this method of assigning oxidation state is more than a mere artifice. Consider the following partial oxidation of methane

$$2CH_4 + \frac{1}{2}O_2 \rightarrow CH_3CH_3 + H_2O$$

Oxygen, the oxidant, has been reduced, and carbon has been oxidized from the -4 to the -3 oxidation state. Now consider

$$CH_3CH_3 + \frac{1}{2}O_2 \rightarrow CH_3CH_2CH_2CH_3 + H_2O$$

Here two carbons (the central ones in the n-butane) have been oxidized from the -3 to the -2 oxidation states, while the end ones have not changed. There exist industrial processes which use these kinds of reactions.

Biological Compounds

Water constitutes about 70% by weight of living organisms, and is central to life processes. Many, many biological processes have to do with the interaction of chemical compounds with water, and thus it

Chapter 23   Introduction to Organic Chemistry

is fitting that the text starts by defining *hydrophilic* (from the Greek, "water loving") and *hydrophobic* ("water hating"). From your past experience it should be clear that gasoline is hydrophobic, whereas salt and sugar are hydrophilic, just based on their solubility behavior. We will see that many biological molecules have a hydrophobic part and a hydrophilic part, and that this has to do with their function.

Completely hydrophobic biological molecules are called *lipids*; the *fats* are examples. *Glycerol* (old-fashioned name: *glycerin*) is the trialcohol related to propane, $CH_2(OH)CH(OH)CH_2(OH)$, and is itself very hydrophilic. However, when each alcohol is esterified to a *fatty acid*, $RCOOH$, where R is a long alkyl chain, the resulting molecule is hydrophobic (Equation 23-18). (You might wish to remind yourself of the esterification reaction, Section 23-6, and two sections back in the Scope.) Esters are not very hydrophilic in general, and the long hydrocarbon tails make fats even less so than simpler esters.

If one end of the glycerol is esterified to phosphoric acid group instead of to a fatty acid, then the compound is called a *phospholipid*. Although we used the term esterification to denote the reaction between a carboxylic acid and an alcohol

$$CR(OH){=}O + R'OH \rightarrow CR(OR'){=}O + H_2O$$

the same sort of reaction can occur between inorganic acids, such as H_3PO_4, and alcohols

$$(OH)_3P{=}O + R'OH \rightarrow (OH)_2(OR')P{=}O$$

The phosphate group (called the *head group*) is hydrophilic, and the fatty *tail* is hydrophobic. The tails tend to avoid water, and the heads are highly stable in water, where they can ionize to form the anion. (More complicated derivatives, such as *phosphatidyl choline*, shown in the text, behave similarly.) Several things can happen in order to satisfy the conflicting demands of the two ends of the molecule.

>*Micelle formation.* Aggregates of phospholipids form little balls with the tails clustered together, excluding water from the interior, while the heads point outward toward the water.
>
>*Bilayer formation.* A bilayer is like a flat micelle two molecules thick. Again, tails in, heads out. These are the bases of biological membranes.
>
>*Detergent action* (not in text). Micelles can dissolve other hydrophobic molecules in their interiors. This is how laundry detergents and soaps work; dirt tends to be hydrophobic. Of

course, laundry detergents are not phospholipids, but they do have a long hydrophobic tail and a hydrophilic head, often a sulfate group.

Surfactant behavior (not in text). At concentrations too low to form bilayers, these sorts of molecules move to the surface of liquid water. The hydrophilic heads remain dissolved and the hydrophobic tails stick up out of the liquid. This reduces the surface tension; i. e., it "makes water wetter."

An *amino acid* has the structure $CR(NH_2)COOH$; it is both an amine and an acid. Since an amine and an acid can react to form an amide (Section 23-6), a long string of amino acids can join together forming a *polyamide*. This is called a *polypeptide*, and a large naturally occurring polypeptide is called a protein. Note that the various R-groups will be hanging off the side of the peptide backbone.

Protein chains range in length from about 50 to about 500 amino acids; however, many chains may form aggregates to give the biologically active protein. The way the protein chain folds is important to its biological function. Note that of the amino acids shown in Table 23.9, some, such as serine, have hydrophilic side-chains, whereas others, such as valine, have hydrophobic side-chains. Certain proteins are soluble in water, and these fold up so that the hydrophobic side-chains are buried, and the hydrophilic side chains are exposed to the water, like a micelle, but all of one big molecule. Other proteins are membrane-bound, and these fold so that hydrophobic side-chains are on the outside, in order to bind well to lipids. What defines a given protein is the specific sequence of amino acids along the polypeptide chain.

Sugars are hydrophilic, as already noted, and polymers of sugars are called *polysaccharides*. These include the *starches*, such as glycogen and amylose (about 300 sugar units) and also the relatively inert *cellulose* (many, many sugar units).

Synthetic Polymers

We have discussed biological polymers; over the past hundred years, but especially over the past forty, synthetic polymers have become part of our lives, to the extent that it is difficult to purchase anything in the Western world that is not either made partly from or packed in containers made partly from synthetic polymers.

We previously defined condensation reactions: ones which eliminate a molecule of water per polymer unit added. *Nylon* is a synthetic polypeptide, and is a *condensation polymer*, because formation of the amide bond eliminates a molecule of water.

On the other hand, *polyvinyl chloride* (pvc) and *polyethylene* are addition polymers. These start with unsaturated *monomers* (single units), *vinyl chloride*, $CH_2=CHCl$ and ethylene, $CH_2=CH_2$, respectively. The free-radical mechanism shown in the text results in a saturated structure, $-CH_2-CHCl-CH_2-CHCl-$ for pvc, and $-CH_2-CH_2-CH_2-CH_2-$ for polyethylene.

Drugs

Drugs are biologically active small molecules. Several examples are given in the text.

Chapter 23 — Introduction to Organic Chemistry

Questions

23.1 Bonding in organic compounds

1. Give the electronic configuration of

    (a) a carbon in the gas phase

    (b) a carbon in ethane

    (c) a carbon in ethylene

    (d) a carbon in acetylene

2. Which of the species in Question 1 has

    (a) the longest C-C bond length?

    (b) the shortest C-C bond length?

3. Which of the species in Question 1 has

    (a) the greatest C-C bond energy?

    (b) the lowest C-C bond energy?

4. (a) Name the most stable conformation of ethane.

    (b) Name the least stable conformation of ethane.

    (c) Account for this difference in stability.

23.2 Geometric isomerism

1. Name and draw the two geometric isomers of 1,2 dibromoethylene. (CHBr=CHBr)

2. There is another isomer of dibromoethylene. What is its name and structural formula?

Chapter 23 — Introduction to Organic Chemistry

3. Which of the following can exhibit geometric isomerism?

   _____(a) $CH_2=CHBr$ bromoethylene

   _____(b) $CH(CH_3)=CH(CH_3)$ 2-butane

   _____(c) $CH_2=C(CH_3)_2$ 1-butane

   _____(d) $C(CH_3)\equiv CCH_3$ 2-butyne

   _____(e) $CH\equiv CCH_2CH_3$ 1-butyne

23.3 Optical isomerism

1. Define the optical activity.

2. Which of the following will exhibit optical activity?

   _____(a) L-alanine

   _____(b) D-alanine

   _____(c) 50% of (a) + 50% of (b)

   _____(d) 25% of (a) + 75% of (b)

   _____(e) CH_3Br bromomethane

   _____(f) CH_2BrCl bromochloromethane

   _____(g) $CHBrClF$ bromochlorofluoromethane

3. A 1:1 mixture of enantiomers is called a _____ mixture.

23.4 Aromaticity

1. The hybrid orbitals exhibited by the carbons in aromatic structures are _____ orbitals.

2. The delocalized π-orbital in an aromatic structure results from the overlap of the atomic _____ orbitals.

Chapter 23 — Introduction to Organic Chemistry

3. Resonance results in (stabilization, destabilization), and the resonance energy is defined as a (positive, negative) quantity.

4. Which shows the greater tendency to undergo addition reactions (such as hydrogenation)

    _____ an isolated double bond

    _____ an aromatic system

5. (a) What is a heterocyclic aromatic system?

    (b) Name one.

23.5 The alkanes

1. Name and give the molecular formula of

    (a) the straight-chain alkane with six carbons

    (b) the straight-chain alkane with three carbons

    (c) the straight-chain alkane with four carbons

2. Which of the above (Question 1) has the highest boiling point?

3. A straight-chain alkane exhibits linear geometry around each carbon (true, false)

4. Give the structural formula for

    (a) n-butane

    (b) isobutane

Chapter 23   Introduction to Organic Chemistry

5. Which of the above (Question 4) has the lower boiling point?

6. Give the standard chemical name (as in text Example 23.3) for

   (a) isobutane

   (b) neopentane $(C(CH_3)_4)$

   (c) isopentane $(CH_3CH_2CH(CH_3)_2)$

23.6 Functional groups

1. Give the name and the structural formula for the compound(s) that will result from addition of HBr to

   (a) ethylene

   (b) 1-butene

   (c) 2-butene

2. Why do alcohols have high boiling points?

3. Give the structural formula for

   (a) 1-propanol

   (b) 2-propanol

23-17

Chapter 23 Introduction to Organic Chemistry

(c) 2-methyl-2-propanol

4. What is the strongest kind of interaction between two ether molecules in the pure liquid?

5. Name and give the structural formulas for two ethers.

6. Tell whether each of the following amines is primary (1°) secondary (2°) or tertiary (3°).

    _____(a) diethylamine

    _____(b) isobutylamine

    _____(c) trimethylamine

    _____(d) aniline

7. Trimethylamine ($N(CH_3)_3$) and methylethylamine ($NH(CH_3)(CH_2CH_3)$) have the same molecular weight. Which would you expect to have higher boiling point?

8. Which of the following are ketones, and which aldehydes?

    _____(a) $(CH_3)_2C=O$

    _____(b) $(CH_3)HC=O$

    _____(c) $H_2C=O$

    _____(d) $(CH_3)(C_6H_5)C=O$

9. The C=O group is called a _____ group

10. The C of a C=O group is _____philic, and is therefore subject to attack by a _____phile, such as _____.

Chapter 23

Introduction to Organic Chemistry

11. Carboxylic acids tend to have (low, high) boiling points.

12. When a carboxylic acid reacts with an alcohol, the result is an _____. One molecule of _____ is also produced.

13. When a carboxylic acid reacts with an amine, the result is an _____. One molecule of _____ is also produced.

23.7 Oxidation states of carbon

1. Give the oxidation state of the starred carbon in each of the following.

    _____ (a) C^*O (carbon monoxide)

    _____ (b) C^*O_2

    _____ (c) CH_3C^*OOH

    _____ (d) $CH_3C^*H_2Br$

    _____ (e) $CH_3C^*H_3$

    _____ (f) $CH_2=C^*H_2$

    _____ (g) $HC\equiv C^*H$

    _____ (h) $H_2C^*=O$

23.8 Brief survey of biologically important compounds

1. Tell whether each of these is hydrophilic or hydrophobic.

    _____ (a) n-hexane

    _____ (b) ethanol

    _____ (c) n-decanol (n-$C_{10}H_{21}OH$)

    _____ (d) acetic acid

    _____ (e) stearic acid (n-$C_{17}H_{35}COOH$)

2. A fat is an ester, where the alcohol is _____, and the acid is any _____ acid.

3. A phospholipid is like a fat, except one _____ has been replaced by a _____ group.

Chapter 23   Introduction to Organic Chemistry

4. Which of the following might you expect to form micelles?

   _____(a) acetic acid

   _____(b) stearic acid

   _____(c) stearin (a fat)

   _____(d) the phospholipid derived from stearin

5. The basic "building blocks" of a protein are _____.

6. The functional group between successive "building blocks" along the protein chain is the _____.

7. What is a zwitterion?

8. (a) All proteins are polypeptides (true, false)

   (b) All polypeptides are proteins (true, false)

9. A compound with the general formula $C_n(H_2O)_n$ is called a _____.

10. (a) Which of these is the largest molecule, and which the smallest?

    _____ amylose

    _____ cellulose

    _____ glucose

    _____ glycogen

    (b) What do they all have in common?

Chapter 23 Introduction to Organic Chemistry

23.9 Synthetic polymers

1. According to the text, what do hair, leather and aardvarks all have in common?

2. Tell whether each of the following monomers would be most likely to polymerize by an addition reaction, a condensation reaction, or not at all.

    _____(a) ethylene

    _____(b) alanine

    _____(c) acetylene

    _____(d) benzene

    _____(e) ethane

23.10 Drugs

1. Drugs are generally (small molecules, polymers).

2. (a) Drugs are generally similar in important ways to compounds occurring naturally in the body. (True, False)

    (b) Drugs are generally different in important ways to compounds occurring naturally in the body. (True, False)

Answers

23.1 Bonding in organic compounds

1. (a) $(2s)^2(2p)^2$
   (b) $(2sp^3)^4$
   (c) $(2sp^2)^3 p$
   (d) $(2sp)^2 p^2$
2. (a) ethane
   (b) acetylene
3. (a) acetylene
   (b) ethane
4. (a) staggered
   (b) eclipsed
   (c) the H's "bump" in the eclipsed form.

23.2 Geometric isomerism

1.
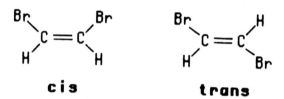

2. 1,1-dibromoethane $CBr_2=CH_2$
3. (b) only. (d) and (e) are linear. In (a), the Br has nothing to be *cis* or *trans* to; similarly for (c).

23.3 Optical isomerism

1. The rotation of the plane of plane-polarized light as it passes through a substance.
2. (a), (b), (d), (g).
3. racemic

23.4 Aromaticity

1. sp^2.
2. p
3. stabilization, positive
4. an isolated double bond
5. (a) A benzene ring with one or more C's replaced by a different kind of atom, such as N.
   (b) pyridine

23.5 The alkanes

1. (a) n-hexane C_6H_{14}
   (b) propane C_3H_8
   (c) n-butane C_4H_{10}
2. n-hexane
3. False (it exhibits the tetrahedral angle, $\approx 109°$)
4. (a) $CH_3CH_2CH_2CH_3$
   (b) $CH_3CH(CH_3)_2$

Chapter 23

5. (b)
6. (a) 2-methylpropane
   (b) 2,3-dimethylpropane
   (c) 2-methylbutane

23.6 Functional groups

1. (a) bromoethane CH_3CH_2Br
   (b) 1-bromobutane $CH_3CH_2CH_2CH_2Br$ and 2-bromobutane $CH_3CH_2CH_2BrCH_3$
   (c) 2-bromobutane $CH_3CH_2CH_2BrCH_3$
2. hydrogen bonding in pure liquid
3. (a) $CH_3CH_2CH_2OH$
   (b) $CH_3CH(OH)CH_3$
   (c) $CH_3C(CH_3)(OH)CH_3$
4. dipole-dipole interactions
5. dimethyl-ether CH_3OCH_3 and methylethyl ether $CH_3CH_2OCH_3$ are two examples.
6. (a) 2°
   (b) 1° (only one isobutyl group)
   (c) 3°
   (d) 1° (only one benzene ring on NH_2)
7. Methylethylamine, since it can hydrogen bond in the pure liquid.
8. (a) ketone
   (b) aldehyde
   (c) aldehyde
   (d) ketone
9. carbonyl
10. electrophilic, nucleophile, NH_3
11. high
12. ester, water
13. amide, water

23.7 Oxidation states of carbon

1. (a) +2
   (b) +4
   (c) +3
   (d) -1
   (e) -3
   (f) -2
   (g) -1
   (h) 0

23.8 Brief survey of biologically important compounds

1. (a) hydrophobic
   (b) hydrophilic
   (c) hydrophobic (long R-group)
   (d) hydrophilic
   (e) hydrophobic (long R-group)
2. glycerol, fatty

3. fatty acid, phosphate
4. (b), (d) Both have hydrophilic heads and long hydrophobic tails.
5. amino acids (L-amino acids, in fact)
6. amide (or peptide group)
7. A single molecule with both a positive and a negative charge on it.
8. (a) true
   (b) false; a protein is a *biologically active* polypeptide.
9. carbohydrate
10. (a) cellulose is largest, glucose smallest
    (b) they are all made up of glucose units

23.9 Synthetic polymers

1. They are all held together by proteins.
2. (a) addition
   (b) condensation
   (c) addition
   (d) not at all (aromaticity makes it inert)
   (e) not at all (no functional groups)

23.10 Drugs

1. Small molecules
2. (a) True
   (b) True
   Consider Figure 23.20 in the text. The aromatic ends of the two molecules are similar; the amino end of dopa has a carboxylate group, enabling it to cross the blood-brain barrier, which the naturally occurring dopamine cannot do.